A. Apostolico M. Crochemore
Z. Galil U. Manber (Eds.)

Combinatorial Pattern Matching

4th Annual Symposium, CPM 93
Padova, Italy, June 2-4, 1993
Proceedings

Springer-Verlag
Berlin Heidelberg New York
London Paris Tokyo
Hong Kong Barcelona
Budapest

Volume Editors

Alberto Apostolico
Computer Science Dept., Purdue University
West Lafayette, IN 47907-1398, USA
and
Dipartimento di Elettronica e Informatica, Università di Padova
I-35131 Padova, Italy

Maxime Crochemore
L.I.T.P., Université Paris VII
F-75251 Paris CX 05, France

Zvi Galil
Columbia University, New York, NY 10027, USA
and
Tel Aviv University, Ramat Aviv, Tel Aviv, Israel

Udi Manber
Computer Science Dept., University of Arizona
Gould-Simpson 721. Tucson, AZ 85721, USA

CR Subject Classification (1991): F.2.2, I.5.4, I.5.0, I.7.3, H.3.3, E.4, G.2.1, J.3

ISBN 3-540-56764-X Springer-Verlag Berlin Heidelberg New York
ISBN 0-387-56764-X Springer-Verlag New York Berlin Heidelberg

© Springer-Verlag Berlin Heidelberg 1993
Printed in Germany

Typesetting: Camera ready by author
Printing and binding: Druckhaus Beltz, Hemsbach/Bergstr.
45/3140-543210 - Printed on acid-free paper

Lecture Notes in Computer Science 684

Foreword

The papers contained in this volume were presented at the fourth annual symposium on Combinatorial Pattern Matching, held June 2-4, 1993 in Padova, Italy. They were selected from 34 abstracts submitted in response to the call for papers.

Combinatorial Pattern Matching addresses issues of searching and matching of strings and more complicated patterns such as trees, regular expressions, extended expressions, etc. The goal is to derive nontrivial combinatorial properties for such structures and then to exploit these properties in order to achieve superior performances for the corresponding computational problems.

In recent years, a steady flow of high-quality scientific study of this subject has changed a sparse set of isolated results into a full-fledged area of algorithmics. This area is expected to grow even further due to the increasing demand for speed and efficiency that comes especially from molecular biology and the Genome project, but also from other diverse areas such as information retrieval (e.g., supporting complicated search queries), pattern recognition (e.g., using strings to represent polygons and string matching to identify them), compilers (e.g., using tree matching), data compression, and program analysis (e.g., program integration efforts). The stated objective of CPM gatherings is to bring together once a year the researchers active in the area for an informal and yet intensive exchange of information about current and future research in the area.

The first three meetings were held at the University of Paris in 1990, at the University of London in 1991, and at the University of Arizona, Tucson, in 1992. The first two meetings were informal and no proceedings were produced. The proceedings of the third meeting appeared as Volume 644 in this series.

Several external referees helped with the selection of papers for CPM 93. The Local Organizing Committee consisted of G. Bilardi, L. Colussi, C. Guerra and L. Toniolo. The Padova Ricerche Consortium provided services and, together with other sponsors, funds for the conference. The efforts of all are gratefully acknowledged. Special thanks are due to F. Bombi for eagerly promoting CPM 93 both within the University of Padova and outside of it.

March 1993

Programm Committee

A. Apostolico, *chair*	G. Gonnet
M. Crochemore	D.S. Hirschberg
A. Ehrenfeucht	U. Manber
A.S. Fraenkel	E.W. Myers
Z. Galil	M.S. Waterman

A Linear Time Pattern Matching Algorithm Between a String and a Tree
  Tetsuo Shibuya

Tight Comparison Bounds for the String Prefix-Matching Problem 1
  Dany Breslauer, Livio Colussi and Laura Toniolo

8-Testing of Pattern Matchings 20
  Daniel Ricchetti, Paola Bonizzoni, Gianluca Della Vedova, and Yuri Pirola

Minimal Separators of Two Words 15
  Emmanuelle Garel

Covering a String 34
  Costas S. Iliopoulos, Dennis W. G. Moore and Kunsoo Park

On the Worst-case Behaviour of Some Approximation Algorithms
  for the Shortest Common Supersequence of k Strings 63
  Robert W. Irving and Campbell B. Fraser

An Algorithm for Locating Non-Overlapping Regions of Maximum
  Alignment Score ....
  Sampath K. Kannan and Eugene W. Myers

Exact and Approximation Algorithms for the Inversion Distance
  between Two Chromosomes 87
  John Kececioglu and David Sankoff

Two Algorithms for the Longest Common Subsequence of Three
  (or More) Strings 105
  Robert W. Irving and Campbell B. Fraser

From Max-trees to Weight-Preserving Pairs to Multiple Sequence Alignment ....
  John Kececioglu

An Algorithm for Approximate Tandem Repeats ....
  Gad M. Landau and Jeanette P. Schmidt

Two-dimensional Pattern Matching in a Sparse Image 174
  Gad M. Landau and Uzi Vishkin

Analysis of a String Pattern in a Probabilistic Framework ....
  Olli Nevalainen and Valeria Simoncini

Detecting False Matches in String Matching Algorithms ....
  S. Muthukrishnan

Table of Contents

A Linear Time Pattern Matching Algorithm Between a String and a Tree
Tatsuya Akutsu
1

Tight Comparison Bounds for the String Prefix-Matching Problem
Dany Breslauer, Livio Colussi and Laura Toniolo
11

3-D Docking of Protein Molecules
Daniel Fischer, Raquel Norel, Ruth Nussinov, and Haim J. Wolfson
20

Minimal Separators of Two Words
Emmanuelle Garel
35

Covering a String
Costas S. Iliopoulos, Dennis W.G. Moore and Kunsoo Park
54

On the Worst-Case Behaviour of Some Approximation Algorithms
for the Shortest Common Supersequence of k Strings
Robert W. Irving and Campbell B. Fraser
63

An Algorithm for Locating Non-Overlapping Regions of Maximum
Alignment Score
Sampath K. Kannan and Eugene W. Myers
74

Exact and Approximation Algorithms for the Inversion Distance
Between Two Chromosomes
John Kececioglu and David Sankoff
87

The Maximum Weight Trace Problem in Multiple Sequence Alignment
John Kececioglu
106

An Algorithm for Approximate Tandem Repeats
Gad M. Landau and Jeanette P. Schmidt
120

Two Dimensional Pattern Matching in a Digitized Image
Gad M. Landau and Uzi Vishkin
134

Analysis of a String Edit Problem in a Probabilistic Framework
Guy Louchard and Wojciech Szpankowski
152

Detecting False Matches in String Matching Algorithms
S. Muthukrishnan
164

On Suboptimal Alignments of Biological Sequences   179
Dalit Naor and Douglas Brutlag

A Fast Filtration Algorithm for the Substring Matching Problem   197
Pavel A. Pevzner and Michael S. Waterman

A Unifying Look at d-Dimensional Periodicities and Space Coverings   215
Mireille Régnier and Ladan Rostami

Approximate String-Matching over Suffix Trees   228
Esko Ukkonen

Multiple Sequence Comparison and n-Dimensional Image Reconstruction   243
Martin Vingron and Pavel A. Pevzner

A New Editing Based Distance Between Unordered Labeled Trees   254
Kaizhong Zhang

A Linear Time Pattern Matching Algorithm Between a String and a Tree

Tatsuya AKUTSU *

Mechanical Engineering Laboratory,
1-2 Namiki, Tsukuba, Ibaraki, 305 Japan.

Abstract. In this paper, we describe a linear time algorithm for testing whether or not there is a path of a tree T ($|V(T)| = n$) that coincides with a string s ($|s| = m$). In the algorithm, $O(n/m)$ vertices are selected from $V(T)$ such that any path of length more than $m - 2$ must contain at least one of the selected vertices. A search is performed using the selected vertices as 'bases.' A suffix tree is used effectively in the algorithm. Although the size of the alphabet is assumed to be bounded by a constant in this paper, the algorithm can be applied to the case of unbounded alphabets by increasing the time complexity to $O(n \log n)$.

Keywords: subtree, subgraph isomorphism, string matching, suffix tree, graph algorithms

1 Introduction

The subgraph isomorphism problem is famous and important in computer science. It is the problem of testing whether or not there is a subgraph of T isomorphic to S when the graphs of S and T are given. It is important for practical applications as well. In particular, many heuristic algorithms have been developed for database systems in chemistry [13, 14].

In general, the problem was proved to be NP-complete [6]. However, polynomial time algorithms have been developed in special cases [10, 12]. When the graphs are simple paths, the problem is reduced to the string matching problem, for which several linear time algorithms have been developed [3, 7]. When the graphs are restricted to trees, the problem is solved in $O(n^{2.5})$ time [4]. Moreover, if the vertex degree of two input trees is bounded by a constant, the problem is solved in $O(n^2)$ time. If the graphs are rooted trees such that the labels of children of each node are

* The author thanks to Prof. Tomio Hirata in Nagoya University for helpful comments. This research was partially supported by the Grand-in-Aid for Scientific Research on Priority Areas, "Genome Informatics", of the Ministry of Education, Science and Culture of Japan.

distinct, whether there is an $o(n^2)$ time algorithm or not had been an open problem for a long time. However, it was solved confirmatively by Kosaraju [8] and the result was improved by Dubiner et.al. [5]. However, as far as we know, there is no $o(n^2)$ time algorithm for undirected trees or rooted trees such that children of a node may have identical labels even if the vertex degree is bounded. In this paper, we show a linear time algorithm for a special case of the problem, that is, the case where S is a path and T is an undirected tree. Moreover, T may have a node such that adjacent vertices have identical labels.

In this paper, T denotes an input undirected tree with labeled vertices and $s = s^1 s^2 \cdots s^m$ denotes an input string of length m. We do not assume that labels of vertices adjacent to the same vertex are different. Although vertices are assumed to be labeled, the result can also be applied to the case of labeled edges. For a graph G, $V(G)$ denotes the set of vertices and $E(G)$ denotes the set of edges. n denotes the number of the vertices of the tree T (i.e., $n = |V(T)|$). For a vertex v, $label(v)$ denotes the label associated with v. We assume that the size of the alphabet is bounded by a constant. The problem is to test whether or not there is a vertex disjoint path $\langle v_1, v_2, \cdots, v_m \rangle$ in T such that $label(v_1) label(v_2) \cdots label(v_m) = s$. This paper describes an $O(n)$ time algorithm for this problem.

Of course, a rooted tree version of the problem, that is, the case where T is a rooted tree and only the paths which do not connect sibling nodes are allowed, is trivially solved in linear time [5] by using a linear time substring matching algorithm [3, 7] with backtracking. However, this method does not seem to work for the problem of this paper. The linear time algorithm which we developed here is based on a different idea: $O(n/m)$ vertices are selected from $V(T)$ such that any path of length more than $m - 2$ must contain at least one of the selected vertices. From each of the selected vertices, a search is performed with traversing the suffix tree associated with s.

2 Suffix Tree

In this section, we give an overview of the well-known data structure, *suffix tree* [2, 11]. The suffix tree is used in on-line string matching for a large fixed text. Moreover, it is applied to a variety of pattern matching problems [1, 5, 9].

Let $s = s^1 s^2 \cdots s^m$ be a string. $|s|$ denotes the length of s. s_i denotes the suffix of s which starts from s^i. s^{-1} denotes the reversed string of s and s_i^{-1} denotes $(s^{-1})_i$. For a string s and a character x, sx denotes the concatenation of s and "x". We assume without loss of generality that the special character '#' does not appear in s. The suffix tree SUF_s associated with s is the rooted tree with m leaves and at most $m - 1$ internal nodes such that

- Each edge is associated with a substring of $s\#$.
- Sibling edges must have labels whose first character is distinct.
- Each leaf is associated with a distinct position of s.
- The concatenation of the labels on the path from the root to a leaf l_i describes $(s\#)_i$.

It is known that the size of a suffix tree is $O(m)$ and a suffix tree can be constructed in $O(m)$ time if the size of the alphabet is bounded by a constant [11]. It is easy to

Suffix Tree

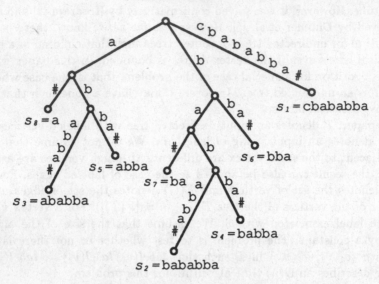

$$PRSX(ab, cbababba) = \{3, 5\}$$
$$PRSX(ba, cbababba) = \{2, 4\}$$

Fig. 1. A suffix tree associated with "cbababba".

see that searching a substring of s that coincides with a given string p can be done in $O(|p|)$ time if SUF_s already exists.

For strings s and t, we define $PRSX(t, s) = \{ i \mid t \text{ is a prefix of } s_i \text{ and } |t| \neq |s_i| \}$ (see Figure 1). If t is a prefix of s_i, we define $rest(t, i) = i + |t|$. Obviously, the concatenation of t and $s_{rest(t,i)}$ becomes s_i. The number of elements in each $PRSX(t, s)$ is at most $m - 1$. Moreover, $PRSX(t, s)$ can be computed in $O(m)$ time as shown below.

Proposition 1. *Assume that a suffix tree SUF_s associated with s already exists. For any string t, $PRSX(t, s)$ is computed in $O(m)$ time.*

Proof. Note that $PRSX(t, s)$ is easily obtained from the set of leaves of the subtree of SUF_s which is rooted at the endpoint of the path spelling t. Since such an endpoint is trivially found in $O(\min\{m, |t|\})$ time and the size of SUF_s is $O(m)$, the elements of $PRSX(t, s)$ can be enumerated in $O(m)$ time. $\square$

Note that $PRSX(t, s)$ can be computed in $O(\min\{m, |t|\})$ time if $PRSX(t, s)$ is represented with an appropriate data structure.

3 Overview of the Algorithm

In the algorithm, $V(T)$ is divided into two disjoint sets V_b (a set of *black* vertices) and V_w (a set of *white* vertices) such that the following conditions are satisfied (see Figure 2):

(A1) $|V_b|$ is $O(n/m)$.

(A2) Any path of T with length (i.e., the number of edges) more than $m - 2$ must contain at least one black vertex.

(A3) If there are two paths such that the endpoints are black and the other vertices are white, such paths have no common white vertex.

In the next section, a linear time algorithm for computing such V_b and V_w is described.

An example for $m = 4$.   A case where condition A3 is violated.

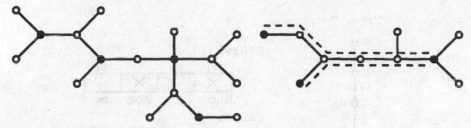

Fig. 2. Black vertices and white vertices.

For each black vertex v of T, two tables $TBL(v, s_i)$ $(1 \leq i \leq m - 1)$ and $TBL(v, s_i^{-1})$ $(1 \leq i \leq m-1)$ are constructed (see Figure 3). Each value in $TBL(v, s_i)$ or $TBL(v, s_i^{-1})$ is either $TRUE$, $FALSE$, or w, where w is one of the adjacent vertices to v.

Let $path(s_i, v)$ denote $\{ w \mid$ there is a path $\langle v = w_1, w = w_2, w_3, \cdots, w_{m-i+1} \rangle$ such that $(i \leq \forall j \leq m)(label(w_{j-i+1}) = s^j) \}$. If $|path(s_i, v)| = 0$, then $TBL(v, s_i) = FALSE$. If $path(s_i, v) = \{w\}$, then $TBL(v, s_i) = w$. Otherwise, $|path(s_i, v)| = 2$, and then $TBL(v, s_i) = TRUE$. $TBL(v, s_i^{-1})$ is defined in the same way.

Once these tables are constructed for all the black vertices, testing whether or not there is a path of T that coincides with s is easy. There is a path if and only if there exists a black vertex v which satisfies the following condition (condition C1):

$$(TBL(v, s_1) \neq FALSE \vee TBL(v, s_1^{-1}) \neq FALSE) \vee$$
$$(2 \leq \exists i < m)(\ (TBL(v, s_i) = TRUE \wedge TBL(v, s_{m+1-i}^{-1}) \neq FALSE) \vee$$
$$(TBL(v, s_i) \neq FALSE \wedge TBL(v, s_{m+1-i}^{-1}) = TRUE) \vee$$
$$(TBL(v, s_i) = w \wedge TBL(v, s_{m+1-i}^{-1}) = u \wedge w \neq u))$$

Note that the condition can be checked in $O(m)$ time for each vertex v.

A procedure for the construction of TBLs is described in section 5. At the beginning of the procedure, all values in TBLs are set to $FALSE$. When a path $P = \langle v = w_1, w = w_2, w_3, \cdots, w_{m-i+1} \rangle$ such that the corresponding label sequence coincides with s_i are found, $TBL(v, s_i)$ is updated using the procedure $UpdateTBL(v, s_i, w)$ listed below. Note that the procedure $UpdateTBL(v, s_i, w)$ can be executed in constant time per call.

Procedure $UpdateTBL(v, s_i, w)$
  begin
    if $TBL(v, s_i) = FALSE \vee TBL(v, s_i) = w$
    then $TBL(v, s_i) \leftarrow w$
    else $TBL(v, s_i) \leftarrow TRUE$
  end

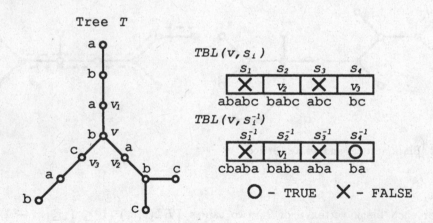

Fig. 3. Examples of $TBL(v, s_i)$s for "ababc".

4 Selection of Vertices

In this section, we describe a procedure for selecting black vertices. It takes T_r and m as inputs where T_r is the rooted tree obtained from T taking an arbitrary vertex $r \in V(T)$ as the root. The procedure consists of the following two phases, where in each phase, vertices are processed in a bottom-up manner.

(Phase 1) For each path from a leaf to the root, black vertices are placed so that no consecutive $\lfloor \frac{m}{2} \rfloor + 1$ white vertices appear.

(Phase 2) Any vertex which has at least two children with black descendants is colored black.

```
Procedure SetColor(T_r, m)
  begin
    M ← ⌊m/2⌋ − 1;
    for all v ∈ V(T) do begin   color(v) ← WHITE;   rank(v) ← 0   end;
    Let W be the list obtained by sorting V(T) topologically in postorder;
    while W is not empty do
      begin
        Remove the first element v from W;
        k ← max{ rank(w) | w is a child of v };
        if k ≥ M then begin   color(v) ← BLACK;   rank(v) ← 0   end
        else rank(v) ← k + 1
      end;
    for all v ∈ V(T) do flag(v) ← FALSE;
    Let W be the list obtained by sorting V(T) topologically in postorder;
    while W is not empty do
      begin
        Remove the first element v from W;
        b ← | { w | flag(w) = TRUE ∧ w is a child of v } |;
        if color(v) = BLACK ∨ b = 1 then flag(v) ← TRUE
        elseif b > 1 then
            begin   flag(v) ← TRUE;   color(v) ← BLACK   end
      end;
    color(r) ← BLACK
  end
```

Proposition 2. *Procedure $SetColor(T_r, m)$ finds a set of black vertices which satisfies conditions A1, A2 and A3 in $O(n)$ time.*

Proof. Since it is easy to see that $SetColor(T_r, m)$ works in $O(n)$ time, we need only consider the correctness of $SetColor(T_r, m)$.

First, let us consider phase 1. Since at least $\lfloor \frac{m}{2} \rfloor - 1$ white vertices are necessary for generating one black vertex, $O(n/m)$ black vertices are generated. Moreover, condition A2 is satisfied after this phase.

Next, let us consider phase 2. To generate a new black vertex, at least two vertices which are already colored black are necessary. Any black vertex used for generating another black vertex can not be used for generating anymore. Thus, $O(n/m)$ black vertices are generated in this phase. Moreover, condition A3 is satisfied after this phase. □

The following proposition is used in the next section.

Proposition 3. *The number of paths such that the endpoints are black and the other vertices are white is $O(n/m)$.*

Proof. Let $P(T)$ be the set of such paths in the tree T where the vertices of T are colored by the procedure $SetColor(T_r, m)$. From condition A3, the following property holds: for any path $p \in P(T)$, one endpoint of p is a descendant of the other endpoint of p in T_r. Thus, for any path $p \in P(T)$, let $e(p)$ denote the endpoint which is a descendant of the other endpoint in T_r. Since T_r is a rooted tree, $e(p_1) \neq e(p_2)$ holds for any two paths p_1 and p_2 in $P(T)$ where $p_1 \neq p_2$. Therefore, the number of paths does not exceed the number of black vertices and the proposition follows. □

5 Body of the Algorithm

In this section, we describe the core part of the algorithm: the construction of $TBLs$. The $TBLs$ are constructed via two phases: $Pass1$ and $Pass2$ (see Figure 4).

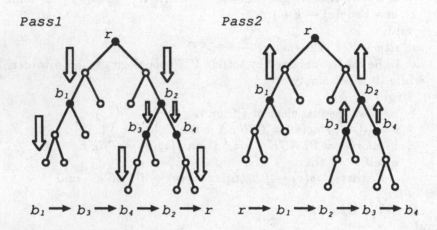

Fig. 4. The order of vertex traversal.

First, we describe $Pass1$. For a black vertex v in T_r, $\hat{V}_r(v)$ denotes the set $\{v\} \bigcup \{ w \mid w$ is a descendant of v and no black vertex except v and w is contained in the path between v and $w \}$. For a black vertex v in T_r, $\hat{T}_r(v)$ denotes the rooted subtree of $\hat{T}_r$ induced by $\hat{V}_r(v)$ (see Figure 5). Globally, the black vertices are visited in postorder. Locally, the depth first search is performed for each subtree $\hat{T}_r(v)$ with traversing the suffix tree. If a leaf of the suffix tree or a black vertex of T_r is encountered, $TBLs$ are updated. The details are described in the procedure $Pass1(T_r, s)$.

$Pass2$ is similar to $Pass1$ with the only difference being the order of vertex traversal. For a black vertex v in T_r, $\tilde{V}_r(v)$ denotes the set $\{v\} \bigcup \{ w \mid w$ is not a descendant of v and no black vertex except v and w is contained in the path between v and $w \}$. For a black vertex v in T_r, $\tilde{T}_r(v)$ denotes the rooted subtree of T such that it is induced by $\tilde{V}_r(v)$ and v is the root (see Figure 5). Globally, the black vertices are visited in preorder. Locally, the depth first search is performed for each subtree $\tilde{T}_r(v)$ with traversing the suffix tree. $Pass2(T, s)$ is obtained from $Pass1(T, s)$ by replacing part (a) with

Let W be the list obtained by sorting V_b topologically in preorder;

and replacing part (b) with

> **while** visiting the vertices of $\tilde{T}_r(v)$ in the depth first way
> with traversing SUF_s **do**   .

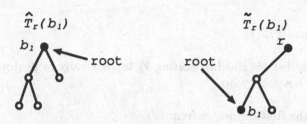

Fig. 5. Examples of $\hat{T}_r(v)$ and $\tilde{T}_r(v)$, using the same tree shown in Figure 4.

The following theorem summarizes the result.

Theorem 4. *Procedure $PatternMatch(T, s)$ tests whether or not there is a path of T that coincides with s in $O(n)$ time.*

Proof. Since it is easy to see that $PatternMatch(T, s)$ is correct, we need only consider the time complexity. Furthermore, it is sufficient to consider only the time for processing each vertex.

First, let us consider the white vertices. Each white vertex is visited at most once per each phase. Since the size of the alphabet is assumed to be bounded, at most constant number of edges of SUF_s are traversed per white vertex. If a leaf of SUF_s is reached, $TBLs$ must be updated. However, it takes constant time. Thus, the total time required for processing white vertices is $O(n)$.

Next, let us consider the black vertices. From Proposition 3, the black vertices are visited at most $O(n/m)$ in total. Per each visit, $O(m)$ time is sufficient for computing $PRSX(t, s)$ and updating a table. And per black vertex, it takes $O(m)$ time for testing condition C1 in $PatternMatch(T, s)$. Thus, the total time required for processing black vertices is $O(n)$. □

```
Procedure PatternMatch(T, s)
  begin
    Select an arbitrary vertex r of T as the root;
    SetColor(T_r, m);
    MakeTable(T_r, s);
    MakeTable(T_r, s^{-1});
    for all black vertices v do
      if condition C1 holds for v then return TRUE;
    return FALSE
  end
```

Procedure $MakeTable(T_r, s)$
  begin
    Make a suffix tree SUF_s;
    Initialize all the TBLs to be $FALSE$;
    $Pass1(T_r, s)$;
    $Pass2(T_r, s)$
  end

Procedure $Pass1(T_r, s)$
  begin
    Let W be the list obtained by sorting V_b topologically in postorder;   -(a)
    while W is not empty **do**
    begin
      Remove the first element v from W;
      while visiting the vertices of $\hat{T}_r(v)$ in the depth first way
      with traversing SUF_s **do**   -(b)
      begin
        Let w be a current vertex;
        Let $\langle v = w_1, w_2, w_3, \cdots, w = w_{m-i+1} \rangle$ be the pass from v to w;
        Let x be the current node of SUF_s;
        if x is the leaf corresponding to s_i **then** $UpdateTBL(v, s_i, w_2)$;
        if $color(w) = BLACK$ **then**
          begin
            Let t be the string: $label(w_1) \cdots label(w_{m-i})$;
            Compute $PRSX(t, s)$;
            for all j in $PRSX(t, s)$ **do**
              if $TBL(w, s_{rest(t,j)}) \neq FALSE \wedge TBL(w, s_{rest(t,j)}) \neq w_{m-i}$
              then $UpdateTBL(v, s_j, w_2)$
          end
      end
    end

6   Concluding Remarks

Although we assumed that the size of the alphabet is bounded by a constant, the algorithm can be directly applied to the case of unbounded alphabets. In such a case, only constructing and searching the suffix tree require additional $O(\log m)$ time. Thus, the time complexity increases to $O(n \log m)$.

The algorithm might also be modified for computing all the matchings where each matching is specified by the positions in the tree corresponding to the endpoints of the path. However, a set of all the matchings can not be computed in $O(n)$ time since, in the worst case, there are $O(n^2)$ matchings. For example, consider a string $s =$"abc" and a tree T such that $V(T) = \{v_1, \cdots, v_k, w_1, \cdots, w_k, u\}$, $E(T) = \{ \{v_i, u\}, \{w_i, u\} \mid 1 \le i \le k \}$, $(1 \le \forall i \le k)(label(v_i) = $ 'a'$)$, $(1 \le \forall i \le k)(label(w_i) = $ 'c'$)$ and $label(u) = $ 'b'. In this case, there are k^2 matchings. If labels of adjacent vertices of each vertex are different, there are at most $O(n)$ matchings. In this case, it would be possible to compute all the matchings in linear or near-linear time.

Although the algorithm in this paper does not seem to have theoretical or practical applications as yet, it is valuable for broadening the class of linear or near-linear time algorithms for pattern matching.

References

1. A. Amir, G. Benson, and M. Farach. "Alphabet independent two dimensional pattern matching". In *Proceedings of ACM Symposium on Theory of Computing*, pp. 59–68, 1992.

2. A. Apostolico, C. Iliopoulos, G. M. Landau, B. Schieber, and U. Vishkin. "Parallel construction of a suffix tree with applications". *Algorithmica*, Vol. 3, pp. 347–365, 1988.

3. R. Boyer and S. Moore. "A fast string searching algorithm". *Communications of the ACM*, Vol. 20, pp. 762–772, 1977.

4. M. J. Chung. "$O(n^{2.5})$ time algorithms for the subgraph homeomorphism on trees". *Journal of Algorithms*, Vol. 8, pp. 106–112, 1987.

5. M. Dubiner, Z. Galil, and E Magen. "Faster tree pattern matching". In *Proceedings of IEEE Symposium on Foundations of Computer Science*, pp. 145–150, 1990.

6. M. R. Garey and D. S. Johnson. *"Computers and Intractability: A Guide to the Theory of NP-completeness"*. Freeman, San Francisco, 1979.

7. D. E. Knuth, J. Morris, and V. Pratt. "Fast pattern matching in strings". *SIAM Journal on Computing*, Vol. 6, pp. 323–350, 1977.

8. S. R. Kosaraju. "Efficient tree pattern matching". In *Proceedings of IEEE Symposium on Foundations of Computer Science*, pp. 178–183, 1989.

9. G. M. Landau and U. Vishkin. "Fast parallel and serial approximate string matching". *Journal of Algorithms*, Vol. 10, pp. 157–169, 1989.

10. A. Lingas. "Subgraph isomorphism for connected graphs of bounded valence and bounded separator is in NC". In *Proceedings of the 1988 International Conference on Parallel Processing*, pp. 304–307, 1988.

11. E. M. McCreight. "A space-efficient suffix tree construction algorithm". *Journal of the ACM*, Vol. 23, pp. 262–272, 1976.

12. S. W. Reyner. "An analysis of a good algorithm for the subtree problems". *SIAM Journal on Computing*, Vol. 6, pp. 730–732, 1977.

13. R. E. Stobaugh. "Chemical substructure searching". *Journal of Chemical Information and Computer Sciences*, Vol. 25, pp. 271–275, 1985.

14. E. H. Sussenguth. "A graph-theoretic algorithm for matching chemical structures". *Journal of Chemical Documentation*, Vol. 5, pp. 36–43, 1965.

Tight Comparison Bounds for the String Prefix-Matching Problem

Dany Breslauer*
Centrum voor Wiskunde en Informatica,
P.O. Box 4079, 1009 AB Amsterdam, The Netherlands.

Livio Colussi†, Laura Toniolo‡
Universitá di Padova,
Dipartimento di Matematica Pura ed Applicata,
Via Belzoni 7, 35131 Padova, Italy.

Abstract

In the *string prefix-matching* problem one is interested in finding the longest prefix of a pattern string of length m that occurs starting at each position of a text string of length n. This is a natural generalization of the string matching problem where only occurrences of the whole pattern are sought. The Knuth-Morris-Pratt string matching algorithm can be easily adapted to solve the string prefix-matching problem without making additional comparisons.

In this paper we study the exact complexity of the string prefix-matching problem in the deterministic sequential comparison model. Our bounds do not account for comparisons made in a pattern preprocessing step. The following results are presented:

1. A family of linear-time string prefix-matching algorithms that make at most $\lfloor \frac{2m-1}{m} n \rfloor$ comparisons.

2. A tight lower bound of $\lfloor \frac{2m-1}{m} n \rfloor$ comparisons for any string prefix-matching algorithm.

We also consider the special case when the pattern and the text strings are the same string and all comparisons are accounted. This problem, which we call the *string self-prefix* problem, is similar to the failure function that is computed in

---

*Partially supported by the European Research Consortium for Informatics and Mathematics post-doctoral fellowship.

†Partially supported by "Progetto Finalizzato Sistemi Informatici e Calcolo Parallelo" of CNR under grant number 89.00026.69.

‡The work was done while this author was visiting at the Department of Computer Science in Columbia University.

the pattern preprocessing of the Knuth-Morris-Pratt string matching algorithm and used in several other comparison efficient algorithms. By using the lower bound for the string prefix-matching problem we are able to show:

3. A lower bound of $2m - \lceil 2\sqrt{m} \rceil$ comparisons for the self-prefix problem.

1 Introduction

In the *string prefix-matching* problem one is interested in finding the longest prefix of a pattern string $\mathcal{P}[1..m]$ that starts at each position of a text string $\mathcal{T}[1..n]$. More formally, the required output of the string prefix-matching problem is an integer array $\Pi[1..n]$ such that for each text position i, $\mathcal{T}[i..i+\Pi[i]-1] = \mathcal{P}[1..\Pi[i]]$ and if $\Pi[i] < m$ and $i+\Pi[i] \leq n$, then $\mathcal{T}[i + \Pi[i]] \neq \mathcal{P}[\Pi[i] + 1]$.

The string prefix-matching problem is a natural generalization of the standard string matching problem where only complete occurrences of the pattern are sought. The classical linear time string matching algorithm of Knuth, Morris and Pratt [9] can be easily adapted to solve the string prefix-matching problem in the same time bounds without making additional comparisons. We assume that the reader is familiar with this algorithm. (Since complete occurrences of the pattern cannot start at text positions larger than $n-m+1$, the string matching algorithm can stop before reaching the end of the text. The prefix-matching algorithm must continue until the end of the text and therefore, it may make at most m extra comparisons.)

In this paper we study the exact number of comparisons performed by algorithms that have access to the input strings by pairwise symbol comparisons that test for equality. This work was motivated by recent results on the exact comparison complexity of the string matching problem [4, 6, 7, 8, 10]: Colussi [6] optimized the Knuth-Morris-Pratt [9] string matching algorithm, which makes $2n - m$ comparisons, using program correctness proof techniques and presented an algorithm that makes $n + \frac{1}{2}(n - m)$ comparisons. His algorithm was later improved by Galil and Giancarlo [8] and further by Breslauer and Galil [4]. Recently, Cole and Hariharan [5] discovered an algorithm that makes only $n + \frac{c}{m}(n - m)$ comparisons, but requires an expensive pattern preprocessing. (All bounds for the string matching algorithms mentioned do not account for the comparisons made in a pattern preprocessing step. The pattern preprocessing step of Cole and Hariharan's algorithm takes $O(m^2)$ time, while the other algorithms use the Knuth-Morris-Pratt pattern preprocessing step that takes linear time.) Cole and Hariharan [5] also improved the lower bounds given by Galil and Giancarlo [7] and Zwick and Paterson [10]. There is still a small gap between the lower and upper bounds for string matching.

The string prefix-matching problem is obviously harder than the standard string matching problem since each text symbol must be either compared directly to the first symbol of the pattern or compared successfully to another symbol, while in the string matching problem some text symbols might not be compared at all, as shown by Boyer and Moore [2]. Interestingly, this "hardness" introduces more structure that makes the analysis of the string prefix-matching problem easier.

This paper presents matching lower and upper bounds for the string prefix matching problem. In particular we give:

1. A family of linear-time string prefix-matching algorithms that make at most $\lfloor \frac{2m-1}{m}n \rfloor$ comparisons. The pattern preprocessing step of these algorithms is almost identical to that of the string matching algorithm of Knuth, Morris and Pratt [9].

   This bound improves on the $2n-1$ comparisons made by the adapted string matching algorithm of Knuth, Morris and Pratt [9].

2. A tight lower bound of $\lfloor \frac{2m-1}{m}n \rfloor$ comparisons for any string prefix-matching algorithm.

These results show that although the string matching and the string prefix-matching problems are closely related, their exact comparison complexities are inherently different:

- When $m \to \infty$ and $n \gg m$ the comparison complexity of the string matching problem approaches n while the comparison complexity of the string prefix-matching problem approaches $2n$.

- The lower bound proofs of the two problems require different arguments: the pattern string that we use for the lower bound is 'ab^{m-1}' while the lower bounds for the string matching problem require patterns with more complex periodicity structures [5, 7, 10].

Finally, we consider the special case when the text and the pattern strings are the same string and all comparisons are accounted. This problem, which we call the *string self-prefix* problem, is similar to the *failure function* that is computed in the pattern preprocessing of the Knuth-Morris-Pratt [9] string matching algorithm using $2m - 4$ comparisons. (These are essentially different representations of the same information: one can be computed from the other in linear time without additional comparisons. Therefore, the lower bound applies also to the computation of the failure function.) The failure function is also used in several other string matching algorithms [4, 6, 8] and in the family of algorithms discusses in this paper. We prove:

3. A lower bound of $2m - \lceil 2\sqrt{m} \rceil$ comparisons for the self-prefix problem.

This paper is organized as follows. Section 2 describes the family of string prefix-matching algorithms and Section 3 gives the matching lower bound. Section 4 uses this lower bound to prove a lower bound on the self-prefix problem.

2 Upper Bounds

In this section we present a family of string prefix-matching algorithms that make at most $\lfloor \frac{2m-1}{m}n \rfloor$ comparisons. The discussion below is in the comparison model where we count only comparisons and all other computation is free. We assume that the algorithms have obtained complete information about the pattern in an unaccounted pattern preprocessing step which may compare even all $\binom{m}{2}$ pairs of pattern symbols. We further assume that the algorithms do not make any comparisons that are implied by the answers to previous comparisons. The algorithms presented can be implemented efficiently in the standard random access machine model [1].

Definition 2.1 *We say that a prefix-matching algorithm is* on-line *if before comparing the text symbol $T[\zeta]$ it has determined if the pattern prefixes that start at text positions l terminate before text position ζ for all text positions l, such that $l < \zeta$.*

Let $\Psi^\zeta = \{\psi_i^\zeta | \zeta - m < \psi_1^\zeta < \psi_2^\zeta < \cdots < \psi_{k_\zeta}^\zeta = \zeta\}$ be the set of all text positions for which $\Pi[\psi_i^\zeta]$ can not be determined without examining $T[\zeta]$. That is, $T[\psi_i^\zeta..\zeta - 1] = \mathcal{P}[1..\zeta - \psi_i^\zeta]$ and $T[\zeta]$ must be compared to check whether $\Pi[\psi_i^\zeta] = \zeta - \psi_i^\zeta$ or $\Pi[\psi_i^\zeta] > \zeta - \psi_i^\zeta$. In this terminology, an on-line prefix-matching algorithm must determine whether $T[\zeta] = \mathcal{P}[\zeta - \psi_i^\zeta + 1]$, for all $\psi_i^\zeta \in \Psi^\zeta$, before examining any text position larger than ζ. Note that $\Psi^{\zeta+1} \subseteq \Psi^\zeta \cup \{\zeta + 1\}$.

Comparison efficient on-line prefix-matching algorithms are somewhat restricted with the choices of comparisons they can make. It is easy to see that they gain no advantage by comparing pairs of text symbols. Furthermore, all comparisons at text position ζ must be between $T[\zeta]$ and some $\mathcal{P}[\zeta - \psi_i^\zeta + 1]$ or otherwise can be answered by an adversary as unequal without giving the algorithm any useful information, provided that the alphabet is large enough. In the rest of this section we consider on-line algorithms that compare $T[\zeta]$ to $\mathcal{P}[\zeta - \psi_i^\zeta + 1]$, for some $\psi_i^\zeta \in \Psi^\zeta$. The only difference between these algorithms is the order in which the pattern symbols $\mathcal{P}[\zeta - \psi_i^\zeta + 1]$ are compared to $T[\zeta]$. These algorithms continue comparing $T[\zeta]$ until $T[\zeta] = \mathcal{P}[\zeta - \psi_i^\zeta + 1]$ for some ψ_i^ζ, or until $T[\zeta] \neq \mathcal{P}[\zeta - \psi_i^\zeta + 1]$ for all ψ_i^ζ, and only then move to the next text position. Note that by the assumption that the algorithms do not to make comparisons which are implied by answers to previous comparisons, and since the algorithms have complete information about the pattern, not all the symbols $\mathcal{P}[\zeta - \psi_i^\zeta + 1]$ have to be compared:

1. If $\mathcal{P}[\zeta - \psi_i^\zeta + 1] = T[\zeta]$, then $\mathcal{P}[\zeta - \psi_\delta^\zeta + 1] = T[\zeta]$, for some $\psi_\delta^\zeta \in \Psi^\zeta$, if and only if $\mathcal{P}[\zeta - \psi_i^\zeta + 1] = \mathcal{P}[\zeta - \psi_\delta^\zeta + 1]$. In this case a comparison model algorithm "knows" which symbol is at text position ζ and it moves to the next text position.

2. If $\mathcal{P}[\zeta - \psi_i^\zeta + 1] \neq T[\zeta]$, then $\mathcal{P}[\zeta - \psi_\delta^\zeta + 1] \neq T[\zeta]$, for all $\psi_\delta^\zeta \in \Psi^\zeta$, such that $\mathcal{P}[\zeta - \psi_i^\zeta + 1] = \mathcal{P}[\zeta - \psi_\delta^\zeta + 1]$. Ideally, a comparison model algorithm should not compare the text symbol $T[\zeta]$ to $\mathcal{P}[\zeta - \psi_\delta^\zeta + 1]$. However, this is not essential for the proofs in this paper as long as the algorithms do not compare some $\mathcal{P}[\zeta - \psi_i^\zeta + 1]$ more than once.

This leads to the definition of a family $\mathcal{F}$ of all on-line comparison model string prefix-matching algorithms that may compare $T[\zeta]$ only to some $\mathcal{P}[\zeta - \psi_i^\zeta + 1]$. The data structures that are used by Breslauer and Galil [4] to implement a family of similar string matching algorithms can be used to implement all algorithms $\mathcal{A} \in \mathcal{F}$ in linear time with a pattern preprocessing step that relies on the Knuth-Morris-Pratt failure function.

Theorem 2.2 *Let $\mathcal{A} \in \mathcal{F}$. Then, except possibly the rule which chooses the order according to which the $\mathcal{P}[\zeta - \psi_i^\zeta + 1]$'s are compared to $T[\zeta]$, $\mathcal{A}$ can be implemented in the standard model in linear time with the Knuth-Morris-Pratt linear time pattern preprocessing step that makes at most $2m - 4$ comparisons.*

The algorithms in the family $\mathcal{F}$ are comparison efficient as we show next:

Lemma 2.3 *Let $\mathcal{A} \in \mathcal{F}$. Then $\mathcal{A}$ makes at most $2n - 1$ comparisons.*

Proof: It is obvious that $\mathcal{A}$ does not need to make more than n comparisons which result in equal answers. In every comparison which results in unequal answer $\mathcal{A}$ determines that at least one prefix of the pattern which starts at some text position ψ_ϵ^ζ terminates at text position ζ. Therefore, $\mathcal{A}$ does not make more than n comparisons which result in unequal answers. However, if all pattern prefixes that start at text positions in Ψ^ζ terminate at text position ζ, then $\mathcal{A}$ moves to the next text position without a comparison that is answered as equal.

Consider the last text position $\zeta = n$. It is clear that if all comparisons at this text position result in unequal answers, then $\mathcal{A}$ got at most $n - 1$ equal answers. On the other hand, if a comparison was answered as equal, then there is at least one pattern prefix which starts at some text position ψ_ϵ^ζ and was not terminated by an inequality answer and, thus, $\mathcal{A}$ got at most $n - 1$ unequal answers. Therefore, $\mathcal{A}$ makes at most $2n - 1$ comparisons. $\square$

The adapted Knuth-Morris-Pratt [9] prefix-matching algorithm is in the family $\mathcal{F}$. There are cases in which it would actually make $2n - 1$ comparisons; e.g. $\mathcal{P}[1..2] = \text{'}ab\text{'}$ and $\mathcal{T}[1..n] = \text{'}a^n\text{'}$. Note that this algorithm compares $\mathcal{T}[\zeta]$ to $\mathcal{P}[\zeta - \psi_\epsilon^\zeta + 1]$ in an increasing order of ψ_ϵ^ζ. This order is the worst possible order as we show in the next theorem.

Define a family of algorithms $\hat{\mathcal{F}}$ of all $\mathcal{A} \in \mathcal{F}$ that compare $\mathcal{P}[\zeta - \psi_1^\zeta + 1]$ only last. Namely, if an algorithm $\mathcal{A} \in \hat{\mathcal{F}}$, then $\mathcal{A}$ compares $\mathcal{T}[\zeta]$ to $\mathcal{P}[\zeta - \psi_1^\zeta + 1]$ only if an unequal answer implies that all pattern prefixes that start at text positions in Ψ^ζ terminate at text position ζ. Note that if $\mathcal{P}[\zeta - \psi_1^\zeta + 1] = \mathcal{P}[\zeta - \psi_\epsilon^\zeta + 1]$, for $\psi_1^\zeta \neq \psi_\epsilon^\zeta$, then $\mathcal{A}$ may compare this pattern symbol at any time.

Theorem 2.4 *Let $\mathcal{A} \in \hat{\mathcal{F}}$. Then $\mathcal{A}$ makes at most $\lfloor \frac{2m-1}{m} n \rfloor$ comparisons.*

Proof: As in Lemma 2.3, every comparison between $\mathcal{T}[\zeta]$ to $\mathcal{P}[\zeta - \psi_\epsilon^\zeta + 1]$ which results in an unequal answer determines that the pattern prefix which starts at text position ψ_ϵ^ζ terminates at text position ζ. We charge such a comparison to text position ψ_ϵ^ζ and charge comparisons that result in equal answers to the text position compared. Using this charging scheme it is obvious that each text position can be charged with at most two comparisons and that comparisons to $\mathcal{T}[\zeta]$ cannot be charged to any text position that is smaller than ψ_1^ζ.

When $\mathcal{A}$ reaches text position ζ, the number of comparisons that are charged to the text positions $\psi_1^\zeta, \cdots, \zeta - 1$ is at most $2(\zeta - \psi_1^\zeta) - (|\Psi^\zeta| - 1)$. This is so since each of these $\zeta - \psi_1^\zeta$ text positions has a comparison that resulted in equal answer charged to it, but at least $|\Psi^\zeta| - 1$ of the text positions do not have a comparison that resulted in unequal answer charged to them.

We prove by induction that the number of comparisons charged to text positions smaller than ψ_1^ζ is at most $\lfloor \frac{2m-1}{m}(\psi_1^\zeta - 1) \rfloor$. This is obviously true at the beginning when $\zeta = 1$. The only concern is when $\mathcal{A}$ advances from ζ to $\zeta + 1$ and $\psi_1^\zeta < \psi_1^{\zeta+1}$.

Let $l = \psi_1^{\zeta+1} - \psi_1^\zeta$. The number of comparisons that were charged to the text positions $\psi_1^\zeta, \cdots, \psi_1^{\zeta+1} - 1$ is at most $2l - 1$ since either at most l text positions were charged with comparisons that resulted in equal answers and ψ_1^ζ was not charged with an unequal answer, or ψ_1^ζ was charged with an unequal answer but then $\psi_1^{\zeta+1} = \zeta + 1$ and text position ζ was not charged with an equal answer. But $l \leq m$ and by simple arithmetic,

$$\lfloor \frac{2m-1}{m}(\psi_1^\zeta - 1) \rfloor + (2l - 1) \leq \lfloor \frac{2m-1}{m}(\psi_1^{\zeta+1} - 1) \rfloor.$$

When $\mathcal{A}$ reaches text position $\zeta = n + 1$, the number of comparisons satisfies,

$$\lfloor \frac{2m-1}{m}(\psi_1^\zeta - 1) \rfloor + 2(\zeta - \psi_1^\zeta) - (|\Psi^\zeta| - 1) \leq \lfloor \frac{2m-1}{m}n \rfloor. \quad \square$$

3 Lower Bounds

In this section we show a lower bound on the number of comparisons required by any string prefix-matching algorithm which may have an unaccounted pattern preprocessing step. We describe an adversary that can force such an algorithm to make at least $\lfloor \frac{2m-1}{m}n \rfloor$ comparisons.

Theorem 3.1 *Any prefix-matching algorithm must make at least $\lfloor \frac{2m-1}{m}n \rfloor$ comparisons.*

Proof: Fix the pattern to $\mathcal{P}[1..m] = `ab^{m-1}`$ and assume that the text alphabet has at least three symbols. We show that an adversary can answer comparisons made by any prefix-matching algorithm in a way that if the algorithm claims to have computed $\Pi[1..n]$ in less than $\lfloor \frac{2m-1}{m}n \rfloor$ comparisons, then it can be fooled.

Consider first algorithms that cannot compare pairs of text symbols. The adversary will maintain each text symbol in one of three states: *unknown*, *potential 'a' or 'b'*, and *fixed 'a' or 'b'*.

Initially the adversary sets all text symbols at positions l, such that $l \equiv 1 \bmod m$, to be potential '*a*'s and all other text symbols to be unknown. A comparison between an unknown text symbol to '*a*' or to '*b*' is answered as unequal and the text symbol is set to be a potential '*b*' or '*a*', respectively. A potential '*a*' or '*b*' is revealed to the algorithm at the cost of one comparison after which it becomes fixed.

If an algorithm claims it has computed $\Pi[1..n]$ before all text symbols are fixed, the adversary has the freedom of setting one of the unknown or potential symbols to an alphabet symbol other than '*a*' and '*b*'. Let u be a text position that is not fixed and assume that all other text symbols become fixed. If $T[u]$ is a potential '*b*', then there exists v such that $u - m < v < u$ and $T[v..u - 1] = `ab^{u-v-1}`$, and the adversary can alter $\Pi[v]$ by fixing $T[u]$ to '*b*' or '*c*'. Similarly, the adversary can alter $\Pi[u]$ if $T[u]$ is unknown or a potential '*a*'. Thus, any algorithm must make two comparisons at each text position except at the text positions that are set initially to be potential '*a*'s, where it has to make only one comparison. The total number of comparisons is at least $\lfloor \frac{2m-1}{m}n \rfloor$.

When pairwise comparisons of text symbols are permitted, the lower bound arguments are slightly more complicated. To keep track of the comparisons the adversary maintains a graph with $n + 2$ vertices that correspond to the n text symbols and the pattern symbols

'a' and 'b'. The edges of the graph correspond to comparisons and are labeled with their outcome ("equal" or "unequal").

The adversary maintains a two-level representation of the edges. This representation satisfies the following invariants:

1. A subgraph that contains the edges that are labeled "unequal" and all vertices.

   We refer to the connected components in this subgraph as *components*. The adversary will maintain the property that components are bipartite graphs.

2. A subgraph that contains the edges that are labeled "equal" and all vertices.

   We refer to the connected components in this subgraph as *super-vertices*. By transitivity, all vertices in a super-vertex correspond to equal symbols. The adversary will maintain the property that vertices which are in the same super-vertex are always in the same side of a single component.

Initially, the graph has $1 + \lceil \frac{n}{m} \rceil$ edges: between the pattern symbol 'a' and the pattern symbol 'b' and between the pattern symbol 'b' and every text position l, such that $l = 1 \bmod m$. These edges are labeled "unequal"; the invariants are clearly satisfied. The adversary answers comparisons as follows:

- A comparison between symbols which correspond to vertices that belong to different components is answered as unequal.

  The two components are merged into a single component which is still bipartite.

- A comparison between symbols which correspond to vertices that belong to the same component is answered as equal if and only if the two vertices are on the same side of the component.

  This may cause two super-vertices to be merged into one. Note that comparisons between vertices that belong to the same component but are on different sides and comparisons between two vertices in the same super-vertex do not contribute anything to the component or super-vertex structure and are practically answered for free.

The invariants are obviously maintained after each comparison is answered. Note that vertices which are in the same super-vertex as one of the pattern symbols correspond to fixed symbols; vertices which are in the same component as the pattern symbols correspond to potential symbols and vertices which are in other components correspond to unknown symbols.

A prefix-matching algorithm can terminate correctly when there is only one component and two super-vertices. Since every connected component with l vertices must have at least $l - 1$ edges, there are at least $n + 1$ edges labeled "unequal" and at least n edges labeled "equal" at termination. Thus, the total number of comparisons is at least $2n + 1 - (1 + \lceil \frac{n}{m} \rceil) = \lfloor \frac{2m-1}{m} n \rfloor$. $\quad \square$

4 Lower Bounds for the Self-Prefix Problem

In this section we consider the special case where the pattern and the text strings are the same string and all comparisons are accounted. This problem is solved in the preprocessing step of the Knuth-Morris-Pratt [9] string matching algorithm in linear time and $2m - 4$ comparisons.

Theorem 4.1 *Fix a positive integer constant k. Then, any self-prefix algorithm that is given an input string of length m, such that $m \geq k$, must make at least $\lfloor \frac{2k-1}{k} m \rfloor - k$ comparisons.*

Proof: The adversary fixes the first k symbols of the string to 'ab^{k-1}' and reveals them to the algorithm for $k - 1$ comparisons. By Theorem 3.1 the algorithm must make at least $\lfloor \frac{2k-1}{k}(m - k) \rfloor$ more comparisons. But, $\lfloor \frac{2k-1}{k}(m - k) \rfloor + k - 1 = \lfloor \frac{2k-1}{k} m \rfloor - k$. $\square$

If the length of the input string is known to the adversary in advance, it can maximize the lower bound as the next corollary shows. In the on-line case, where the string is given a symbol at a time and its length not known in advance, there seems to be a tradeoff between maximizing the number of comparisons in the short term and in the long term.

Corollary 4.2 *The lower bound in Theorem 4.1 has a maximal value of $2m - \lceil 2\sqrt{m} \rceil$.*

Proof: It is easy to verify that the maximum is achieved for $k = \lfloor \sqrt{m} \rfloor$ and also for $k = \lceil \sqrt{m} \rceil$. $\square$

5 Concluding Remarks

The lower and upper bounds presented in this paper are shown to be tight only for the pattern string 'ab^{m-1}'. Recently, we have been able to obtain bounds that depend on the given pattern string [3].

6 Acknowledgments

We thank Matt Franklin, Raffaele "The Great Raf" Giancarlo and Moti Yung for comments on early versions of this paper.

References

[1] A. V. Aho, J. E. Hopcroft, and J. D. Ullman. *The Design and Analysis of Computer Algorithms.* Addison-Wesley, Reading, MA., 1974.

[2] R. S. Boyer and J. S. Moore. A fast string searching algorithm. *Comm. of the ACM*, 20:762–772, 1977.

[3] D. Breslauer, L. Colussi, and L. Toniolo. On the Exact Complexity of the String Prefix-Matching Problem. In preparation, 1993.

[4] D. Breslauer and Z. Galil. Efficient Comparison Based String Matching. *J. Complexity*, 1993. To appear.

[5] R. Cole and R. Hariharan. Tighter Bounds on The Exact Complexity of String Matching. In *Proc. 33rd IEEE Symp. on Foundations of Computer Science*, pages 600–609, 1992.

[6] L. Colussi. Correctness and efficiency of string matching algorithms. *Inform. and Control*, 95:225–251, 1991.

[7] Z. Galil and R. Giancarlo. On the exact complexity of string matching: lower bounds. *SIAM J. Comput.*, 20(6):1008–1020, 1991.

[8] Z. Galil and R. Giancarlo. The exact complexity of string matching: upper bounds. *SIAM J. Comput.*, 21(3):407–437, 1992.

[9] D. E. Knuth, J. H. Morris, and V. R. Pratt. Fast pattern matching in strings. *SIAM J. Comput.*, 6:322–350, 1977.

[10] U. Zwick and M. S. Paterson. Lower bounds for string matching in the sequential comparison model. Manuscript, 1991.

3-D Docking of Protein Molecules

Daniel Fischer[1,2], Raquel Norel[1], Ruth Nussinov[2,3]*, Haim J. Wolfson[1]**

[1] Computer Science Department, Raymond and Beverly Sackler Faculty of Exact Sciences, Tel Aviv University
[2] Sackler Inst. of Molecular Medicine, Faculty of Medicine, Tel Aviv University
[3] Lab of Math. Biology, PRI - Dynacor, NCI-FCRF, NIH

Abstract. We present geometric algorithms which tackle the docking problem in Molecular Biology. This problem is a central research topic both for synthetic drug design and for biomolecular recognition and interaction of proteins in general. Our algorithms have been implemented and experimented on several 'real world' biological examples. The preliminary experimental results show an order of magnitude improvement in the actual run-time of the algorithms in comparison to previously published techniques. The matching part of our algorithm is based on the *Geometric Hashing* technique, originally developed for Computer Vision applications. The algorithmic similarity between the problems emphasizes the importance of interdisciplinary research in Computational Molecular Biology. Future research directions are outlined as well.

1 Introduction

The problem of receptor-ligand recognition and binding is encountered in a very large number of biological processes. It is a prerequisite in cell-cell recognition, in enzyme catalysis (and inhibition) and in regulation of development and growth. The type of molecules involved is very diverse as well: proteins, nucleic acids (DNA and RNA), carbohydrates and lipids. Whereas the receptor molecules are usually fairly large (often proteins), the ligands can be large (e.g. proteins) or small, such as cofactors, small chemicals or drugs. Regardless of the type of the ligand molecules which are involved, the principles of recognition and binding between the receptor and the ligand are similar: geometry and chemistry.

Since proteins interact at their surfaces, the geometric aspect of the protein docking problem can be modeled as a 3-D motion invariant surface matching task, namely, *given two 3-D compact bodies (proteins), determine the rotation and translation of one of them relative to the other so that the there is a large fit between the respective surfaces, and no penetration of one body into the other*. It can be viewed as an attempt to fit two 3-D puzzle pieces to achieve a good contact. One should note, however, that the fit cannot be perfect as in toy puzzle fitting, and a

* Work on this paper was supported by grant No. 91-00219 from the US-Israel Binational Science Foundation (BSF), Jerusalem, Israel.

** Work on this paper was supported by grant No. 89-00481 from the US-Israel Binational Science Foundation (BSF), Jerusalem, Israel and by a grant from the Basic Research Foundation of Tel Aviv University.

considerable tolerance should be allowed between the corresponding surface points. The respective *docking sites* on both protein surfaces are *a-priori* unknown, and the detection of a partial matching area is required. Also, we do not necessarily look for the solution which maximizes the contact area, since the optimal solution should satisfy both geometric and chemical criteria. The resulting solutions satisfying the geometric criterion should be later evaluated chemically, and the energies of the docked complexes should be minimized to find the best fit. Thus, we are interested in all the solutions which have a 'large enough' fit.

In this paper we treat the first, geometrical, aspect. We assume that both the receptor and the ligand can be modeled as rigid objects. [4] Several key issues in protein docking are

1. Protein surface shape representation.
2. Docking site prediction.
3. Efficient partial surface matching algorithms.

The protein docking problem is usually treated in two separate cases: protein-protein recognition, and protein-ligand docking. In the first case both proteins to be docked are usually large (about thousand atoms). In the second case the receptor is a large protein molecule, whereas the ligand is a relatively small molecule of few hundred atoms . Although from a geometrical standpoint both problems are similar, since the complexity of the matching algorithms is input dependent, techniques which are appropriate for protein-ligand docking will usually not be practical for protein-protein docking. A sparse shape representation, which still preserves enough informative features for docking is required in this case.

We present in detail an algorithm for the ligand-protein docking problem and also sketch the main features of our new protein-protein docking algorithm, which is currently being tested. We present experimental results of both algorithms. Ligand - protein docking is a key problem in rational drug design, which is, potentially, one of the most useful industrial applications of Computational Molecular Biology. Traditionally, new drugs have been discovered by screening a large number of synthetic and natural compounds, and by examining their activity. This procedure is extremely slow, is of an inherently random nature, and also suffers from many limitations, such as the difference between the action of candidate drugs on animals (to whom it is administered) and on humans (see [29]). Thus, computer aided drug design has the potential to be a most powerful and rapid tool.

A major difficulty in the development of a rational drug design scheme is having a simple and fast procedure for docking potential drug candidates to a target molecule. The targets can be enzymes, cellular receptors or nucleic acids ([29]). For example, if a disease is a result of the overproduction of a certain compound, then some of the enzymes involved in the production of this compound can be blocked by docking a drug molecule to it (see an example in Section 4). A reasonable docking scheme should employ automated methods for database search.

This problem is closely related to assembly and object recognition problems in robotics and computer vision. There one is frequently faced with questions such as

---

[4] Currently we are investigating new techniques which allow restricted flexibility of the molecules.

the automatic assembly of fitting parts by a robot (a challenging application is an assembly of a jigsaw puzzle ([3])), or the detection of a known part in a cluttered scene by fitting the surfaces of the known parts to the surface of the observed scene (for surveys see [2, 4]). Only a partial match can be required here, since the parts may partially occlude each other. The close analogy between the types of problems addressed, brought about this interdisciplinary research endeavour, developing and adapting the Geometric Hashing techniques borrowed from Computer vision discipline [21, 23, 22, 24] and applying them to central problems in molecular biology [27, 11, 12]. Here, we further develop this technique, adapt and apply it to the problem of receptor-ligand recognition and docking. The application of these computer-vision tools brought about a reduction in the complexity of the proposed algorithm compared to previously reported results.

Kuhl et al. ([18]) formulate the docking problem as a maximum clique search problem in the, so called, docking graph. The vertices of the docking graph are all pairs of points from the ligand and the receptor respectively. An edge exists between two vertices, if the distance between the ligand points does not differ more than a predefined threshold from the distance between the corresponding receptor points. A maximum clique represents a maximum matching set, where all the inter-atomic distances are consistent. Although maximum clique search is NP-complete, they claim practical complexity of $(nm)^{2.8}$ (n and m are the numbers of atoms on the receptor and the ligand respectively) on randomly generated graphs of sizes from 16 to 1024 (note that the size of the graph is already quadratic in the number of atoms involved). No real experimental results are reported.

In real applications the first step is to model the molecular surface. Most of the methods use Connolly's MS algorithm ([6, 5]), which is briefly discussed in section 2.

In a seminal work, Kuntz et al. ([19]) suggested to focus the search of a docking site in the invaginations of the receptor surface. This follows the assumption that active sites of receptors are usually located in such invaginations (see [26]). They describe a receptor using spheres which are locally complementary to invaginations and protrusions of the molecular surface (see the SPHGEN procedure described in [19] or section 2 for details). Connected clusters of these spheres focus on local invaginations. The centers of the spheres may be thought as pseudo-atoms. A docking graph of these pseudo-atoms is considered as described above. 4-cliques are searched in this graph, using heuristic tree search techniques. For each such 4-clique a corresponding rotation and translation is computed, the ligand is mapped onto the receptor and overlapping solutions are discarded. Our protein-ligand docking algorithm follows the same basic steps, however, its matching part is based on *Geometric Hashing* rather than on 4-clique search. Our preliminary experimental results (section 4) show an order of magnitude improvement in the actual run-time of the algorithm in comparison to the DOCK2 program of [31].

Other geometrical methods (e.g. [15]) have made a brute-force search over all the (quantized) values of 3-D rotations, while employing some heuristics to detect the translation parameters. A recent paper ([17]) utilizes the Fast Fourier Transform for translation detection.

A promising docking approach was suggested by Connolly ([7]). In order to reduce the search he looks for special 'interest points' on the surface, which are local knobs an holes. A shape function is suggested whose local maxima and minima model these knobs and holes. The docking procedure is performed for four-tuples of knobs and holes with complementary shape. With this method Connolly succeeded in docking the α-subunit and β-subunit of hemoglobin, but failed in docking the trypsin-trypsin inhibitor complex, which is characterized by a relatively flat contact surface. Our protein-protein docking method uses a similar sparse 'interest point' detection step, however instead of matching four-tuples, we efficiently match pairs of interest points together with their associated normals. This improves both the complexity of the method, and also allows matching in cases where one cannot find matching knob/hole four-tuples, such as the trypsin-trypsin inhibitor complex, which has been succesfully matched by our algorithm.

This paper is organized as follows. Sections 2 and 3 give an outline of the algorithms that we have applied, Section 4 discusses few representative experimental results. Finally, in section 5 we summarize our results and discuss future research directions.

2   The Ligand-Protein Docking Algorithm

Since the docking is a complex problem, matching being just one part of it, we describe in detail the main steps of our algorithm.

(i) Construct the molecular surface, using Connolly's (MS) surface representation procedure [6, 5].

(ii) Use Kuntz's clustered spheres ([19]) approach (SPHGEN) to represent the negative image of the receptor (and, in most cases, as described below, the positive image of the ligand) to detect candidate invaginations and protrusions.

(iii) Apply *Geometric Hashing* (GH) ([12]) to get candidate receptor/ligand seed matches.

(iv) Filter the potentially geometrically acceptable solutions, by checking for overlaps of the ligand and receptor atoms. This is done using a 3-D grid.

An additional *focusing* step is applied after step (ii) if the number of clustered spheres is larger than a specified size. It follows the practically tested assumption, that atoms of a docked substructure are in spatial proximity. The focusing step divides the cluster into several overlapping regions (we have divided it into nine regions, which approximately fit the eight octants and a central region). Matching structures are sought only within these regions. This heuristic restricts the search, while the overlap between the regions prevents missing 'correct' matches on the borders between the regions.

Now, we describe our algorithm in more detail, following the steps outlined above.

(i) A major problem in docking is the representation of the molecular surface. Connolly's method [6, 5], which is based on the Richard's definition [30], is often used. A 'water' sphere (with 1.4Å diameter) is rolled over both the receptor and the ligand van der Waals surfaces. The molecular surface is defined by the points where the probe sphere 'touches' these surfaces. In practice, the surface is described by dots (sampled on the contact surface) with their associated surface normals.

(ii) Next we employ Kuntz's [19] clustered spheres approach. The surface dots generated by Connolly's algorithm, along with their associated normals, are used to generate the spheres. The spheres do not intersect the surface, being located outside the receptor surface and inside the ligand. Initially a sphere is generated for every two dots with its center along one of the normals. Subsequently, the number of spheres is reduced, and only about one sphere per surface atom is kept. Spheres intersecting each other constitute a cluster. The largest clusters are the ones of interest. The clustered spheres accomplish three goals. First, they reduce the number of surface points produced in the dot description. Second, they focus on the concave/convex regions in the receptor/ligand, that is, on the local invaginations/protrusions (see also [26]). These are likely to constitute the binding sites of the molecules. Third, constructing the sphere clusters outside the receptor surface has the effect of bringing the centers of the spheres closer than their corresponding receptor atoms. This representation enables comparisons of the distances between the receptor spheres and the ligand spheres (or atoms) found inside the molecular surface. Nevertheless, the clustered sphere representation also has some drawbacks. First, if the binding of the two molecules occurs on a wide, flat, surface, it might not be described by an appreciable cluster. Second, the construction of the spheres depends critically on the choice of the surface dots, and their associated normals. Even for the two very similar alcohol dehydrogenase chains (PDB code: 6ADH), with an RMS of 1.23Å and with the same number of atoms, the sizes of the clusters generated differ by about a factor of two.

(iii) The clustered spheres surface representation is used as an input to the geometric hashing algorithm ([12]), which has been originally developed for Computer Vision applications ([23]). Here, we have further developed this technique, and adapted it to the problem of matching two complementary surfaces. We give a short reminder of the technique. For details the reader is referred to [12].

Our surface comparison problem can be stated as follows: *Given the 3-D coordinates of the atoms (or clustered sphere centers) of the receptor and ligand surfaces, find a rigid transformation (rotation and translation) in space, so that a 'sufficient' number of atoms of one molecule matches the atoms of the other molecule.* The matching is done in two major steps which we shall nickname 'detection of seed matches', and 'extension of the match'.

Detection of the seed matches is the core of the algorithm. It allows simultaneous matching of several ligand surfaces to a receptor surface (or several receptor surfaces to a ligand). A seed match is represented by a list of matching pairs of atoms (from one ligand and the receptor) and by a 3-D rigid transformation (rotation and translation) which superimposes these pairs. The number of the matching pairs should be above a threshold (minimal score) which is either a static or dynamic parameter of the algorithm.

Assume that we match several ligands to a receptor region (restricted by the focusing heuristic). In a preprocessing stage which is done off-line and only once, the ligand interatomic distance information is arranged in a hash-table as follows. For each ligand in the database, all possible pairs of atoms are chosen. The line segment defined by them is called a basis. For each additional atom on the ligand surface its distance to the two basis endpoints is computed. These 3 distances serve as an address to the hash table where the information on the ligand and the underlying basis is stored as a pair (ligand,basis). Each hash-table entry contains information regarding all triangles (from all ligands) having the same 3 distances. The basis length and the lengths of the triangle sides are restricted between minima and maxima thresholds, which are set by the user. In the actual recognition stage a similar procedure is done for the receptor surface. First, a pair of points (clustered sphere centers) satisfying the distance constraints is chosen as a basis. Then, triangles are built by adding third points. For each receptor triangle, the lengths of the triangle sides are taken as an address to the hash-table, where all matching ligand triangles are stored. Votes are accumulated for each pair (ligand, basis) that appeared in the appropriate hash-table entry. If a particular pair scored relatively high, then we found a possible match between the corresponding ligand/receptor points. The list of these points is called a 'seed match' and it is built by retrieving the pairs of third points of the triangles that voted for this particular (ligand, basis) match. The inter-distances between all pairs of points in the list are verified. Pairs of points that deviate from a specified threshold are removed from the list.

In order to allow for 'inexact' matches, when the hash-table is accessed for a receptor triangle, several entries of the hash-table are accessed according to error thresholds.

The complexity of the first step is of the order of n^3, where n is the number of 'interest atoms/spheres' of the receptor[5]. One should note that if the hash-table bins are not heavily occupied, the simultaneous processing of several ligands does not increase linearly the recognition time.

The seed matches obtained are extended and refined using the following iterative procedure. Using a least squares technique [14] the best rotation and translation is computed to accommodate the matching pairs of each seed match. Then, the ligand is transformed accordingly and superimposed on the receptor. Pairs of atoms that lie 'close enough' after the transformation are candidate additional matching pairs ('close enough' is a parameter usually set at 2.5$\mathring{A}$). If there are several candidate ligand atoms within a small neighborhood of a receptor atom, the closest one is selected. The new matching pairs are added to the seed match, a new optimal (in the least squares sense) transformation is computed. Then, the ligand is transformed again and if there is a pair for which the distance between its two matched points is above a specified threshold, the pair is removed from the match.

This procedure is done in few iterations, till no significant improvement is obtained. Typically, about 2 iterations are enough due to the good quality of the initial seed matches.

(iv) In the last step of the algorithm, we prune those candidate matches which result in the penetration of the ligand into the receptor. A seed match implies existence of complementarity between receptor-ligand surface patches. Other regions of the two molecules can, however, overlap. The following scheme has been adopted in order to check for overlaps. The atomic coordinates of the receptor

---

[5] Assuming $O(1)$ processing for each hash-table access. Also, the distance restrictions between the triangle vertices contribute to the reduction of the practical run time

are mapped onto a 3-D orthogonal grid. All grid cubes containing receptor atoms are defined as interior cubes. Using the same procedure, the dots generated by the MS method (with a water sphere probe size 1.4Å, and density $1/Å^2$) are also mapped onto the grid. Cubes containing either only surface dots or both receptor atoms and molecular surface dots are defined as surface cubes. We next consider the ligand. All ligand atoms are transformed by the rotation and translation as described above, and mapped onto the same grid. If a ligand atom falls within an interior cube, the corresponding seed match is discarded. The remaining potential solutions are further weeded according to two criteria - the number of matching pairs and their RMS. The thresholds are set above the average size and below the average RMS of all the candidate seed matches. The remaining candidate matchings are the output of the program.

The method described here is a geometrical one. It searches for optimal fit of patches of surfaces between the receptor and the ligand, discarding potential solutions in which there are overlaps between ligand and receptor atoms. Following these procedures, we are still often left with several geometrically acceptable docked solutions. These should be the input to routines examining the chemical interactions between the receptor - ligand atoms at the interface. Minimizing the energies of the geometrically optimally docked structures, is clearly an essential next step which will enable discrimination between the favorable and unfavorable solutions.

3   The Protein-Protein Docking Algorithm

In this section we shortly outline our protein-protein matching algorithm, which has been recently implemented and is currently being tested.

In protein-protein matching one cannot usually rely on the SPHGEN procedure to segment the candidate binding site, since the docking occurs over a relatively large (and possibly flat) surface area. Also using atom centers is quite impractical due to the large number of atoms involved. Thus our approach is a purely geometric one. It can be summarized in the following steps :

1. Compute the Molecular Surface of the proteins using the MS algorithm ([6, 5]), which gives a dense representation of surface dots with their associated normals.
2. Reduce the dense surface representation to a sparse representation by geometrically significant *interest points*. In our case we follow Connolly ([7]) by extracting points which represent 'knobs' and 'holes'. These are detected by computing at each surface point (output of the previous step) the surface volume intersected by a ball of a prespecified radius, centered at the point. This shape descriptor achieves local minima at 'knobs' and local maxima at 'holes'. The averaged normals at the appropriate points are computed as well.
3. Apply a geometric matching algorithm to the suitable interest points with their associated normals. In our case 'knobs' of the receptor may match 'holes' of the ligand and vice versa. Note that a pair of points, together with their associated normals can uniquely define a 3-D rigid transformation. Thus, if the number of interest points extracted is sufficient, one can apply a Geometric Hashing type

matching algorithm for these points. If the number of points extracted is relatively sparse, as has been in our experiments, one could use a straightforward and less efficient technique, as outlined below. For each pair of points (which are not far away from each other) on the ligand a shape signature is computed. This signature includes the point labels (knobs/holes), the distance between the points, and the angles between the normals at the points and between the normals and the line segment connecting the two points. For each pair of receptor interest points, one creates the appropriate complementary signature, and checks, which of the ligand interest point pairs are consistent with it. A pair of receptor points (with their normals) and a pair of ligand points (with their normals) with consistent shape signatures define a 3-D translation and rotation between the ligand and the receptor. If this transformation implies penetration of one protein into the other, it is discarded. When no penetration occurs, it is scored according to the size of the matching surface area.

Different *interest point* extraction methods can be employed. If the molecular surface data were accurate and smooth, a natural choice would be to consider maxima and minima of principal curvatures. For noisy data as ours robust shape functions are required.

4 Experimental Results

4.1 Protein-Ligand Docking

Figures 1-3 and Table I show the results obtained for three protein-ligand docking experiments of complexed structures (i.e. both receptor and ligand coordinates appear in the already docked orientation). These complexes are: (i) heme bound to myoglobin, (ii) NADPH bound to dihydrofolate reductase and (iii) tyrosinyl adenylate bound to tyrosyl-tRNA synthetase. In one of the examples a factor of over 29 improvement in time over the DOCK2 ([32, 31]) program of Kuntz et al. ([32, 31]) has been obtained. Also the number of potential solutions which have passed our initial filtering stages is significantly lower than the number given by DOCK2. Although in all cases the ligand has been treated separately from the receptor, its coordinates have not been rotated or translated with respect to the original ones in the complexed crystal. Correct docking solutions should, therefore, display rotations and translations close to zero. Clearly, however, in the calculations it is assumed that the rotations and translations, docking the ligand onto the receptor, are unknown. From figures 1-3, it can be seen that correct candidate matches have been obtained. These figures show plots of the potential geometrical solutions obtained. Each diamond plotted represents a seed match which passed the grid overlap test, the score (i.e. number of matching pairs) and RMS tests. The X-axis is the l_2-norm of the translation vector in Å. The Y-axis is the total angular distance (in radians) [6]. A summary of these three experiments are shown in Table I.

---

[6] Each rotation matrix can be represented as a single rotation around an equivalent axis (the equivalent angle-axis representation [8]). The angular distance definition coincides with this angle and is computed as $arccos\frac{tr(R)-1}{2}$, where R is the rotation matrix and tr is the trace of M

Myoglobin The oxygen carriers in vertebrates are the proteins *hemoglobin* and *myoglobin*. Myoglobin, which is located in muscle, serves as a reserve supply of oxygen and facilitates the movement of oxygen within muscle [33]. In order to carry the oxygen, myoglobin binds the heme molecule.

The three dimensional structure of myoglobin from sperm whale (with a bound heme molecule) was resolved in 1977 [34] (PDB code: 4MBN). Figure 1 shows a plot of the results of our docking of the ligand (heme) into the receptor (myogloblin). The receptor and ligand coordinates have been separately taken from their complexed structures. The number of points used for the matching procedure was 325 for the receptor and 43 for the ligand. This example required 2.3 cpu minutes on a Silicon Graphics Workstation. Four good solutions (out of 134) were obtained. DOCK2 required 16 minutes for the same experiment.

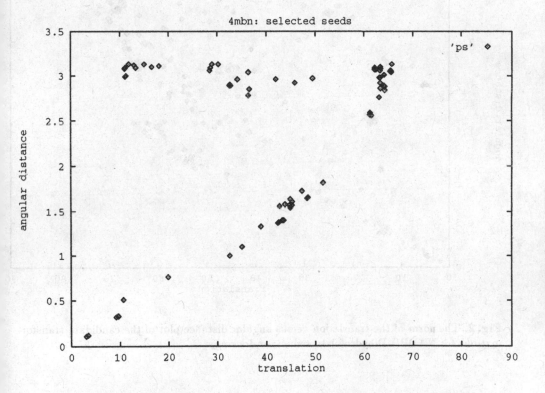

Fig. 1. The norm of the translation versus angular distance plot of the candidate transformations for heme-myoglobin docking.

Dihydrofolate Reductase The usefulness of several drugs important in the treatment of cancer, bacterial infections, and malaria stems from the inhibition of the enzyme dihydrofolate reductase ([20]). Here we present an example of docking the NADPH with dihydrofolate reductase.

The crystal structure of dihydrofolate reductase complexed with NADPH and methotrexate was resolved in 1982 ([16]).

Figure 2 shows the results of this experiment. Nine solutions (out of 359) are very close to the original orientation. The receptor was represented by 479 spheres and the ligand by 45. This example required 5.9 minutes of cpu. DOCK2 could not complete this example (after a few cpu hours and a very large output file, the program stopped).

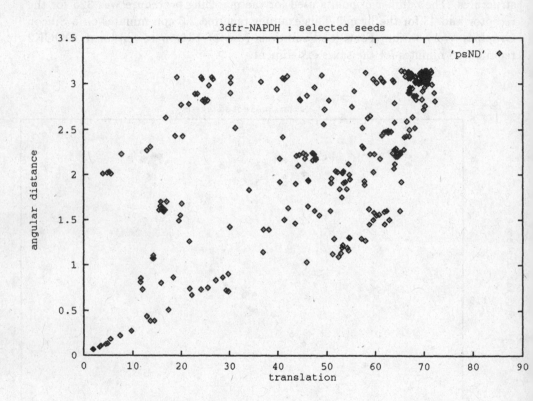

Fig. 2. The norm of the translation versus angular distance plot of the candidate transformations for NADPH-Dihydrofolate reductase docking.

Tyrosyl-Transfer RNA Synthetase Transfer RNAs (tRNA) are relatively small molecules that act as carriers of specific amino acids during protein synthesis on the ribosome ([25]). The 3-D structure of tyrosyl-transfer RNA synthetase complexed with tyrosinyl adenylate (PDB code: 3TS1), was resolved in 1989 ([28]). The most striking feature of the subunit surface is a deep cleft which forms the adenylate binding site. Opening from the bottom of this cleft is a narrow pocket which is the tyrosine binding site. The bound tyrosine substrate lies almost at the centroid of the subunit [28].

Figure 3 shows the results of this experiment. More than 50 solutions (out of 496) are very close to the original orientation. The receptor was represented by 184 spheres and the ligand by 34. This example required 1.3 minutes of CPU, whereas DOCK2 required 38 minutes.

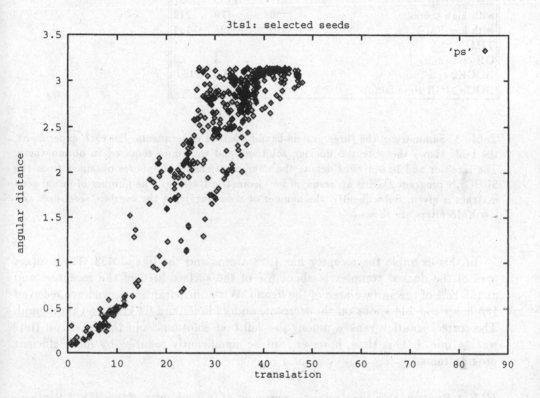

Fig. 3. The norm of the translation versus angular distance plot of the candidate transformations for tyrosinyl adenylate - tyrosyl-tRNA synthetase docking.

4.2 Protein-Protein Docking

We have run successfully about 10 examples of protein-protein docking experiments. Almost in all the experiments the correct solution was obtained and ranked relatively high among the candidate solutions. To get the flavor of the results we give two representative examples.

Trypsin - Trypsin inhibitor This is considered a difficult example, since the contact surface at the docking site is relatively flat, and one cannot find a sufficient number of matching knob/hole pairs. Connolly's method ([7]), which required four such pairs failed to solve it.

PDB code:	3dfr	3ts1	4mbn
Ligand:	NAPH	adenylate	HEME
receptor size:	479	184	325
ligand size:	45	34	43
GH # seed matches before overlap	5340	2604	944
after overlap	1417	1135	337
with high scores	730	770	212
with low RMS	359	496	134
TIMES			
GH cpu (min)	5.9	1.3	2.3
DOCK2 cpu (min)		38.0	16.0
DOCK2/GH time factor		29.2	6.9

Table 1. Summary of the three protein-ligand docking experiments. For each experiment, the table shows the potential docking solutions and the times required to obtain them. The receptor and ligand sizes denote the number of clustered spheres obtained from the SPHGEN program . GH is an acronym for Geometric Hashing. The number of initial seed matches is given. Subsequently, the number of seeds surviving the overlap, seed size, and low RMS filters are shown.

In this example the receptor has 1628 atoms and the ligand 453. The contact area of the docked complex is about 6% of the surface area of the receptor, and about 17% of the surface area of the ligand. With our parameters, we have received 185 holes and 144 knobs on the receptor, and 23 holes and 64 knobs on the ligand. The correct solution ranked among the 250 best solutions, and the total cpu time was 44 min. (This time, however, can be significantly reduced by more efficient programming.)

HIV-1 Protease In the docking experiment of two subunits of the HIV-1 Protease the correct solution was achieved with the highest score. Here both proteins had 745 atoms. The contact area was about 25% of the surface area of each protein. The number of holes and knobs on both subunits was about 50 and 100 respectively. The algorithm completed in 28 min cpu time on a Silicon Graphics workstation.

5 Discussion and Future Research

In this paper we have presented an algorithm for docking protein molecules. This problem is central for biomolecular recognition and binding and for synthetic drug design . We described experimental results, which have been obtained by a fully automated procedure. These results have been obtained in a significantly shorter time (typically, in few minutes of cpu time) than in previous methods. Our main algorithmic tool is the Geometric Hashing technique, which was originally developed for partially occluded object recognition in Computer Vision. The algorithmic similarity between problems which appear in such apparently distinct fields as Computer Vision/Robotics and Molecular Biology emphasizes the importance of interdisciplinary research in this area.

The results reported represent our ongoing research effort in pattern matching in Molecular Biology. Currently, we are exploring several research directions aimed at reducing the complexity of the Geometric Hashing application. In order to reduce complexity, one may look for either biological or geometric relations, which allow a unique determination of a full 3-D basis, thus making the number of explored bases linear in the number of interest points considered, and the overall worst case complexity - quadratic. One can, theoretically, define a unique 3-D coordinate frame at a given surface point. From Differential Geometry we know that such a frame can be defined by the two orthogonal directions of the principal curvatures, and by the normal at the point. We shall investigate the numerical stability of such an approach, for some special critical points, such as the 'knobs' and 'holes' we were using in the protein-protein docking.

An additional significant speed up in our algorithm can be obtained by chemical classification of the 'interest points'. By labeling each point according to some chemical property (e.g. electrostatic potential), and by allowing only chemically consistent matches, many candidate matches can be rapidly pruned.

We shall also investigate additional directions in order to improve the other stages of the docking algorithm, which we have currently borrowed from other methods. Connolly's MS algorithm is the most extensively used algorithm for molecular surface modeling. However, other approaches can be explored as well. Halperin ([13]) recently suggested to model the molecular surface generation as a motion planning problem, where the atoms of the molecule are considered as obstacles, and the water molecule as the moving robot. Certain 'fatness' properties of the atoms involved imply a favorable complexity of such an approach. (This observation shows again the potential benefit from Robotics techniques in Molecular Biology.) Also the invagination detection which is currently performed using Kuntz et. al. SPHGEN ([19]) procedure could be potentially improved by techniques of Computational Geometry, such as 3-D convex hull computation, or Voronoi diagram based approaches ([26]).

The methods described in this paper all deal with the matching of rigid three-dimensional structures. However, recent studies ([10]) suggest that conformational transitions in proteins can involve the relative movement of relatively rigid structural elements. These transitions can be visualized as 'hinge bending', or, using the language of roboticists, the existence of joints with rotational degrees of freedom. Docking of such flexible structures has been investigated by DesJarlais et al. ([9]). In our Computer Vision research we have recently developed a new technique for efficient recognition of partially occluded objects having rotary or sliding joints ([1, 35]). This technique has been successfully tested in the recognition of two dimensional objects. An adaptation of this technique for the three dimensional case would make it directly applicable to the problems of structural comparison that we are facing in Molecular Biology.

References

1. A. Beinglass and H. J. Wolfson. Articulated Object Recognition, or, How to Generalize the Generalized Hough Transform. In *Proceedings of the IEEE Computer Vision and Pattern Recognition Conf.*, pages 461 – 466, Maui, Hawaii, June 1991.

2. P. J. Besl and R. C. Jain. Three-Dimensional Object Recognition. *ACM Computing Surveys*, 17(1):75–154, 1985.

3. G. Burdea and H. J. Wolfson. Solving Jigsaw Puzzles by a Robot. *IEEE Trans. on Robotics and Automation*, 5(6):752–764, 1989.

4. R. T. Chin and C. R. Dyer. Model-Based Recognition in Robot Vision. *ACM Computing Surveys*, 18(1):67–108, 1986.

5. M.L. Connolly. Analytical molecular surface calculation. *J. Appl. Cryst.*, 16:548–558, 1983.

6. M.L. Connolly. Solvent-accessible surfaces of proteins and nucleic acids. *Science*, 221:709–713, 1983.

7. M.L. Connolly. Shape complementarity at the hemoglobin $\alpha_1\beta_1$ subunit interface. *Biopolymers*, 25:1229–1247, 1986.

8. J. J. Craig. *Introduction to Robotics*. Addison-Wesley, Readings, MA., 1986.

9. R.L. DesJarlais, R.P. Sheridan, J.S. Dixon, I.D. Kuntz, and R. Venkataraghavan. Docking Flexible Ligands to Macromolecular Receptors by Molecular Shape. *J. Med. Chem.*, 29:2149–2153, 1986.

10. H.R. Faber and B.W. Matthews. *Nature*, 348:263–266, 1990.

11. D. Fischer, O. Bachar, R. Nussinov, and H.J. Wolfson. An Efficient Computer Vision based technique for detection of three dimensional structural motifs in Proteins. *J. Biomolec. Str. and Dyn.*, 9(4):769–789, 1992.

12. D. Fischer, R. Nussinov, and H. J. Wolfson. 3-d substructure matching in protein molecules. In *Third Symposium on Combinatorial Pattern Matching*, Tucson, Arizona, USA, April 1992. Springer Verlag. in press.

13. D. Halperin. A data structure for representing molecules and efficient computation of molecular surfaces. 1992. personal communication.

14. B. K. P. Horn. Closed-form solution of absolute orientation using unit quaternions. *J. Opt. Soc. Amer. A*, 4(4):629–642, 1987.

15. F. Jiang and S.H. Kim. Soft docking : Matching of molecular surface cubes. *J. Mol. Biol.*, 219:79–102, 1991.

16. Bolin J.T., Filman D. J., Matthews D. A., Hamlin R. C., and Kraut J. Crystal structures of eschorichia coli and lactobacillus casei dihydrofolate reductase refined at 1.7å resolution. *J.Biol.Chem*, 257:13650–13662, 1982.

17. E. Katchalski-Katzir, I. Shariv, M. Eisenstein, A.A. Friesem, C. Aflalo, and I.A. Vakser. Molecular Surface Recognition: Determination of Geometric Fit between Protein and their Ligands by Correlation Techniques. *Proc. Natl. Acad. Sci. USA*, 89:2195–2199, 1992.

18. F.S. Kuhl, G.M. Crippen, and D. K. Friesen. A Combinatorial Algorithm for Calculating Ligand Binding. *J. Comp. Chem.*, 5(1):24–34, 1984.

19. I.D. Kuntz, J.M. Blaney, S.J. Oatley, R. Langridge, and T.E. Ferrin. A Geometric Approach to Macromolecule-Ligand Interactions. *J. Mol. Biol.*, 161:269–288, 1982.

20. L. F. Kuyper. Inhibitors of Dihydrofolate Reductase. In T. J. Perun and C. L Propst, editors, *Computer-Aided Drug Design*, pages 327–369. Marcel Dekker, New York, 1989.

21. Y. Lamdan, J. T. Schwartz, and H. J. Wolfson. On Recognition of 3-D Objects from 2-D Images. In *Proceedings of IEEE Int. Conf. on Robotics and Automation*, pages 1407–1413, Philadelphia, Pa., April 1988.

22. Y. Lamdan, J. T. Schwartz, and H. J. Wolfson. Affine Invariant Model-Based Object Recognition. *IEEE Trans. on Robotics and Automation*, 6(5):578–589, 1990.

23. Y. Lamdan and H. J. Wolfson. Geometric Hashing: A General and Efficient Model-Based Recognition Scheme. In *Proceedings of the IEEE Int. Conf. on Computer Vision*, pages 238–249, Tampa, Florida, December 1988.

24. Y. Lamdan and H. J. Wolfson. On the Error Analysis of Geometric Hashing. In *Proceedings of the IEEE Computer Vision and Pattern Recognition Conf.*, pages 22–27, Maui, Hawaii, June 1991.

25. A.L. Lehninger. *Principles of Biochemistry*. Worth Publishers, Inc., 1982.

26. R. A. Lewis. Clefts and Binding Sites in Protein Receptors. In Langone J. J., editor, *Methods in Enzymology, Vol. 202*, pages 126–156. Academic Press, San Diego, 1991.

27. R. Nussinov and H.J. Wolfson. Efficient detection of three-dimensional motifs in biological macromolecules by computer vision techniques. *Proc. Natl. Acad. Sci. USA*, 88:10495–10499, 1991.

28. Brick P., Bhat T.N., and Blow D.M. Structure of tyrosil-trna synthetase refined at 2.3å resolution. *J. Mol. Biol.*, 208:83–98, 1989.

29. C. L. Propst and T. J. Perun. Introduction to Computer-Aided Drug Design. In T. J. Perun and C. L Propst, editors, *Computer-Aided Drug Design*, pages 1–16. Marcel Dekker, New York, 1989.

30. F.M. Richards. Areas, volumes, packing and protein structure. *Ann. Rev. Biophys. Bioeng.*, 6:151–176, 1977.

31. B.K. Shoichet, D.L. Bodian, and I.D. Kuntz. Molecular docking using shape descriptors. *J. Comp. Chem.*, 13:380–397, 1992.

32. B.K. Shoichet and I.D. Kuntz. Protein Docking and Complementarity. *J. Mol. Biol.*, 221:327–346, 1991.

33. L. Stryer. *Biochemistry*. W.H. Freeman, San Francisco, 1981.

34. T. Takano. Structure of myoglobin refined at 2.0å resolution. *J. Mol. Biol.*, 110:569–584, 1977.

35. H. J. Wolfson. Generalizing the Generalized Hough Transform. *Pattern Recognition Letters*, 12(9):565 – 573, 1991.

MINIMAL SEPARATORS OF TWO WORDS

E. Garel, L.I.T.P
2, Place Jussieu
75251 PARIS Cedex 05, France

Abstract

Let u and v be two distinct words on the alphabet A. Let us call sep-min(u, v) the set of minimal length elements in A^* which are subwords (subsequences) of one and only one of u and v. This article aims at defining an automaton recognizing sep-min(u, v).

In the case where $u = va$, $a \in A$, the complexity in time of the algorithm for constructing the automaton is $O(|u|\mathrm{Card}A)$. In the general case, the complexity is proportional to the product of the size of the automaton by the size of the alphabet.

Keywords : Algorithms, Combinatorics on words, Automata and Formal Languages.

1.Introduction

There has been an increasing advance in combinatoric on words during these last few years. It was motivated by theoretical considerations (such as the theories of automata and languages) and by pratical considerations (such as the treatment of molecular sequences in biology).

Our topic is related to the comparison of character sequences. In this area the study of subwords (subsequences) of words plays an important role. The strategy commonly adopted to compare several sequences is to search for the maximum of analogies between these sequences. This leads to the notion of common subsequences. That becomes a basic principle and a lot of algorithms shores up detection of common subwords (see the algorithms used in the comparison of files or in the molecular sequences alignement problems).

Imre Simon' s approach is quite different. The algorithm that he proposes in [4] points at the positions where two sequences are dissimilar. We first give some basic terminology before a more explicit description. Let us consider a finite alphabet A and two words u and v in A^*. We say that $w \in A^*$ *distinguishes* u and v if w is a subword of one and only one of u and v. The set of the words which distinguish u and v is denoted by sep(u, v) and its elements are called the *separators* of u and v.

Imre Simon algorithm computes one shortest word among the separators of two words u and v. His program, based on the theorem 6.2.6 of [3], reads simultaneously each of the two words, from the left to the right. Then it needs to compute a family of separators that distinguishes particular prefixes of u and v.

The aim of this article is to study the same problem and to present the construction of an automaton which recognizes the set of minimal length separators of two words u and v. We denote sep-min(u, v) this set. But in contrast of Simon algorithm our program provides directly the construction of an automaton recognizing sep-min(u, v).

We first give a characterization of sep-min(u, v). It is proved that this set is equal to the subset of the minimal length elements in the union $\bigcup_{a \in A}$ sep$(ua, u) \triangle$ sep(va, v), where $\triangle$ denotes the symmetric difference operator.

According to this characterization, we show that the automaton can be calculated by using a dynamic programming algorithm. The complexity of the algorithm is proportional to the product of the size of the alphabet by the size of the automaton that has been built.

In the case where $u = va$, $(a \in A)$ a slightly different algorithm can be used. Its complexity is $O(|u| \text{Card } A)$.

In the two algorithms proposed, we borrow from Simon' s work the idea of locating the letters appearing in the separators of u and v by certain lists of integers corresponding to special positions in u and v.

In section 2, we precise definitions and notations. In section 3, we prove the fundamental characterization of scp-min(u, v), u and $v \in A^*$. In sections 4 and 5 we present respectively the construction of sep-min(u, v) for u and $v \in A^*$ and $u \neq v$ and the construction of sep-min(ua, u) for $u \in A^*$, $a \in A$. We finish with some examples and the comparison of the two given algorithms.

2. Definitions

Let A be a finite alphabet. We denote by A^+ the set of non-empty words, by ε the empty word and note $A^* = A^+ \cup \{\varepsilon\}$. The cardinal of A is denoted by $|A|$.

Let $u \in A^+$, then u is a sequence of letters which is written $u = u[1] \cdots u[n]$, $u[i] \in A$, $1 \leq i \leq n$ or $u = u[1 \cdots n]$. The **length** n of u is denoted by $|u|$, $|u| = n$.

We also denote by $u_{\rightarrow i}$ the prefix $u[1 \cdots i]$, $1 \leq i \leq n$ of a word $u = u[1 \cdots n] \in A^+$.

The set letters of A which appear in u is called the alphabet of u and is written $Alph(u)$.

A word $w \in A^+$ is a **subword** of u if there exists a sequence of indices $(l_1, \cdots, l_k)$, such that $1 \leq l_1 < \cdots < l_k \leq |u|$ and $w = u[l_1] \cdots u[l_k]$. A word w is a **factor** of u if there exist two indices i and j such as $1 \leq i < j \leq |u|$ and $w = u[i \cdots j]$.

Note that this terminology is not universal.

Note also that A^* is partially ordered by the divisibility relation "is a subword of".

We say that $w \in A^*$ is a **separator** of two given words u and v in A^*, or that w **distinguishes** u and v, if w is a subword of one and only one of u and v.

The set of separators of u and v is denoted by $sep(u,v)$, the subset of its minimal length elements by $sep\text{-}min(u,v)$ and the length of the minimal separators is denoted by $lm(u,v)$.

example 1

Let $A = \{a, b, c\}$, $u = bacbc$, $v = bacb$. Then abc and cc are both in $sep(u,v)$ and both minimal separators of u and v for the relation "is a subword of". However cc is shorter than abc and in this particular case we have $sep\text{-}min(u,v) = \{cc\}$.

Computing the set of all separators of u and v which are minimal for the relation "is a subword of" would be another problem.

We define the **truncature map** $\tau : A^+ \longrightarrow A^*$ by setting $u^\tau = u[1 \cdots n - 1]$ when $u \in A^+$ is of length n. Notice that $|u| = 1$ implies $u^\tau = \epsilon$.

Let w and u be two words in A^+ and j an integer ($j \in N$). We define the integers $l_j(w, u)$ in the following way :

$$l_j(w, u) = \begin{cases} r \leq |u| & \text{if } w_{\to j} \in sep(u_{\to r}, u_{\to r-1}) \\ \infty & \text{if } w_{\to j} \text{ is not a subword of } u. \end{cases}$$

The sequence $(l_1(w, u), \cdots, l_{|w|}(w, u))$ is denoted by $PL(w, u)$.

Let us assume that $w_{\to j}$ is a subword of u. In this case there exists an unique index r of u verifying $w_{\to j} \in sep(u_{\to r}, u_{\to r-1})$. Let $r = l_j(w, u)$ and $r' = l_{j-1}(w, u)$, we have

- $w[j] = u[r]$,
- $w_{\to j}$ is a subword of $u_{\to r}$, and is not a subword of $u_{\to r-1}$, in particular $w[j] \notin Alph(u[r' + 1 \cdots r - 1])$.
- $w[1 \cdots j] = u[l_1(w, u)] \cdots u[l_j(w, u)]$.

Note also that if $w \in sep(ua, u)$, with $a \in A$, the last letter of w is necessarily a.

We denote by $SM(u, v)$ the automaton recognizing $sep\text{-}min(u, v)$ which will be constructed by the algorithms 1 or 2 for the words u and v.

3. A Characterization of sep-min(u, v), u and $v \in A^*$

In this section, for two given words u and v on an alphabet A, we will denote by $S(u, v)$ the set

$$S(u, v) = \bigcup_{a \in A} \text{sep}(ua, u) \; \Delta \; \text{sep}(va, v).$$

(Δ is the symmetric difference).

Proposition 3.1

Let u and v be two distinct words on an alphabet A. The set sep-min(u, v) is equal to the subset of the minimal length elements of $S(u, v)$.

Proof :

1) - We first verify the inclusion sep-min$(u, v) \subset S(u, v)$.

Let $w \in$ sep-min(u, v). Suppose that w is a subword of u and is not a subword of v. Let r be the length of the longest prefix of w which is a subword of v. Hence we have $r < |w|$, $w_{\rightarrow r}w[r + 1] \in \text{sep}(vw[r + 1], v)$ and $w_{\rightarrow r+1}$ is a subword of u. In particular that implies $w_{\rightarrow r+1} \in \text{sep}(u, v)$.

Because of the minimality of the length of w in sep(u, v) we have $r + 1 = |w|$ and $w = w_{\rightarrow r+1} \in \text{sep}(vw[r + 1], v)$.

Similarly, if w is a subword of v, (and is not subword of u), then $w \in$ sep-min(ua, u) with a the last letter of w. It follows, for any shortest separator w,

$$w \in \bigcup_{a \in A} \text{sep}(ua, u) \cup \text{sep}(va, v).$$

Since all words in sep$(ua, u) \cup$ sep(va, v) end with an a, the sets sep$(ua, u) \cup$ sep(va, v) are pairwise disjoint when a runs through A. Also, since a word in sep(ua, u) cannot be a subword of u, we have sep$(u, v) \cap$ sep$(ua, u) \cap$ sep$(va, v) = \emptyset$. Therefore we can that conclude $w \in S(u, v)$.

2) - Conversely, let w be a minimal length element in $S(u, v)$. By the above inclusion we have $|w| \leq lm(u, v)$.

Suppose now $w \in \text{sep}(ua, u)$ with a is the last letter of w. Let r be the maximum index such that $w[1 \ldots r]$ is a subword of v, (in particular $r \leq |w|$).

Suppose $r < |w|$ and note $w' = w[1] \ldots w[r]w[r + 1]$. The word w' is an element of sep$(vw[r + 1], v)$ and it verifies $r + 1 \neq |w|$. (In fact, if $r + 1 = |w|$ then $w' = w$ and $w \in \text{sep}(ua, u) \cap \text{sep}(va, v)$).

Suppose now $r + 1 < |w|$. Then w' is a subword of u ; thus we have $w' \in \text{sep}(u, v)$ and also $w' \in S(u, v)$. This implies $|w| \leq |w'| = r + 1$, in contradiction with the hypothesis.

The only possible situation is $r = |w|$, w is a subword of v and $w \in \text{sep}(u, v)$. The minimality of the length of w in S implies that w belongs to sep-min(u, v). Thus the proposition holds.

4. Construction of sep-min(u, v) for u and $v \in A^*$ and $u \neq v$

The algorithm which computes sep-min(u, v) is based on the following proposition :

Conventions : Let u and $v \in A^*$, $u \neq v$ and let $w \in$ sep-min(u, v). Let us note $PL(w, u) = (l_1, \cdots, l_{|w|})$ and $PL(w, v) = (h_1, \cdots, h_{|w|})$.

In the case where w is not a subword of u, (and then w^τ is a subword of u), we will implicitly set $l_{|w|} = |u|+1$ and will consider that $u[|u|+1] = w[|w|]$; in the opposite, if w is not a subword of v, we will set $h_{|w|} = |v|+1$ and will consider that $v[|v|+1] = w[|w|]$.

Proposition 4.1

Using the conventions above, we have for every index j with $1 \leq j \leq |w|$, the prefix $w[1 \cdots j]$ has a minimal length in $\text{sep}(u_{\to l_j}, u_{\to l_j -1}) \cap \text{sep}(v_{\to h_j}, v_{\to h_j -1})$.

Proof

Let $w \in$ sep-min(u, v) be the given word above and let j be an integer, $1 \leq j \leq |w|$; suppose $w_{\to j}$ does not have a minimal length in $\text{sep}(u_{\to l_j}, u_{\to l_j -1}) \cap \text{sep}(v_{\to h_j}, v_{\to h_j -1})$. Choose $z \in \text{sep}(u_{\to l_j}, u_{\to l_j -1}) \cap \text{sep}(v_{\to h_j}, v_{\to h_j -1})$ such that $|z| < j$.

The word $w' = zw[j + 1 \cdots |w|]$ is still a separator of u and v. We have $|w'| = |z| + |w| - j < |w|$ and $w \notin$ sep-min(u, v), which leads to a contradiction.

●

Now, we can compute an automaton that recognizes sep-min(u, v) by using a dynamic programming algorithm.

First, we need to specify the different variables which will be used.

- In the set of ASCII code, **A** is an interval corresponding to the alphabet A. The elements belonging to **A** are letters and the type letter is char.

- The words u and v are memorized in arrays $[1 \cdots |u|]$ of letter denoted by **u** and $[1 \cdots |v|]$ of letter denoted by **v** respectively.

- a, c are type letter variables,

- i, i', j, j', d, k are integer variables where i and i' are indices of u, j and j' indices of v. The variable d will contain the value $lm(u, v)$ and is initialized by $\inf(|u|, |v|)+1$ because we always have $lm(u, v) \leq \inf(|u|, |v|) + 1$.

- The set of the automaton's states include the state $(0, 0)$, some couples of integers (i, j) verifying $1 \leq i \leq |u|$, $1 \leq j \leq |v|$ and $u[i] = v[j]$ and some states denoted by $sa, a \in A$.

- The initial state is $(0, 0)$ and the terminal states are sa, $a \in A$.

- The transitions are always labelled with the letters in A. If $((i,j),(i',j'))$ is an edge its label is necessarily the letter $a = u[i'] = v[j']$.

A letter $a \in A$ is free for a state (i,j) if it is still possible to create a labelled a transition from the state (i,j).

It is necessary to forbid a transition on a particular symbol a from a state (i,j) when the separators obtained by concatenation with a are not minimal.

When a new state (i,j) is created every letter $a \in A$ is free for this state.

- We define two dimensional arrays of integer, SP_u and SP_v, $[0 \cdots |u|, |A|]$ and $[0 \cdots |v|], |A|]$ respectively, as follows :

For every integer $n \in N$, $0 \leq n \leq |u|$, and for every $a \in A$, we set :

$$SP_u[n, a] = \begin{cases} m & \text{if } n < m \leq |u|, \ a \notin Alph(u[n+1 \cdots m-1]) \text{ and } u[m] = a. \\ \infty & \text{if } a \notin Alph(u[n+1 \cdots |u|]). \end{cases}$$

We define $SP_v[n, a]$ in a similar way replacing u by v.

- H is a two dimensional integer array, $[0 \cdots |u|, 0 \cdots |v|]$, initialized by 0. Through the algorithm, $H[i,j]$ takes the first value k so that a word $w \in sep(u_{\rightarrow i}, u_{\rightarrow i-1}) \cap sep(v_{\rightarrow j}, v_{\rightarrow j-1})$ is found and $|w| = k$.

In fact the minimal length of the elements of $sep(u_{\rightarrow i}, u_{\rightarrow i-1}) \cap sep(v_{\rightarrow j}, v_{\rightarrow j-1})$ is k in that case.

ALGORITHM 1

Input. Two words u and $v \in A^*$.

Output. The automaton that recognizes sep-min(u, v).

Method.

1) Initialize the array H by 0 ;
   Calculate the values contained in SP_u and SP_v ;
   Initialize d with inf $(|u|, |v|) + 1$ if $u \neq v$ else 0 ;
   if $(d = 0)$ then the automaton is empty - stop -
   else create the initial state $(0, 0)$, (every label a is free for $(0, 0)$),
        and go to 2) ;

2) $k \leftarrow 1$;

   while $k \leq d$ do
   begin

        for each state (i,j) such that $H[i,j] = k - 1$ do
           for each letter $a \in A$ do

begin

- Set $i' \leftarrow SP_u[i,a], j' \leftarrow SP_v[j,a]$;

- Create the state (i',j') and the labelled a edge from (i,j) to (i',j') if $(\ (i' \leq |u|), (j' \leq |v|), (H[i',j'] = 0)$ and $(k < d))$, Set $H[i',j'] \leftarrow k$;

- Create the labelled a edge $((i,j),(i',j')\ )$ if $(\ (i' \leq |u|), (j' \leq |v|), (H[i',j'] = k)$ and $(k < d)\ )$;

- Forbid the label a to the state (i,j)
  if $(\ (H[i',j'] \neq 0)$ and $(H[i',j'] < k)\ )$;

- Forbid the label a to the state (i,j) if $((i' = \infty)$ and $(j' = \infty))$ or if $((d = k$ and $(i' \leq |u|)$ and $(j' \leq |v|))$;

- If $(\ (i' = \infty$ and $j' \leq |v|)$ or $(\ i' \leq |u|$ and $j' = \infty)\ )$
  then create a labelled a edge from (i,j) to the terminal state sa
  and set $d \leftarrow k$;

end ;

Set $k \leftarrow k + 1$;

end ;

Before proving that the automaton built by the algorithm 1 recognizes sep-min(u,v), we give an example.

example 2 Let $A = \{a,b\}$, $u = abababab$, $v = babababa$.

The algorithm provides

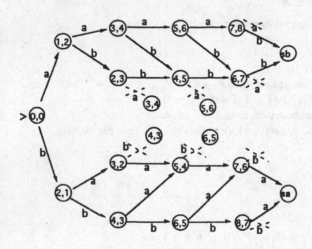

fig.1 $SM(u,v), u = abababab, v = babababa$.

Note

$$u = \begin{array}{cccccccc} a & b & a & b & a & b & a & b \\ 1 & 2 & 3 & 4 & 5 & 6 & 7 & 8 \end{array}$$

$$v = \begin{array}{cccccccc} b & a & b & a & b & a & b & a \\ 1 & 2 & 3 & 4 & 5 & 6 & 7 & 8 \end{array}$$

We obtain two terminal states sa and sb,
sep-min$(u, v) = \{aaaab, aaabb, aabbb, abbbb, baaaa, bbaaa, bbbaa, bbbba\}$.

Note, for example, the state $(2,3)$ is examined at step $k = 3$ and here ab is the only word recognized by $(2,3)$. Letter a gives $(i', j') = (3,4)$. Thus the word aba is an element of sep$(u[1\cdots3], u[1\cdots2]) \cap$ sep$(v[1\cdots4], v[1\cdots3])$ corresponding to the path $((0,0),(1,2),(2,3),(3,4))$.

But $(3,4)$ is already a state of the automaton, $H[3,4] = 2$. The word aa is minimal in sep$(aba, ab) \cap$ sep$(baba, bab)$. It corresponds to the path $((0,0),(1,2),(3,4))$. We have $|aba| = |aa| + 1$, the word aba is not minimal and the label a must be forbidden to $(2,3)$.

The program provides 30 times calculations of the values $i' = SP_u[i, a]$ and $j' = SP_v[j, a]$ for particular indices (i, j) and letter a. Note than $30 < |A|.|uv| = 16 \times 2 = 32$, and $|u|.|v|.|A| = 128$, $|A|^d = 2^5 = 32$.

The automaton, which is minimal here, countains 17 states and $17 = |u| + |v| + 1$.

We now prove that the automaton recognizes sep-min(u, v).

First, denote by $Int(i, j)$, $1 \le i \le |u|$ and $1 \le j \le |v|$, the set of the minimal elements in sep$(u_{\to i}, u_{\to i-1}) \cap$ sep$(v_{\to j}, v_{\to j-1})$.

We denote by $E(i', j')$ the set of states (i, j) in the automaton which verify $i' = SP_u[i, a]$, $j' = SP_v[j, a]$ and $H[i, j] = H[i', j'] - 1$.

We denote by $P(i, j)$ the set of paths in the automaton which start in $(0, 0)$ and end in (i, j).

We will say that a state (i, j) recognizes a word $w \in A^+$ if there exists a path starting at $(0,0)$ and ending at (i, j) such that w can be obtained by the concatenation of the labels of the successive edges of this path.

Proposition 4.2

Let u and v be two distinct words of A^+. Let i and j be two indices, , $1 \le i \le |u|$, $1 \le j \le |v|$ and denote by k the length of the elements of $Int(i, j)$.

1) If $1 \le k < lm(u, v)$ the couple (i, j) is a state of the automaton built and $H[i, j] = k$.

2) Conversely, if (i, j) is a state of the automaton built such that $H[i, j] = k$, then the length of the elements of $Int(i, j)$ is k.

Moreover, there exists a biunivoque correspondence between the words w of $Int(i,j)$ and the paths of the automaton starting in $(0, 0)$ and ending in (i,j). Every w is obtained by concatenation of the labels of the successive edges in the corresponding path.

Proof

Proof is obtained by reccurrence on k and first examine the case $k = 1$ $(1 < \overset{\bullet}{lm}(u,v))$.

remark

If one of u or v is equal to ϵ then sep-min(u,v) is exactly the alphabet of the other one. In fact if $Alph(u) \neq Alph(v)$ we obtain sep-min$(u,v) = \text{Alph}(u) \triangle \text{Alph}(v)$.
The condition $(1 < lm(u,v))$ implies $Alph(u) = Alph(v)$.

On enterring the while-statement the only initial state $(0,0)$ has been created. By execution of the contained statement, for all letters $a \in A$, the values $SP_u[0,a]$ and $SP_v[0,a]$ are calculated. A state (i,j) and a labelled a edge $((0,0),(i,j))$ are created if and only if $a \in Alph(u) \cap Alph(v)$, $i = SP_u[0,a]$ and $j = SP_v[0,a]$. In other words $a \notin Alph(u[1\cdots i-1]) \cup Alph(v[1\cdots j-1])$ and $u[i] = v[j] = a$. That exactly implies than $Int(i,j) = \{a\}$.

Hence, $H[i,j]$ takes the value 1 which is the length of the unique element belonging to $Int(i,j)$. Proposition 4.2 holds for $k = 1$.

Suppose now that the proposition 4.2 holds for every integer $l, 1 \leq l \leq k-1 < lm(u,v) - 1$.

In particular, at the end of step $k-1$ of while, by the reccurrence hypothesis, the set of the states in the automaton describes exactly the set of indices couples (i,j) such that $1 \leq i \leq |u|$, $1 \leq j \leq |v|$, $u[i] = v[j]$ and the length l of the elements of $Int(i,j)$ verifies $1 \leq l \leq k-1$. Further we have $H[i,j] = l$ and $Int(i,j)$ is recognized by (i,j).

As a consequence, for all other couples of indices (i,j), $H[i,j] = 0$ means than either $Int(i,j) = \emptyset$ or the length of the elements of $Int(i,j)$ is $\geq k$.

Let (i,j) be a state such as $H[i,j] = k-1$ and let $a \in A$. Note $i' = SP_u[i,a]$, $j' = SP_v[j,a]$.

If $(i' = \infty$ and $j' = \infty)$ then $a \notin Alph(u[i\cdots|u|]) \cup Alph(v[j\cdots|v|])$. Thus for every $w \in Int(i,j)$ we have $wa \in \text{sep}(ua,u) \cap \text{sep}(va,v)$ and, as the program does, the transition a must be forbidden to the state (i,j).

We obtain $(i' \leq |u|$ and $j' \leq |v|)$ if and only if

$a \notin Alph(u[i+1\cdots i'-1]) \cup Alph(v[j+1\cdots j'-1])$ and $u[i'] = v[j'] = a$. This implies in particular that for every $w \in Int(i,j)$

    1, $wa \in \text{sep}(u_{\to i'}, u_{\to i'-1}) \cap \text{sep}(v_{\to j'}, v_{\to j'-1})$,
    2, $|wa| = k$.

As a consequence, as the program does,

- a new state (i', j') and a labelled a edge $((i, j), (i', j'))$ must be created if and only if $H[i', j'] = 0$ at the enter of step k. (In this case, all the words wa, $w \in Int(i, j)$, belong to $Int(i', j')$ and $H[i', j']$ must be set k).

- a labelled by a transition must be forbidden to (i, j) if and only if $1 \leq H[i', j'] \leq k - 1$. (In this case $Int(i', j') \neq \emptyset$, the length of its elements is $\leq k - 1$ and the words wa, for any $w \in Int(i, j)$, are not minimal in $sep(u_{\to i'}, u_{\to i'-1}) \cap sep(v_{\to j'}, v_{\to j'-1})$).

Now prove than there exists a biunivoque correspondence between $P(i', j')$ and $Int(i', j')$. The reccurrence hypothesis implies there exists a bijection between $P(i, j)$ and $Int(i, j)$ if $H[i, j] \leq k - 1$. Further the sets $P(i, j)$ are clearly pairwise disjoint when (i, j) runs through $E(i', j')$.

If the state (i', j') is created at step k then every path $s \in P(i, j)$ is extended by a labelled a edge $v = ((i, j), (i', j'))$. The new path sv ends in (i', j'). If $w \in Int(i, j)$ is the word which corresponds to s, the word wa will be the word which corresponds to sa and moreover $wa \in Int(i', j')$.

Thus at the end of step k the set $P(i', j')$ is equal to the following union

$$P(i', j') = \cup\{sv,\ s \in P(i, j),\ (i, j) \in E(i', j'),\ v \text{ a labelled } a \text{ edge } v = ((i, j), (i', j'))\}.$$

Clearly there exists a biunivoque correspondence between $P(i', j')$ and the union $\cup\{wa,\ w \in Int(i, j),\ (i, j) \in E(i', j')\}$.

(Remark : For all paths s and s' extended by an edge v as these defined above, $s \neq s'$ implies $sv \neq s'v$. For all words w and w' and for all $a \in A$, $w \neq w'$ implies $wa \neq w'a$.)

Since the wa above are of minimal length in $sep(u_{\to i'}, u_{\to i'-1}) \cap scp(v_{\to j'}, v_{\to j'-1})$ we have

$$\cup\{wa, w \in Int(i, j),\ (i, j) \in E(i', j'), \} \subset Int(i', j')$$

It is not very difficult to see that, in fact, we have the equality.

In order to verify it, choose w' an element in $Int(i', j')$ and let $a = u[i'] = v[j']$. Using the reccurrence hypothesis, there exist two indices i and j such as the prefix $w = w'^r$ is recognized by (i, j). The couple (i, j) verifies $H[i, j] = k - 1$, $i' = SP_u[i, a]$ and $j' = SP_v[j, a]$. Hence we have $(i, j) \in E(i', j')$ and w' belongs to the union above.

This also proves that, at the end of step k, every couple (i', j'), whose length of the elements of $Int(i', j')$ is k, has become a state in the automaton such as $H[i', j'] = k$.

•

Proposition 4.3

The automaton constructed by algorithm 1 recognizes sep-min(u, v).

Proof

Let again u and v be two distinct words. By proposition 4.2, at the end of step $k = lm(u,v) - 1$ of while, the automaton constructed contains exactly all the indices couples (i,j) such as $Int(i,j) \neq \emptyset$ and the length of the elements of $Int(i,j)$ is $\leq lm(u,v) - 1$.

Thus for any $w \in$ sep-min(u,v) the prefix w^τ is recognized by a state in the automaton.

Using similar notations as those defined in proposition 4.2, let
$$E(sa) = \{(i,j),\ H[i,j] = lm(u,v) - 1,\quad a \in Alph(u[i+1\cdots|u|])\Delta Alph(v[j+1\cdots|v|]),$$
and
$$P(sa) = \cup\{sv,\ s \in P(i,j),\ (i,j) \in E(sa),\ v \text{ a labelled by } a \text{ edge } v = ((i,j),(sa))\}.$$

In step $k = lm(u,v)$ the program examines every state (i,j) such as $H[i,j] = k - 1$. Let $a \in A$, let (i,j) a state such as $H[i,j] = lm(u,v) - 1$ and note i' and j' the values $i' = SP_u[i,a]$ and $j' = SP_v[j,a]$ which are associated to (i,j) and a.

If $(\ (i' = \infty$ and $j' = \infty)$ then for every $w \in Int(i,j)$ we have $wa \in$ sep$(ua,u) \cap$ sep(va,v) and, as the program does, the transition a must be forbidden to the state (i,j).

Before finding a first separator $w \in$ sep(u,v), the program continues as described in proposition 4.2 and constructs the states $(i',j'), i' \leq |u|, j' \leq |v|$, such as the length of the words in $Int(i',j')$ is $lm(u,v)$.

After having set $d \leftarrow lm(u,v)$, program forbids the transition a to every state $(i,j) \in E(i',j')$ such as $i' \leq |u|$ and $j' \leq |v|$).

This last instruction is not absolutely necessary . At the end of the while statement it is sufficient to delete all the states (i',j') such as $H[i',j'] = lm(u,v)$, $i' \leq |u|, j' \leq |v|$, and forbid the transition $a = u[i'] = v[j']$ from every state $(i,j) \in E(i',j')$.

When the condition $(\ (i' = \infty$ and $j' \leq |v|)$ or $(\ i' \leq |u|$ and $j' = \infty)\ )$ is satisfied, (i.e. $a \in Alph(u[i+1\cdots|u|])\Delta Alph(v[j+1\cdots|v|])\ )$ the program creates a labelled a edge $(\ (i,j),\ sa)$, ending in the terminal state sa and sets $d \leftarrow lm(u,v)$. Thus the state sa will recognize every word za, $z \in Int(i,j)$.

Further as in proposition 4.2, at the end of step $lm(u,v)$ the set $P(sa)$ will be in bijection with $Sa = \cup\{wa,\ w \in Int(i,j),\ (i,j) \in E(sa)\}$,
and Sa recognizes all words $w \in$ sep-min(u,v) whose last letter is a.

In fact, suppose there exists a word $wa \in$ sep-min(u,v). As $|w| = lm(u,v) - 1$ there exists a state (i,j) in the automaton recognizing w. Further $H[i,j] = lm(u,v) - 1$, $a \in Alph(u[i+1\cdots|u|])\Delta Alph(v[j+1\cdots|v|])$, and $(i,j) \in E(sa)$. Thus for every $z \in Int(i,j)$ (i.e. z recognized by (i,j)) the word za is also in sep-min(u,v) and is recognized by Sa. As sep-min$(u,v) = \bigcup_{a\in A} Sa$, at the end of while, the automaton constructed recognizes sep-min(u,v).

Proposition 4.4

Complexity in time of the algorithm 1 is proportional to the product of the size of the alphabet by the size of the automaton constructed.

Proof

Every instruction in the principal loop is in constant time. At each step either a new edge is created, labelled with some letter $a \in A$, or the label a is forbidden to the courant state. The program studies the existence of a new transition from a given state for every letter $a \in A$. The complexity in time is proportional to the product of the number of states which appear in the automaton by the size of the alphabet.

•

Remark

1) - Let $d = lm(u, v)$ be the length of the elements belonging to $sepmin(u, v)$. Note n_k the the number of the states (i, j) such as $H[i, j] = k$.

We have $n_1 \leq |A|$. If $Alph(u) = Alph(v) = A$ then we obtain exactly $n_1 = |A|$.

Let (i, j) be a state such as $H[i, j] = k$. At most (i, j) is the origin of $|A|$ edges labelled by different letters of A.

Thus we can deduce $n_{k+1} \leq n_k.|A|$, $1 \leq k < d$.

Let n be the number of the states contained in the automaton. We have $n \leq 1 + |A| + \cdots + |A|^d = \frac{|A|^{d+1}-1}{|A|-1}$. The size of the automaton is $O(|A|^d)$.

2) - At most one couple (i, j) such as $u[i] = u[j]$ is a state in the automaton and, at most, (i, j) is the origin of $|A|$ vertices. The size of the automaton is $\leq |u|.|v|$, $(u \neq \epsilon, v \neq \epsilon)$ and the numbers of edges is $\leq |u|.|v|.|A|$.

We obtain a complexity in $O(\inf (|u|.|v|.|A|, |A|^d))$, where $d = lm(u, v)$, (d always verifies $d \leq \inf (|u|, |v|) + 1$).

conjecture : Complexity is $O(|uv|.|A|)$.

The number of states in the built automaton is $\leq |u| + |v| + 1$.

5. Construction of sep-min(u, u^r) for $u \in A^+$

A slightly different algorithm can be used to construct an automaton which recognizes sep-min(u, v) when $v = u^r$.

It is based on the following description of sep-min(u, u^r) :

Proposition 5.1

Let $u \in A^+$ and $w \in$ sep-min(u, u^r). Note $PL(w, u) = (l_1, \cdots, l_{|w|})$. For every index j with $1 \leq j \leq |w|$ we have

$$w[1 \cdots j] \in \text{sep-min}(u[1 \cdots l_j], u[1 \cdots l_j]^r).$$

proof :

Let $u \in A^+$, $w \in$ sep-min(u, u^r) and $PL(w, u) = (l_1, \cdots, l_{|w|})$. By definition $PL(w, u)$ verifies, for every j with $1 \leq j \leq |w|$,

$$w[1 \cdots j] \in \text{sep}(u_{\to l_j}, u_{\to l_j - 1}).$$

Suppose the length of $w[1 \cdots j]$ is not minimal in sep$(u_{\to l_j}, u_{\to l_j - 1})$ and let z be a shortest length separator in sep$(u_{\to l_j}, u_{\to l_j - 1})$. As $w[j+1] \notin Alph(u[l_j + 1 \cdots l_{j+1} - 1])$ the word $y = zw[j+1] \cdots w[\|w\|]$ is an element of sep(u, u^r). We have $|z| < j$ and $|y| = |z| + |w| - j < |w|$, and this leads to a contradiction.

●

We can now compute the automaton that recognizes sep-min(u, u^r) using a dynamic programming algorithm.

We present first the different variables which are used :

- **A** is again an interval corresponding to the alphabet A and $a \in A$ is of type letter.

- The given word u is memorized in an array$[1 \cdots |u|]$ of letter, noted **u**.

- a, c are type letter variables,

- i and j are integer variables and are indices of u.

- The automaton built contains at most $|u|+1$ states, numbered from 0 to $|u|$. The initial state is 0, the terminal state is $|u|$.

The transitions are labelled by the letters in A. If (i, j) is an edge then $i < j$ and its label is $u[j]$.

A letter $a \in A$ is **free** for a state i if no transition labelled by a has its origin in i and if it is still possible to create a labelled by a transition from i.

It is necessary to **forbid** a transition on a symbol a from a state i when separators which are obtained by concatenation with a are not minimal.

- L and G are integer arrays with dimension $[0 \cdots |u|]$ and $[\|A\|]$ respectively. L and G are initialized by 0.

For every i, $1 \leq i \leq |u|$, L[i] would be equal to $lm(u_{\to i}, u_{\to i-1})$.

At the end of the step i we set $G[u[i]] \leftarrow L[i]$, and for any letter $c \in A$, we set $G[c] \leftarrow lm(u[1 \cdots i]c, u[1 \cdots i])$ - 1.

ALGORITHM 2

Input. A word $u \in A^+$.

Output. The automaton that recognizes sep-min(u, u^r).

Method.

1, Initialize the arrays G, L by 0 ; the initial state 0 is created and every letter a is free for state 0.

2, for $i \leftarrow 1$ to $|u|$ do

begin

  • Set $a \leftarrow u[i]$, $L[i] \leftarrow G[a] +1$;

  • for every state j such as $(j < i$ and $L[j] = G[a])$ do
  begin
      Create the edge (j, i) labelled with a if a is free in j ;
  end ;

  • Set $G[a] \leftarrow L[i]$;

  • for every state j such as $(j < i$ and $L[j] \geq L[i])$ do
  begin
      Suppress every edge (j, h), $j < h < i$ and forbid $u[h]$ to j ;

      Forbid a to j and suppress the states which are become inacessible ;

  end ;

  • for every letter $c \in A$ such as $(G[c] > G[a])$ Set $G[c] \leftarrow G[a]$;

end ;

First, the complexity of algorithm 2 is given by the following proposition :

Proposition 5.2

  Complexity in time of the algorithm 2 is $O(|u||A|)$.

Proof

The variable i goes from 1 to $|u|$.

If (i, j) is a labelled a edge then $i < j$, $a \notin Alph(u[i + 1 \cdots j - 1])$, and $u[j] = a$.

We deduce than, for every letter $a \in A$ there will be no more than $|u|$ creations and interdictions of edges labelled by a produced by the algorithm.

At most every state i is the origin of $|A|$ vertices (i, h), $i < h$ corresponding to distinct letters. The complexity is $O(|u|.|A|)$.

Proposition 5.3

  Algorithm 2 constructs an automaton recognizing sep-min(u, u^r).

Proof

Proof of proposition 5.3 can be obtained by reccurrence on the index i of u, $1 \leq i \leq |u|$.

The hypothesis of the reccurrence is

At the end of each step i the following conditions are satisfied

1) $L[i] = lm(u_{\to i}, u_{\to i-1})$;

2) For every letter $c \in A$ $G[c]$ is equal to $lm(u_{\to i}c, u_{\to i})$ - 1 ;

3) If a letter a verifies : for any $w \in$ sep-min$(u_{\to i}, u_{\to i-1})$, wa is never the prefix of any minimal separator in sep-min$(u_{\to h}, u_{\to h-1})$, for any h, $i < h$ then a is forbidden to the state i.

4) The automaton, whose initial state is 0 and terminal state is i, recognizes sep-min$(u_{\to i}, u_{\to i-1})$. Every $w \in$ sep-min$(u_{\to i}, u_{\to i-1})$ corresponds to an unique path starting at 0 and ending at i and $PL(w, u)$ is exactly the states values sequence in this path.

The proof needs very technical lemma. We give the lemma and let the real proof to the reader.

lemma 1

Let u in A^+. For any $a \in A$ we have
$\{w = w'a \text{ with } w' \in$ sep-min$(u, u^r)\} \subset$ sep (ua, u) and $lm(ua, u) \leq lm(u, u^r) + 1$.

Proof

If $w' \in$ sep-min(u, u^r), we clearly have $w'a \in$ sep(ua, u) and $|w'a| = |w'|+1 \geq lm(ua, u)$.

●

lemma 2

Let $u \in A^+$ and a the last letter of u. We have $lm(ua, u) = lm(u, u^r) + 1$ and sep-min$(ua, u) = \{wa \in A^*, \text{ with } w \in sepmin(u, u^r)\}$.

Proof

Suppose $u = u^r a$, $a \in A$, and let $w \in$ sep-min(ua, u). We have $w = w^r a$ with w^r subword of u.

Let u' be the prefix of u, given by the proposition 3.1, such as $w^r \in$ sep-min(u', u'^r).

If $|u'| < |u|$ then w^r is a subword of u^r, $w^r a$ becomes subword of u and $w \notin$ sep-min(ua, u). That leads to a contradiction. Thus the equality
$\quad$ sep-min$(ua, u) = \{wa \in A^*, \text{ with } w \in$ sep-min$(u, u^r)\}$ is hold.

●

lemma 3

Let $u \in A^+$, a the last letter of u, $u = u^r a$, and let $c \in A$, $c \neq a$, we have
sep-min$(u^r c, u^r) \subset$ sep(uc, u) and $lm(uc, u) \leq lm(u^r c, u^r)$.

Proof

Let $z \in$ sep-min$(u^r c, u^r)$ and u' the prefix of u^r, given by the proposition 5.1, such as $z^r \in$ sep-min(u', u'^r).

Moreover $c \notin Alph(u[|u'| + 1 \cdots |u| - 1])$ and $a \neq c$, we have $c \notin Alph(u[|u'| + 1 \cdots |u|])$ and $z = z^r c \in$ sep(uc, u).

We deduce sep-min$(u^r c, u^r) \subset$ sep(uc, u) and $lm(uc, u) \leq lm(u^r c, u^r)$.

●

lemma 4

Let $u \in A^+$, a the last letter of u and let c be a letter, $c \neq a$. We have

i) If $lm(u^r c, u^r) \leq lm(u, u^r)$ then sep-min$(uc, u) =$ sep-min$(u^r c, u^r)$ and $lm(uc, u) = lm(u^r c, u^r)$;

ii) If $lm(u^r c, u^r) = lm(u, u^r) + 1$ then sep-min$(uc, u) =$ sep-min$(u^r c, u^r) \cup \{w \in A^*, w = w'c$ with $w' \in$ sep-min$(u, u^r)\}$ and $lm(uc, u) = lm(u^r c, u^r)$.

iii) If $lm(u^r c, u^r) > lm(u, u^r) + 1$ then sep-min$(uc, u) = \{w \in A^*, w = w'c$ with $w' \in$ sep-min$(u, u^r)\}$ and $lm(uc, u) = lm(u, u^r) + 1$;

Proof

i) - First, suppose that $lm(u^r c, u^r) \leq lm(u, u^r)$; using lemma 3 we can assert that sep-min$(u^r c, u^r) \subset$ sep(uc, u) and thus $lm(uc, u) \leq lm(u^r c, u^r) \leq lm(u, u^r)$.

Let $w \in$ sep-min(uc, u), $w = w^r c$; let u' be the prefix of u, given by proposition 3.1, such that $w^r \in$ sep-min(u', u'^r).

If $|u'| = |u|$ then $w^r \in$ sep-min(u, u^r) and $|w| = |w^r c| = lm(uc, u) = lm(u, u^r) + 1$; and $lm(u, u^r) > lm(u, u^r)$; and that is contrary to the hypothesis.

Necessarily, we have $|u'| < |u|$; w^r is a subword of u^r and $w = w^r c$ is subword of uc. By hypothesis $w = w^r c$ is not a subword of u, w is not more a subword of u^r ; we have $w \in$ sep $(u^r c, u^r)$ and $lm(uc, u) \geq lm(u^r c, u^r)$. We can conclude $lm(uc, u) = lm(u^r c, u^r)$ and $w \in$ sep-min$(u^r c, u^r)$.

ii) - Suppose $lm(u^r c, u^r) = lm(u, u^r) + 1$. In this case we have
$$lm(uc, u) \leq lm(u^r c, u^r) = lm(u, u^r) + 1.$$
Let $w \in$ sep-min(uc, u) and let u' be the prefix of u, given by the proposition 3.1, such that $w^r \in$ sep-min(u', u'^r).

If $|u'| < |u|$, as $c \neq a$, we have $w \in$ sep-min$(u^r c, u^r)$ and $lm(u^r c, u^r) \leq lm(uc, u)$, that implies $|w| = lm(u, u^r) + 1$.

If $|u'| = |u|$ then we have $w \in$ sep-min(uc, u) and $|w| = lm(u, u^r) + 1$. In conclusion we obtain
sep-min$(uc, u) =$ sep-min$(u^r c, u^r) \cup \{w \in A^*, w = w'c$ with $w' \in$ sep-min$(u, u^r)\}$.

iii) - Suppose now $lm(u^r c, u^r) > lm(u, u^r) + 1$. Using the lemma 1 we have
$lm(uc, u) \leq lm(u, u^r) + 1 < lm(u^r c, u^r)$ and, if $w \in$ sep-min$(u^r c, u^r)$ then $w \notin$ sep-min(uc, u) ;

If $w \in$ sep-min(uc, u), the only possible case is $w \in \{z \in A^*, z = z'c$ with $z' \in$ sep min $(u, u^r)\}$ and iii, is hold.

lemma 5

Let $u \in A^+$, and let i and j, $j < i$ be two integer.

Note $a = u[i]$, $lm(u_{\to i}, u_{\to i-1}) = k$ and choose some $z \in$ sep-min$(u_{\to j}, u_{\to j-1})$.

1) If $lm(u_{\to j}, u_{\to j-1}) \geq k$ then za is not the prefix of any element belonging to sep-min$(u_{\to h}, u_{\to h-1})$ with $h > i$.

2) If $lm(u_{\to j}, u_{\to j-1}) > k$ then z is not the prefix of any element belonging to sep-min$(u_{\to h}, u_{\to h-1})$ with $h \geq i$.

Proof

1) - Suppose $lm(u_{\to j}, u_{\to j-1}) \geq k$, $z \in$ sep-min$(u_{\to j}, u_{\to j-1})$, $h > i$, and let $w \in$ sep$(u_{\to h}, u_{\to h-1})$ such that za is a prefix of w, in particular za is a subword of $u[1 \cdots i]$.

Let w' be the longest prefix of w such as w' is a subword of $u[1 \cdots i]$; w is factorized by $w = w'w''$. We also have za is a prefix of w' and $|za| \leq |w'|$. Note l the index of u such as $w' \in$ sep $(u_{\to l}, u_{\to l-1})$, b the first letter of w'' and l' the index of u such as $w'b \in$ sep$(u[1 \cdots l'], u[1 \cdots l' - 1])$. We have $l \leq i < l'$ and $b \notin$ Alph $(u[l + 1 \cdots l' - 1])$.

Choose some v in sep-min$(u_{\to i}, u_{\to i-1})$, vb is not a subword of $u_{\to i}$ and thus $vw'' \in$ sep$(u_{\to h}, u_{\to h-1})$.

We have $|vw''| = k + |w''| < |w'| + |w''| = |w|$ and w is not minimal in sep $(u_{\to h}, u_{\to h-1})$.

2) - Suppose $lm(u_{\to j}, u_{\to j-1}) > k$, $z \in$ sep-min$(u_{\to j}, u_{\to j-1})$.

First, if $h = i$ then we have $lm(u_{\to i}, u_{\to i-1}) = k < |z|$; z can't be the prefix of one element in sep-min$(u_{\to i}, u_{\to i-1})$.

Suppose $h > i$ and let $w \in$ sep $(u_{\to h}, u_{\to h-1})$ such as z is a prefix of w. We have $k < |z| \leq |w|$.

Let w' be the longest prefix of w such as w' is a subword of $u[1 \cdots i]$; w is factorized by $w = w'w''$; z is also a prefix of w' and $|z| \leq |w'|$.

Note l the index of u such as $w' \in$ sep$(u_{\to l}, u_{\to l-1})$, b the first letter of w'' and l' the index of u such as $w'b \in$ sep$(u_{\to l'}, u_{\to l'-1})$.

We have $l \leq i < l'$ and $b \notin Alph(u[l + 1 \cdots l' - 1])$.

Choose some $v \in$ sep-min$(u[1 \cdots i], u[1 \cdots i - 1])$, vb is not a subword of $u[1 \cdots i]$ and thus $vw'' \in$ sep$(u_{\to h}, u_{\to h-1})$.

We have $|vw''| = k + |w''| < |w'| + |w''| = |w|$ and w is not minimal in sep$(u_{\to h}, u_{\to h-1})$.

$\bullet$

6. examples

remark

Let u and $v \in A^+$ be two distinct words. It is not true that

$$\text{sep-min}(u, v) \subset \bigcup_{a \in A} \text{sep-min}(ua, u) \,\triangle\, \text{sep-min}(va, v),$$

as it is illustrated by the following example :

example 3

Let $u = acaba$ and $v = aacaba$. A quick computation shows that sep-min$(ux, u) =$ sep-min(vx, v) for every letter x in A. We obtain :

sep-min(ua, u)	$=$	sep-min(va, v)	$=$	$\{baa\}$
sep-min(ub, u)	$=$	sep-min(vb, v)	$=$	$\{bb\}$
sep-min(uc, u)	$=$	sep-min(vc, v)	$=$	$\{bc, cc\}$

and sep-min$(u, v) = \{aac\}$.

example 4

This example compares the efficiency of the two algorithms. Let $A = \{a, b\}$, and consider $u = caaababac$, $v = u^{\tau} = caaababa$.

we can see that sep-min$(u, u^{\tau}) = \{ac, bc, cc\}$ and $|u| + |u^{\tau}| + 1 = 20$.

$$u = \begin{array}{cccccccccc} c & b & a & a & a & b & a & b & a & c \\ 1 & 2 & 3 & 4 & 5 & 6 & 7 & 8 & 9 & 10 \end{array}$$

The algorithm 1 provides

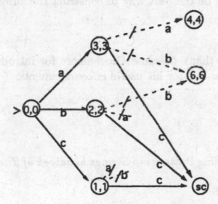

fig.2 $SM(u, u^{\tau}), u = cbaaababac.$

At most 7 states and 9 edges are created ; the transition a is forbidden from state (2,2) and (1,1), the transition b is forbidden from state (1,1). At the end the edges $((3,3),(4,4))$, $((3,3),(6,6))$, $((2,2),(6,6))$ are suppressed.

The algorithm 2 provides

53

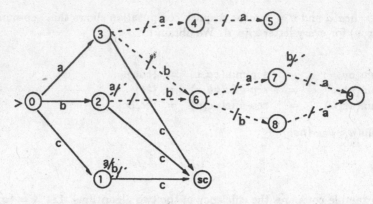

fig.3 $SM(u, u^r), u = cbaaababac.$

$|u|.|A| = 27$; exactly 11 states and 14 edges are created; 4 transitions must be forbidden, a and b from state 1, a from state 2 and b from state 7 ; 6 states (the states 4, 5, 6, 7, 8, 9) and 8 edges must be suppressed.

Conclusion :

For calculating sep-min(ua, u), $u \in A^*$, $a \in A$ algorithm 2 is more natural to consider. But algorithm 1 seems to be the best way to construct the automaton $SM(u, v)$ in any case.

Acknowledgements : I wish to thank Maxime Crochemore for introducing me to the study of Imre Simon's algorithm and for his useful encouragements.

Bibliographie

[1] AHO, A.V. Algorithms for Finding Patterns in Strings *handbook of Theorical Computer Science, Vol A* (1990) 257–300.

[2] EILENBERG, S., *Automata, Languages and Machine, Vol A* (Academic Press 1974).

[3] LOTHAIRE, M., *Combinatorics on words* (Addison-Weysley, Reading, MA, 1983).

[4] SIMON, I., An Algorithm to Distinguish Words efficiently by their Subwords, *Personal Communication.*

[5] WATERMAN, M.S., *Mathemetical Methods for DNA Sequences*, (CRC Press 1989).

Covering a String

Costas S. Iliopoulos[*]
King's College London and
Curtin University

Dennis W. G. Moore[†]
Curtin University

Kunsoo Park[‡]
King's College London

Abstract

We consider the problem of finding the repetitive structures of a given string x. The period u of the string x grasps the repetitiveness of x, since x is a prefix of a string constructed by concatenations of u. We generalize the concept of repetitiveness as follows: A string w *covers* a string x if there exists a string constructed by concatenations and superpositions of w of which x is a substring. A substring w of x is called a *seed* of x if w covers x. We present an $O(n \log n)$ time algorithm for finding all the seeds of a given string of length n.

Keywords: Combinatorial algorithms on words, string algorithms, periodicity of strings, covering of strings, partitioning.

1. Introduction

Regularities in strings arise in many areas of science: combinatorics, coding and automata theory, molecular biology, formal language theory, system theory, etc. Here we study string problems focused on finding the repetitive structures of a given string x. A typical regularity, the period u of the string x, grasps the repetitiveness of x, since x is a prefix of a string constructed by concatenations of u. We consider a problem derived by generalizing this concept of repetitiveness by allowing overlaps between the repeated segments.

Apostolico, Farach and Iliopoulos [3] considered the *Aligned String Covering* (ASC) problem: Given a string x, find a substring w of x such that x can be constructed by

---

[*] Department of Computer Science, King's College London, Strand, London, U.K. and School of Computing, Curtin University, Perth, WA, Australia. Email: csi@dcs.kcl.ac.uk. Partially supported by the SERC grant GR/F 00898, the NATO grant CRG 900293, the ESPRIT BRA grant for ALCOM II, and the MRC grant G 9115730.

[†] School of Computing, Curtin University, Perth, WA, Australia. Email: moore@marsh.cs.curtin.edu.au.

[‡] Department of Computer Science, King's College London, Strand, London, U.K. Email: kpark@dcs.kcl.ac.uk. Partially supported by the MRC grant G 9115730.

concatenations and superpositions of w; the string w is said to provide an *aligned cover* for x. The shortest such w ($\neq x$) is called the *quasiperiod* of x, and x is said to be *quasiperiodic* if x has a quasiperiod. For example, if $x = abaabababaaba$, then x is quasiperiodic with the quasiperiod $w = aba$. A linear-time algorithm for computing the quasiperiod was given in [3]. Moreover, Apostolico and Ehrenfeucht [2] introduced *maximal quasiperiodic* substrings of a string x, and presented an $O(n \log^2 n)$ time algorithm for computing all maximal quasiperiodic substrings of x.

We focus on the *General String Covering* (GSC) problem. We say that a string w *covers* a string x if there exists a string constructed by concatenations and superpositions of w of which x is a substring. For example, $abca$ covers $abcabcaabc$. A substring w of a string x is called a *seed* of x if w covers x. The GSC problem is as follows: Given a string x of length n, compute all the seeds of x. Note that there may be more than one shortest seeds (e.g., for $abababa$, both ab and ba are the shortest seeds).

The Aligned String Covering problem considered in [3] is restricted in the following sense. The computational focus of the ASC problem is on finding the quasiperiod of x rather than finding all possible strings which provide aligned covers for x. More importantly, the first and the last occurrences of the quasiperiod w in x must align exactly with the given string x. Therefore, the period of x may not provide an aligned cover for x, though a periodic string is quasiperiodic. Consider the string $abcabcabca$, which is quasiperiodic with the quasiperiod $abca$. However, the period abc does not provide an aligned cover for the string.

We present an $O(n \log n)$ time algorithm for finding all the seeds of a given string of length n. Since the possible number of seeds of a string of length n can be as large as $O(n^2)$, this is achieved by reporting a group of seeds in one step.

In Section 2 we classify the seeds of a given string into two kinds: easy seeds and hard seeds. Then we describe how to find all easy seeds in $O(n)$ time. In Sections 3 and 4 we describe how to find all hard seeds in $O(n \log n)$ time. In Section 5 we mention open problems related to the General String Covering problem.

2. Preliminaries

A *string* is a sequence of zero or more symbols in an alphabet Σ. The set of all strings over the alphabet Σ is denoted by Σ^*. A string x of length n is represented by $x_1 \cdots x_n$, where $x_i \in \Sigma$ for $1 \leq i \leq n$. A string $w = w_1 \cdots w_m$ is a *substring* of $x = x_1 \cdots x_n$ if $w_1 \cdots w_m = x_i \cdots x_{i+m-1}$ for some i. A string w is a *prefix* of x if $x = wu$ for $u \in \Sigma^*$. Similarly, w is a *suffix* of x if $x = uw$ for $u \in \Sigma^*$.

The string xy is a *concatenation* of two strings x and y. The concatenations of k copies of x is denoted by x^k. For two strings $x = x_1 \cdots x_n$ and $y = y_1 \cdots y_m$ such that $x_{n-i+1} \cdots x_n = y_1 \cdots y_i$ for some $i \geq 1$, the string $x_1 \cdots x_n y_{i+1} \cdots y_m$ is a *superposition of x and y with i overlap*. A string $w = w_1 \cdots w_n$ is a *cyclic rotation* of $x = x_1 \cdots x_n$ if $w_1 \cdots w_n = x_i \cdots x_n x_1 \cdots x_{i-1}$ for some $1 \leq i \leq n$ (for $i = 1$, $w = x$). For a substring w of x, uwv for $u, v \in \Sigma^*$ is an *extension* of w in x if uwv is a substring of x; wv for $v \in \Sigma^*$ is a *right extension* of w in x if wv is a substring of x; uw for $u \in \Sigma^*$ is a *left extension* of w in x if uw is a substring of x.

A string u is a *period* of x if x is a prefix of u^k for some k, or equivalently if x is a prefix of ux. The *period* of a string x is the shortest period of x. A string w

semi-covers a string x, if x can be constructed by concatenations and superpositions of w. A string w *covers* a string x, if there exists a string z constructed by concatenations and superpositions of w of which x is a substring. Such a string z is called a *cover* of x by w, and the shortest cover is called *the cover* of x by w. A *seed* of a string x is a substring of x that covers x.

In the definition of the seed, we do not consider strings that are longer than the given string x, because we are interested in the repetitive structures of x. We also restrict ourselves to substrings of x, because any other string w that covers x can be easily derived from a substring of x as Theorem 1 shows.

Lemma 1. Let w be a seed of x. If w covers x only by concatenations, all the cyclic rotations of w cover x by concatenations.

Proof. Since w covers x by concatenations, there is a cover w^k of x by w. Let $\hat{w}$ be a cyclic rotation of w. Since w^k is a substring of $\hat{w}^{k+2}$, $\hat{w}^{k+2}$ is a cover of x and therefore $\hat{w}$ covers x by concatenations. □

Lemma 2. Let w be a seed of x. If w covers x only by concatenations, all the extensions of w in x cover x.

Proof. Let $\hat{w} = uwv$ be an extension of w in x. Since w covers x by concatenations, there is a cover w^k of x by w and u and v are a suffix and a prefix of w, respectively. Thus $uw^k v$ is constructed by superpositions of k copies of $\hat{w}$. Since $uw^k v$ is a cover of x by $\hat{w}$, $\hat{w}$ covers x. □

Theorem 1. Let y be a string such that $|y| \leq |x|$ and y is not a substring of x. If y covers x, then there is a seed of x that is either a substring of y or a cyclic rotation of y or a cyclic rotation of a substring of y.

Proof. Let $m = |y|$ and $n = |x|$. Since $m \leq n$ and y is not a substring of x, y covers x with exactly two copies. Let z be the cover of x by y, and i be the integer such that x is a prefix of $z_i \cdots z_{|z|}$.

1. $z = y^2$: Let $w = y_i \cdots y_m y_1 \cdots y_{i-1}$, which is a cyclic rotation of y. Since $|w| = m$, w is a prefix of x. Since w^2 is a cover of x, w is a seed of x. By Lemma 1 y is derived from w.
2. $z = yy_j \cdots y_m$ for $j > 1$. There are two cases.
   2.1. $i \leq j$: Let $w = y_i \cdots y_{m-j+i}$, which is a substring of y. Since $|w| = m - j + 1 < m$, w is a prefix of x. Since w^3 is a cover of x, w is a seed of x. By Lemma 2 y is derived from w.
   2.2. $i > j$: Let $w = y_i \cdots y_m y_j \cdots y_{i-1}$, which is a cyclic rotation of $y_1 \cdots y_{m-j+1}$. Since $|w| = m - j + 1 < m$, w is a prefix of x. Since w^2 is a cover of x, w is a seed of x. By Lemmas 1 and 2 y is derived from w. □

For a string of length n there are $O(n^2)$ different substrings, each of which can possibly be a seed. In fact, the string $(a_1 a_2 \cdots a_m)^4$, where $a_i \neq a_j$ for $i \neq j$, has $O(m^2)$ seeds. Our algorithm finds all seeds in $O(n \log n)$ time by reporting several seeds at once.

Each seed w of x is reported by a pair (i, j), where i and j are the start position and the end position, respectively, of an occurrence of w in x. If $(i, j), (i, j+1), \ldots, (i, j+k)$ are seeds, we report them by a triple (i, j, k). If $(i, j), (i-1, j), \ldots, (i-k, j)$ are seeds, we report them by a triple $(i, j, -k)$.

We do the following preprocessing on the given string x. For each position i of x, let $F[i]$ be the length of the maximal proper prefix of $x_1 \cdots x_i$ which is also a suffix of $x_1 \cdots x_i$, and let $B[i]$ be the length of the maximal proper suffix of $x_i \cdots x_n$ which is a prefix of $x_i \cdots x_n$. The arrays $F[1..n]$ and $B[1..n]$ can be computed in $O(n)$ time by the preprocessing of the Knuth-Morris-Pratt algorithm [9] for string matching.

For $1 \leq i \leq n$, let $F_i^+ = i - F[i]$. Then $F_i^+ > 0$ for all i, because i is the length of $x_1 \cdots x_i$, and $F[i]$ is the length of a proper prefix of $x_1 \cdots x_i$. Consider the values of F from left to right. Let $F[i] = r$. Then $F[i+1] = r+1$ if $x_{i+1} = x_{r+1}$, and $F[i+1] \leq r$ otherwise. That is, the F values increase by 1 or are nonincreasing. Therefore, the F^+ values are nondecreasing (monotonically increasing) from left to right. Since the F^+ values are between 1 and n, and they are nondecreasing, we can define the break points of the F^+ values: For $1 \leq i \leq n$, let $MF^+[i]$ be the largest j such that $F_j^+ \leq i$. We do not store the F^+ values into a separate array, because they are easily computed from the array F. The array MF^+ is computed in $O(n)$ time by scanning the F^+ values from left to right.

Similarly, let $B_i^+ = (n - i + 1) - B[i]$. Then $B_i^+ > 0$ for all i, because $n - i + 1$ is the length of $x_i \cdots x_n$, and $B[i]$ is the length of a proper suffix of $x_i \cdots x_n$. The B^+ values are nondecreasing from right to left. For $1 \leq i \leq n$, let $MB^+[i]$ be the smallest j such that $B_j^+ \leq i$. The array MB^+ is computed in $O(n)$ time by scanning the B^+ values from right to left.

We classify the seeds of x into two kinds: A seed w is an *easy seed* if there is a substring of w which covers x only by concatenations; w is a *hard seed* otherwise. Let $u = x_1 \cdots x_p$ be the period of x. It is easy to see that u covers x by concatenations. The following lemmas characterize easy seeds of x.

Lemma 3. A seed w is an easy seed if and only if $|w| \geq p$.

Proof. (if) Let w be a seed such that $|w| \geq p$. Since w is a substring of x and $|w| \geq p$, w contains a cyclic rotation $\hat{u}$ of u as a substring. Since $\hat{u}$ covers x by concatenations by Lemma 1, w is an easy seed.

(only if) Let w be an easy seed. Suppose that $|w| < p$. Since w is an easy seed, there is a substring $\hat{w}$ of w that covers x by concatenations; i.e., x is a substring of w^k. Let w' be the cyclic rotation of $\hat{w}$ which is a prefix of x. Then w' covers x by concatenations by Lemma 1, and therefore w' is a period of x. Since the length of w' is less than p, we have a contradiction. Therefore, $|w| \geq p$. $\square$

Lemma 4. A substring w of x is an easy seed if and only if w is a right extension of a cyclic rotation of u.

Proof. (if) Let w be a right extension of a cyclic rotation $\hat{u}$ of u. Since $\hat{u}$ covers x by concatenations by Lemma 1, w is an easy seed.

(only if) Let w be an easy seed. By Lemma 3, $|w| \geq p$. Let $\hat{w} = w_1 \cdots w_p$. Since $\hat{w}$ is a substring of x whose length is p, it is a cyclic rotation of u. Thus w is a right extension of a cyclic rotation of $\hat{x}$. □

We first find all the easy seeds of x. Let $p = n - F[n]$, the length of the period of x. By Lemma 4, all cyclic rotations of $x_1 \cdots x_p$ and their right extensions are seeds of x. Therefore, all easy seeds can be reported in $O(p)$ time. Including the preprocessing, we find easy seeds in $O(n)$ time. In the following sections we will find all the hard seeds in $O(n \log n)$ time.

3. Finding hard seeds

A substring w of the given string x is a *candidate* for a hard seed if there exists a substring x' of $x = ux'v$ such that w semi-covers x' and $|u|, |v| < |w|$. The string u (v) is called the *head* (*tail*) of x with respect to w. In order for the candidate w to be a seed, the head u and the tail v must be covered by w. If w covers the head u (the tail v), then the overlap between the w covering u (v) and the first (last) occurrence of w in x' is called the *left overlap* (*right overlap*) of w. The length of the left (right) overlap of w is called *l-size* (*r-size*) of w. We divide hard seeds into two types:

(1) A hard seed is a type-A seed if its l-size is larger than or equal to its r-size.
(2) A hard seed is a type-B seed if its l-size is smaller than its r-size.

We describe our algorithm for finding all type-A seeds. The algorithm for type-B seeds is analogous, and will be discussed at the end of Section 4. Assume that we have a set S of candidates w_i for $f \leq i \leq g$ such that

1. $|w_i| = i$, and
2. w_i has t occurrences in x and the j-th occurrence of w_i is $x_{e_j} \cdots x_{e_j+i-1}$ for $1 \leq j \leq t$ ($e_1 = e_t$ if there is only one occurrence).

That is, the set of start positions of the occurrences are the same for every candidate in S. The set S will be called a *candidate set* of x. By the definition of candidates, the head and the tail of x with respect to w_f are shorter than w; i.e., $e_1 \leq f$ and $n - e_t + 1 < 2f$. In Section 4 we will show how to find all the candidate sets of x.

We now describe how to find the type-A seeds among the candidates of S in constant time. Let $k = (n - e_t + 1) - B[e_t]$; i.e., w_k (if any) covers the tail of x with respect to w_k by concatenation.

(i) If $k > g$, then no candidate w_i for $f \leq i \leq g$ covers the tail of x with respect to w_i. Therefore, none of the candidates are seeds.
(ii) If $f \leq k \leq g$, then w_k is the shortest candidate which covers the tail of x with respect to itself.
(iii) If $k < f$, then w_f is the shortest candidate which covers the tail of x with respect to itself.

In Cases (ii) and (iii), let $h = \max(k, f)$ (i.e., w_h is the shortest candidate which covers the tail of x). We consider candidates $w_h, \ldots, w_g$, and find type-A seeds among them. For $h \leq i \leq g$, the r-size of w_i is $i - k$. Note that $e_1 + i - 1$ is the end position of the first occurrence of w_i in x for $h \leq i \leq g$. For $h \leq i \leq g$, the l-size of w_i (if any) is $F[e_1 + i - 1] - e_1 + 1$. Since we are looking for type-A seeds, we want to find candidates whose l-size is greater than or equal to its r-size; i.e., w_i for $h \leq i \leq g$ such that $F[e_1 + i - 1] - e_1 + 1 \geq i - k$ (or $(e_1 + i - 1) - F[e_1 + i - 1] = F^+_{e_1 + i - 1} \leq k$). Let s be the integer such that $e_1 + s - 1 = MF^+[k]$; i.e., s is the largest integer such that the l-size of w_s is larger than or equal to its r-size.

1. Case $s < h$: For $h \leq i \leq g$, the l-size of w_i is less than its r-size, since F^+ is nonincreasing. Therefore, there are no type-A seeds in S.
2. Case $s \geq h$: For $h \leq i \leq \min(s, g)$, the l-size of w_i is greater than or equal to its r-size. Since the r-size of w_i is greater than or equal to 0, so is its l-size. Thus w_i covers the head of x. Therefore, $w_h, \ldots, w_{\min(s,g)}$ are type-A seeds.

4. Finding candidate sets

In this section we find all the candidate sets of the given string x. Since the length of a hard seed is less than p by Lemma 3, we find candidates whose length is less than p. A candidate set S is $\{w_i \mid f \leq i \leq g\}$ such that

1. $|w_i| = i$, and
2. w_i has t occurrences in x and the j-th occurrence of w_i is $x_{e_j} \cdots x_{e_j + i - 1}$ for $1 \leq j \leq t$.

For each substring w of x, the *start-set* of w is the set of start positions of all occurrences of w in x. For each start-set, we maintain its elements in ascending order. This sorted list will be used to find a candidate set from a start-set. An *equi-set* is a set of substrings of x whose start-sets are the same. Note that a start-set is associated with an equi-set and vice versa. Although there are $O(n^2)$ substrings of x, there are only $O(n)$ distinct start-sets.

Lemma 5. [5,7] There are $O(n)$ distinct start-sets (i.e., $O(n)$ equi-sets).

To find the start-sets, we define equivalence relations E_l for $1 \leq l < n$. E_l is defined on the positions $\{1, 2, \ldots, n - l + 1\}$ of x: iE_lj if $x_i \cdots x_{i+l-1} = x_j \cdots x_{j+l-1}$. Now we maintain equivalence classes for each E_l. Note that the start-set of a substring of length l is an equivalence class of E_l. If a start-set A is an equivalence class of $E_l, \ldots, E_{l'}$, the equi-set associated with A is the set of strings of length l to l' whose start positions are the elements of A.

We now describe how to find all the start-sets. It is easy to see that E_{l+1} is a refinement of E_l, excluding the position $n - l + 1$. The equivalence classes of E_1 can be computed by scanning x in $O(n \log n)$ time when the alphabet is general. Then we compute $E_2, E_3, \ldots$ successively until all classes are singleton sets or $l = p$. At stage l of the refinements we compute E_{l+1} from E_l. The refinement is based on:

$$iE_{l+1}j \quad \text{if and only if} \quad iE_lj \text{ and } (i+1)E_l(j+1).$$

That is, i and j in an equivalence class of E_l belong to the same equivalence class of E_{l+1} if and only if $(i+1)E_l(j+1)$. An easy solution is: (1) take each class C of E_l; (2)

procedure PARTITION
  compute E_1;
  SMALL ← all classes of E_1;
  $l ← 1$;
  while $l < p$ and there is a non-singleton class of E_l **do**
    copy classes in SMALL into QUEUE;
    empty SMALL;
    $l ← l + 1$;
    while QUEUE not empty **do**
      extract a class C from QUEUE;
      partition with respect to C;
      for each split class D, maintain its new subclasses;
    end do
    for each split class D (into r subclasses) **do**
      put $r - 1$ small subclasses of D into SMALL;
    end do
  end do
end

Fig. 1. Procedure PARTITION

partition C so that $i, j \in C$ go to the same class of E_{l+1} if and only if $(i+1)E_l(j+1)$. This method leads to $O(n^2)$ time, since each refinement requires $O(n)$ time and there can be $O(n)$ stages of refinements.

We do the refinement more efficiently as follows.

(1) Take a class C of E_l.
(2) Instead of partitioning C, we partition with respect to C those classes D of E_l which has at least one i such that $i + 1 \in C$, and one j such that $j + 1 \notin C$. That is, each D is partitioned into classes $\{i \in D \mid i+1 \in C\}$ and $\{i \in D \mid i+1 \notin C\}$.

Note that at the end of stage l, for each class D of E_{l+1} there exists one class A of E_l such that for all $i \in D$, $i + 1 \in A$. This fact can be more easily observed in terms of strings: if aw for $a \in \Sigma$ and $w \in \Sigma^+$ is the string whose start-set is D, then w is the string whose start-set is A. We call D a *preimage class* of A.

(i) If a class A of E_l is not split at stage l, we need not partition the preimage classes of A with respect to A at stage $l + 1$.
(ii) If A is split into $C_1, \ldots, C_r$ at stage l, we need to partition the preimage classes of A at stage $l+1$. For a preimage class D, let $D_s = \{i \in D \mid i+1 \in C_s\}$ for $1 \le s \le r$. We can partition the preimage class D with respect to any $r-1$ classes of $C_1, \ldots, C_r$, and the result will be the same because $D_s = D - (D_1 + \cdots + D_{s-1} + D_{s+1} + \cdots + D_r)$. Since we can choose any $r - 1$ classes, we partition with respect to $r - 1$ small ones except the largest at stage $l + 1$.

Procedure PARTITION in Fig. 1 shows the partitioning algorithm. Since there are $O(n)$ classes, we represent each class by a number k for $1 \le k \le cn$. Each class is implemented by a doubly linked list of its elements in ascending order. When we partition with respect to C in Fig. 1, the classes D which are partitioned with respect to C can be easily identified because D is a class that contains i such that $i+1$ is in C. See [6] for more details of implementation, where the time complexity of PARTITION is shown to be proportional to the sum of the sizes of classes C with respect to which the partitioning is made. Initially ($l = 1$), all classes of E_1 are in SMALL; i.e., all positions in x belong to SMALL. Consider a position i in a class D of E_l. Suppose that D is split at stage l and the subclass D' containing i is put in SMALL. Then $|D'| \leq |D|/2$. Therefore, one position cannot belong to SMALL more than $\log n$ times. Since there are n positions, Procedure PARTITION takes $O(n \log n)$ time.

This partitioning is similar to the single function partitioning due to Hopcroft [8,1]. The former can be viewed as a special case of the latter in which $f(i) = i + 1$ with the following two exceptions.

1. One position is excluded at each stage.
2. Each stage must be separated from another, because each stage deals with equivalence relation E_l. (Thus the algorithm in [1] cannot be applied directly to our partitioning.)

Crochemore [6] used this partitioning for computing all repetitions in a string. Although the approach of Apostolico and Preparata [4] computes the same equivalence classes with their elements sorted, the partitioning method in Fig. 1 is simpler and more elegant ([4] uses quite complicated data structures).

We compute equi-sets from the start-sets as follows. When a new class A (which is a start-set) is formed at stage l, we record the number $l + 1$ with the class A (i.e., A is a class of E_{l+1}). The class A will be either split later or remain unchanged until PARTITION stops (if A is a singleton set or stage p is reached). If the class A is split into smaller classes at stage l' for $l' > l$, the equi-set associated with A is the set of strings of length $l + 1$ to l' whose start-set is A. If A remains unchanged, let e_t be the largest element in A. The equi-set associated with A is the set of strings of length $l + 1$ to $\min(p - 1, n - e_t + 1)$ whose start-set is A.

We now obtain candidate sets from the equi-sets. Let $S = \{w_l, w_{l+1}, \ldots, w_g\}$ be an equi-set such that $|w_i| = i$ for $l \leq i \leq g$, and its associated start-set is A. From the start-set A, the differences of consecutive positions can be computed. For example, if $A = \{3, 5, 9, 12\}$, the differences are $2, 4, 3$. During the partitioning, we maintain only the maximum difference between elements for each equivalent class that has at least two elements. Since each class is implemented by a doubly linked list of increasing elements, and elements are only deleted once a class is formed (there are no insertions), computing maximum differences does not take more time than the partitioning. Let d be the maximum difference in A; $d = 0$ if A is a singleton set. Let $h = \max(d, l)$, and let e_1 and e_t be the first and last start positions of w_h in the given string x, respectively. If $h \leq g$, then w_i for $h \leq i \leq g$ semi-covers the substring x' of x, where $x' = x_{e_1} \cdots x_{e_t + i - 1}$. We now check the lengths of the head and tail of x. The length of the head (tail) of x with respect to w_i is $e_1 - 1$ ($n - e_t + 1 - i$). A candidate w_i must satisfy $i \geq e_1$ and $i > n - e_t + 1 - i$, since the head and tail must be shorter than w_i. Let $f = \max(h, e_1, \lceil \frac{n - e_t + 2}{2} \rceil)$. If $f \leq g$, then $\{w_f, \ldots, w_g\}$ is a candidate set.

For the type-B seeds, we compute the sets of end positions of substrings of x. Checking the heads and tails is symmetric to the case of type-A seeds. From the discussion so far, we have the following theorem.

Theorem 2. All the hard seeds of the given string x can be found in $O(n \log n)$ time.

5. Conclusion

We have presented an $O(n \log n)$ time algorithm for finding all the seeds of a given string of length n. An interesting open problem is to find segments which approximately cover a given string, together with the problem of computing the approximate period of a given string. Here we considered the exact version of this problem, where all segments must be the same. Another open problem is to design an efficient parallel algorithm for the General String Covering problem.

References

[1] A. V. Aho, J. E. Hopcroft and J. D. Ullman, "The Design and Analysis of Computer Algorithms," Addison-Wesley, 1974.

[2] A. Apostolico and A. Ehrenfeucht, Efficient detection of quasiperiodicities in strings, to appear in *Theoret. Comput. Sci.*

[3] A. Apostolico, M. Farach and C. S. Iliopoulos, Optimal superprimitivity testing for strings, *Inform. Process. Lett.* **39** (1991), 17–20.

[4] A. Apostolico and F. P. Preparata, Optimal off-line detection of repetitions in a string, *Theoret. Comput. Sci.* **22** (1983), 297–315.

[5] A. Blumer, J. Blumer, D. Haussler, A. Ehrenfeucht, M. T. Chen and J. Seiferas, The smallest automaton recognizing the subwords of a text, *Theoret. Comput. Sci.* **40** (1985), 31–55.

[6] M. Crochemore, An optimal algorithm for computing the repetitions in a word, *Inform. Process. Lett.* **12** (1981), 244–250.

[7] M. Crochemore, Transducers and repetitions, *Theoret. Comput. Sci.* **45** (1986), 63–86.

[8] J. E. Hopcroft, An $n \log n$ algorithm for minimizing states in a finite automaton, in Kohavi and Paz, ed., "Theory of Machines and Computations," Academic Press, New York, 1971, pp.189–196.

[9] D. E. Knuth, J. H. Morris and V. R. Pratt, Fast pattern matching in strings, *SIAM J. Comput.* **6** (1977), 323–350.

On the worst-case behaviour of some approximation algorithms for the shortest common supersequence of k strings

Robert W. Irving and Campbell B. Fraser*

Computing Science Department,
University of Glasgow,
Glasgow G12 8QQ,
Scotland.

Abstract. Two natural polynomial-time approximation algorithms for the shortest common supersequence (SCS) of k strings are analysed from the point of view of worst-case performance guarantee. Both algorithms behave badly in the worst case, whether the underlying alphabet is unbounded or of fixed size. For a Tournament style algorithm proposed by Timkovskii, we show that the length of the SCS found is between $k/4$ and $(3k+2)/8$ times the length of the optimal in the worst case. The corresponding bounds proved for the obvious Greedy algorithm are $(4k+17)/27$ and $(k-1)/e$. Even for a binary alphabet, no constant performance guarantee is possible for either algorithm, in contrast with the guarantee of 2 provided by a trivial algorithm in that case.

1 Introduction

Many algorithms have been proposed for finding a longest common subsequence (LCS), and equivalently a shortest common supersequence (SCS) of two strings — see, for example, [1] [4], [6], [11], [12], [15], [16], [17]. In the case of $k > 2$ strings, the LCS and SCS problems are no longer equivalent. Each problem can, however, be solved by a dynamic programming approach [8], and in the case of the LCS problem at least, some attempts have been made to improve upon the basic dynamic programming approach [5], [7].

However, both the LCS problem and the SCS problem are known to be NP-hard for general k [10], and remain so even when subject to quite severe restrictions. For example, they remain NP-hard if the strings are over an alphabet of size 2[14]. The problem therefore arises of designing fast approximation algorithms capable of finding a common subsequence or supersequence close to optimal in all cases, or perhaps in some special cases. However, Jiang and Li [9] have shown that, at least over a general alphabet, there can be no polynomial-time approximation algorithm for either LCS or SCS with a constant worst-case performance guarantee (unless P = NP). In this paper, we focus on the SCS problem, and in particular on the worst-case behaviour of two natural approximation algorithms for the SCS of k strings.

---

* Supported by a postgraduate research studentship from the Science and Engineering Research Council

Timkovskii [14] proposed a tournament-style approximation algorithm for SCS, and asked for a characterisation of its worst-case behaviour. Bradford and Jenkyns [3] gave an example to show that the Tournament algorithm does not always find a shortest common supersequence (though Timkovskii had not suggested that it would do so). A second obvious approximation algorithm involves a greedy strategy, in which we start with the first string in the set, and repeatedly find a SCS of the current common supersequence and the next string in the original set.

It is our aim in this paper to characterise the worst-case behaviour of both the Tournament algorithm and the Greedy algorithm. Our results show that both of the algorithms perform badly in the worst case, even for a small fixed alphabet. In the case of the Tournament algorithm, we show that the worst-case ratio of the length of the supersequence found to the length of the optimal lies between $k/4$ and $(3k + 2)/8$. In the case of the Greedy algorithm, this ratio is between $(4k + 17)/27$ and $(k - 1)/e$, and this lower bound can be attained even over a fixed size alphabet. The Tournament algorithm may not be quite as badly behaved in this case, but the ratio is still unbounded, and this can be contrasted with a trivial approximation algorithm which achieves a worst-case ratio of 2 for a binary alphabet.

2   Worst-case behaviour of the Tournament algorithm

The Tournament algorithm applied to $k = 2^p$ strings $\alpha_1^0, \ldots, \alpha_k^0$ of length n proceeds as follows. The tournament has p 'rounds'. In the rth round there are 2^{p-r} 'matches'. The ith match generates an arbitrary shortest common supersequence α_i^r of the two strings α_{2i-1}^{r-1} and α_{2i}^{r-1}, for $i = 1, \ldots, 2^{p-r}$.

It follows that α_i^r is a common supersequence of the 2^r strings $\alpha_{(i-1)2^r+1}^0, \ldots, \alpha_{i2^r}^0$, and in particular that α_1^p is a common supersequence of all of the 2^p original strings.

It is easy to verify that, using a standard $O(mn)$ algorithm for the SCS of 2 strings of lengths m and n, the Tournament algorithm for k strings has $O(k^2n^2)$ complexity.

Theorem 1 *For a given set of $k = 2^p$ strings, denote by s the length of an SCS, and by t the length of a CS returned by the Tournament algorithm. Then*

$$\frac{t}{s} \le \frac{3k + 2}{8}.$$

Proof Consider an SCS α, of length s say, of the strings $\alpha_1^0, \ldots, \alpha_k^0$, together with a particular embedding of each of the k individual strings in α. For $i = 1, 2, \ldots, k$, let x_i denote the number of positions in α that correspond to symbols in exactly i of the individual strings in the chosen embedding. Then

$$\sum_{i=1}^{k} x_i = s \tag{1}$$

and

$$\sum_{i=1}^{k} i x_i = kn \tag{2}$$

Also, it is easy to see that

$$t = kn - \sum_{r=1}^{p} \sum_{i=1}^{2^{p-r}} l_{2i-1,2i}^{r-1} \tag{3}$$

where $l_{2i-1,2i}^{r-1}$ is the length of the longest common subsequence of the strings α_{2i-1}^{r-1} and α_{2i}^{r-1}.

Let m_{ij} $(1 \le i < j \le k)$ be the number of positions in α that correspond to symbols in both α_i^0 and α_j^0 for the chosen embeddings.

Since α_{2i-1}^{r-1} is a supersequence of the 2^{r-1} strings $\alpha_{(i-1)2^r+1}^0, \ldots, \alpha_{(i-1)2^r+2^{r-1}}^0$, and α_{2i}^{r-1} is a supersequence of the 2^{r-1} strings $\alpha_{(i-1)2^r+2^{r-1}+1}^0, \ldots, \alpha_{i2^r}^0$, it follows that

$$l_{2i-1,2i}^{r-1} \ge \max_{\substack{1 \le u \le 2^{r-1} \\ 1 \le v \le 2^{r-1}}} m_{(i-1)2^r+u,(i-1)2^r+2^{r-1}+v}.$$

and so

$$l_{2i-1,2i}^{r-1} \ge \frac{1}{2^{2r-2}} \sum_{u=1}^{2^{r-1}} \sum_{v=1}^{2^{r-1}} m_{(i-1)2^r+u,(i-1)2^r+2^{r-1}+v}. \tag{4}$$

Combining (3) and (4), we get

$$t \le kn - X \tag{5}$$

where

$$X = \sum_{r=1}^{p} \frac{1}{2^{2r-2}} \sum_{i=1}^{2^{p-r}} \sum_{u=1}^{2^{r-1}} \sum_{v=1}^{2^{r-1}} m_{(i-1)2^r+u,(i-1)2^r+2^{r-1}+v}. \tag{6}$$

Manipulation of (6) gives

$$X = \frac{1}{2^{2p-2}} \sum_{1 \le u < v \le 2^p} m_{uv} + \sum_{r=1}^{p-1} \frac{3}{2^{2p-2r}} \sum_{i=0}^{2^r-1} \sum_{i2^{p-r}+1 \le u < v \le (i+1)2^{p-r}} m_{uv}. \tag{7}$$

This may be seen by noting that, in both (6) and (7) the coefficient of m_{uv} is $1/2^{2x}$, where x is such that 2^x divides some number in the closed interval $[u, v-1]$ but 2^{x+1} does not.

We now seek a lower bound, in terms of $x_1, \ldots, x_{2^p}$, for the two expressions involving $\sum m_{uv}$ in (7). The first is easy, namely

$$\sum_{1 \le u < v \le 2^p} m_{uv} = \sum_{j=2}^{2^p} \binom{j}{2} x_j. \tag{8}$$

This is because x_j counts the number of positions in α that represent symbols in exactly j of the strings for the chosen embeddings. For any particular choice of exactly j strings, there are $\binom{j}{2}$ choices of m_{uv} that contribute.

The lower bound for

$$\sum_{i=0}^{2^r-1} \sum_{i2^{p-r}+1 \le u < v \le (i+1)2^{p-r}} m_{uv} \tag{9}$$

in terms of the x_j is less obvious. Recall that x_j counts the number of positions in the SCS α that correspond to symbols in exactly j of the original strings. So we ask, for any particular j strings, what is the smallest number of terms in (9) that count the positions in α at which just these j strings are represented? This smallest number will arise when the indexes of the j strings are as evenly distributed as possible in the 2^r intervals $[i2^{p-r} + 1, (i+1)2^{p-r}]$ $(i = 0, \ldots, 2^r - 1)$, and will be precisely

$$\sum_{w=0}^{2^r-1} \binom{y_w}{2} \tag{10}$$

where y_w is the number of indexes in the wth interval.

But in a typical case of even distribution, there will be $\lfloor \frac{j}{2^r} \rfloor$ indices in the first interval, $\lfloor \frac{j+1}{2^r} \rfloor$ indices in the second interval, ..., and $\lfloor \frac{j+2^r-1}{2^r} \rfloor$ indices in the 2^rth interval, and these values of y_w, together with (10), establish

$$\sum_{i=0}^{2^r-1} \sum_{i2^{p-r}+1 \leq u < v \leq (i+1)2^{p-r}} m_{uv} \geq \sum_{j=1}^{2^p} \sum_{w=0}^{2^r-1} \binom{\lfloor \frac{j+w}{2^r} \rfloor}{2} x_j. \tag{11}$$

From (7), (8), (9) and (11) we obtain

$$X \geq \sum_{i=1}^{2^p} \left\{ \frac{1}{2^{2p-2}} \binom{i}{2} + \sum_{r=1}^{p-1} \frac{3}{2^{2p-2r}} \sum_{w=0}^{2^r-1} \binom{\lfloor \frac{i+w}{2^r} \rfloor}{2} \right\} x_i. \tag{12}$$

So, combining (1), (2), (3), (5) and (12)

$$\frac{t}{s} \leq \frac{\sum_{i=1}^{k} c_i x_i}{\sum_{i=1}^{k} x_i}. \tag{13}$$

where

$$c_i = i - \frac{1}{2^{2p-2}} \binom{i}{2} - \sum_{r=1}^{p-1} \frac{3}{2^{2p-2r}} \sum_{w=0}^{2^r-1} \binom{\lfloor \frac{i+w}{2^r} \rfloor}{2}. \tag{14}$$

It is immediate from (13) that

$$\frac{t}{s} \leq \max_{1 \leq i \leq k} c_i. \tag{15}$$

It is not hard to show that $c_i < c_{i+1}$ if $i \leq 2^{p-1} - 1$, and $c_i > c_{i+1}$ if $i \geq 2^{p-1}$, so that the maximum value of c_i occurs when $i = 2^{p-1} (= \frac{k}{2})$. Substituting in (14) and simplifying gives $\frac{t}{s} \leq \frac{3k+2}{8}$, and the proof of the theorem is complete. $\square$

3   Bad examples for the Tournament algorithm

We give an example involving $k = 2^p$ strings of length $2^{k/2-1}$ over an alphabet of size $2^{k/2}$. We denote the elements of the alphabet by the binary representations of the integers $0, 1, \ldots, 2^{k/2} - 1$, i.e., by strings of length $k/2$ over the alphabet $\{0,1\}$, and we let α be the string of length $2^{k/2}$ comprising the symbols of the alphabet in natural order.

For $i = 1, 2, \ldots, k/2$, we define

α^0_{2i-1} = the subsequence of α consisting of elements whose ith bit is 0;

α^0_{2i} = the subsequence of α consisting of elements whose ith bit is 1.

So every string is a substring of α, and therefore the SCS of $\alpha_1, \ldots, \alpha_k$ has length $2^{k/2}$.

In the first round of the tournament, α^0_{2i-1} is matched with α^0_{2i} ($i = 1, \ldots, k/2$), and a feasible outcome is

α^1_i = the permutation of α obtained from α by flipping the ith bit.

Theorem 2 *The length of the SCS of* $\alpha^1_1, \ldots, \alpha^1_{k/2}$ *is at least*

$$f(k) = k2^{k/2-2} + 3k2^{k/4-3}.$$

Proof Consider an SCS β of $\alpha^1_1, \ldots, \alpha^1_{k/2}$, and a fixed embedding of each α^1_i in β. Let S_{ij} ($1 \leq i < j \leq k/2$) be the set of positions in β representing a symbol in both α^1_i and α^1_j in the chosen embeddings, and let γ_{ij} be the subsequence of β corresponding to the positions in S_{ij}. Then the symbols in γ_{ij} have fixed values in the ith and jth bits.

We can partition the set $T_i = \{1, \ldots, k/2\} \setminus \{i\}$ into two subsets, $T_i = T_{i0} \cup T_{i1}$ where $j \in T_{i0}$ if the elements in γ_{ij} have ith bit 0, and $j \in T_{i1}$ otherwise. Suppose that $|T_{i0}| = x$ and $|T_{i1}| = y$, with $x + y = k/2 - 1$.

The symbols of α^1_i that are in γ_{ij} for some $j \in T_{i0}$ must have ith bit 0 and a fixed value in one of x of the other bits. There can be at most $2^{k/2-1} - 2^{k/2-x-1}$ such symbols. Likewise, there can be at most $2^{k/2-1} - 2^{k/2-y-1}$ symbols of α^1_i that are in γ_{ij} for some $j \in T_{i1}$. So in total, at most $2^{k/2} - 2^{k/2-1}/(2^{-x} + 2^{-y})$ symbols of β are part of the embedding of α^1_i and of α^1_j for some $j \neq i$, with $x + y = k/2 - 1$. This quantity is maximised when x and y are $k/4$ and $k/4 - 1$ in some order, giving a maximum value of $2^{k/2} - 3.2^{k/4-1}$. If this maximum degree of overlap is attained for every i and j, then the length of the string β will be given by $(k/2)2^{k/2} - (1/2)(k/2)\{2^{k/2} - 3.2^{k/4-1}\}$, which simplifies to $k2^{k/2-2} + 3k2^{k/4-3}$. □

From this theorem, and the fact that the SCS of the original strings $\alpha^0_1, \ldots, \alpha^0_k$ have an SCS of length $2^{k/2}$, we get the following immediate corollary.

Corollary 1 *For k an arbitrary power of 2, there is a set of k strings for which*

$$\frac{t}{s} \geq \frac{k}{4}\left(1 + \frac{3}{2^{k/4+1}}\right)$$

where s is the length of an SCS and t the length of a CS found by applying the Tournament algorithm.

4 The Tournament algorithm on an alphabet of fixed size

The bad examples for the Tournament algorithm given in the previous section were constructed using an unbounded alphabet. In the case where the alphabet is of fixed size, we might hope that the Tournament algorithm will behave in a more reliable way. However, in this section, we show that, even in the extreme case of alphabet size 2, the algorithm cannot guarantee to find a common supersequence of length within a constant factor of optimal. In this respect, therefore, it is not even as good as the trivial algorithm that always returns the string $(ab)^n$ as a supersequence of any number of strings of length n over the alphabet $\{a, b\}$, and which trivially guarantees to be within a factor of 2 of optimal in all cases.

Theorem 3 *For k an arbitrary power of 2, there is a set of k strings over an alphabet of size 2 for which*

$$\frac{t}{s} \geq \frac{1}{2}(1 + \log_2 k)$$

where s is the length of an SCS and t the length of a CS found by applying the Tournament algorithm.

Proof Consider the instance involving $k = 2^p$ strings all of length 1 defined by $\alpha_{2i-1}^0 = a$, $\alpha_{2i}^0 = b$ for $1 \leq i \leq 2^{p-1}$. It is easy to see that a feasible application of the tournament algorithm leads to the following:

$$\alpha_{2i-1}^{2j} = (ab)^j a$$

$$\alpha_{2i-1}^{2j-1} = (ab)^j$$

$$\alpha_{2i}^{2j} = (ba)^j b$$

$$\alpha_{2i}^{2j-1} = (ba)^j$$

Hence, after the rth round of the tournament, all of the strings have length $r + 1$, and, in particular, the common supersequence generated by the algorithm has length $p + 1 = 1 + \log_2 k$. But the shortest common supersequence has length 2, and the proof of the theorem is complete. $\square$

5 Worst case behaviour of the Greedy algorithm

The Greedy algorithm applied to k strings $\alpha_1, \ldots, \alpha_k$ of length n proceeds as follows. There are $k-1$ steps. Define $\beta_1 = \alpha_1$. In the ith step, an arbitrary SCS of the strings α_{i+1} and β_i is found and labelled β_{i+1}. Hence β_i is a common supersequence of the strings $\alpha_1, \ldots, \alpha_i$, and in particular β_k is a supersequence of all k strings.

It is easy to verify that, using a standard $O(mn)$ algorithm for the SCS of two strings of lengths m and n, the Greedy algorithm for k strings has $O(k^2 n^2)$ complexity.

Theorem 4 *For a given set of k strings, denote by s the length of an SCS, and by g the length of a CS returned by the Greedy algorithm. Then*

$$\frac{g}{s} \leq \frac{k-1}{e}.$$

Proof As before, consider an SCS α of the strings $\alpha_1, \ldots, \alpha_k$, together with a particular embedding of each of the k individual strings in α. For $i = 1, 2, \ldots, k$, let x_i denote the number of positions in α that correspond to symbols in exactly i of the individual strings in the chosen embedding. Then equations (1) and (2) hold as before.

Also, it is easy to see that

$$g = kn - \sum_{r=1}^{k-1} l_r \tag{16}$$

where l_r is the length of a longest common subsequence of the strings α_{r+1} and β_r.

Let m_{ij} $(1 \leq i < j \leq k)$ be the number of positions in α that correspond to symbols in both α_i and α_j for the chosen embeddings.

Since β_r is a supersequence of the strings $\alpha_1, \ldots, \alpha_r$, it follows that

$$l_r \geq \max_{1 \leq i \leq r} m_{i,r+1}$$

and so,

$$l_r \geq \frac{1}{r} \sum_{i=1}^{r} m_{i,r+1}. \tag{17}$$

Combining (16) and (17) we get

$$g \leq kn - X \tag{18}$$

where

$$X = \sum_{r=1}^{k-1} \frac{1}{r} \sum_{i=1}^{r} m_{i,r+1}. \tag{19}$$

We now seek a lower bound, in terms of $x_1, \ldots, x_k$, for the expression on the right hand side of (19). Recall that x_j counts the number of positions in the SCS α that correspond to symbols in exactly j of the original strings. So we ask, for any choice of j strings, how small the sum of coefficients of terms in (19) can be, terms that count the positions in α at which just these j strings are represented. This smallest sum will clearly arise for the particular j strings $\alpha_{k-j+1}, \alpha_{k-j+2}, \ldots, \alpha_k$, and will have value

$$\sum_{l=k-j+1}^{k-1} \frac{l-k+j}{l} = j - \sum_{l=k-j+1}^{k-1} \frac{k-j}{l}.$$

So

$$X \geq \sum_{j=1}^{k} \left(j - \sum_{l=k-j+1}^{k-1} \frac{k-j}{l} \right) x_j. \tag{20}$$

Combining (1), (2), (18) and (20) gives

$$\frac{g}{s} \leq \frac{\sum_{i=1}^{k} \sum_{l=k-i+1}^{k-1} \frac{k-i}{l} x_i}{\sum_{i=1}^{k} x_i}. \tag{21}$$

It is immediate from (21) that

$$\frac{g}{s} \leq \max_{1 \leq i \leq k} \sum_{l=k-i+1}^{k-1} \frac{k-i}{l}$$

$$= \max_{1 \leq j \leq k-1} j \sum_{l=j+1}^{k-1} \frac{1}{l}$$

$$= \max_{1 \leq j \leq k-1} j(H_{k-1} - H_j)$$

$$\leq \max_{1 \leq j \leq k-1} j(\log(k-1) - \log j).$$

By calculus, the maximum value of $x(K - \log x)$ occurs when $\log x = K - 1$, i.e., when $x = e^{K-1} = \frac{k-1}{e}$. Hence

$$\frac{g}{s} \leq \frac{k-1}{e}(\log(k-1) - \log(k-1) + 1) = \frac{k-1}{e}.$$

This completes the proof of the theorem. □

6 Bad examples for the Greedy algorithm

In the case of the Tournament algorithm we gave separate examples to establish lower bounds on any performance guarantee for unbounded and fixed-size alphabets. However for the Greedy algorithm it is sufficient to provide one construction for the extreme case of alphabet size 2. This demonstrates that, in the worst case, the Greedy algorithm is not as good as the trivial algorithm for a fixed size alphabet, indeed that the worst-case ratio of the Greedy solution to the optimal solution is at least a constant times k for any alphabet.

Theorem 5 *There is a set of k strings over an alphabet of size 2 for which*

$$\frac{g}{s} \geq \frac{k+7}{8}$$

where s is the length of an SCS and g the length of a CS found by applying the Greedy algorithm.

Proof Consider the set of k strings of length $n = 2^{k-2}$ over the alphabet $\{a, b\}$ defined by $\alpha_1 = a^n, \alpha_2 = b^n$, and for $3 \leq i \leq k$, $\alpha_i = (b^{2^{k-i}} a^{2^{k-i}})^{2^{i-3}}$. The length of the LCS of α_1 and α_2 is clearly 0 so we can take $\beta_2 = a^n b^n$. The length of the LCS of β_2 and α_3 is clearly $n/2$ so we can take $\beta_3 = a^n b^n a^{n/2}$. We now show by induction that the Greedy algorithm can choose

$$\beta_i = a^n b^n a^{2^{k-i}} \bigoplus_{j=1}^{2^{i-3}-1} b^{\mathcal{F}(i,j)} a^{2^{k-i}} \text{ for } 3 \leq i \leq k$$

where

$$\mathcal{F}(i,j) = (max \ 2^p : 2^p | j).2^{k-i}$$

and $\oplus$ denotes concatenation.

The base case is $\beta_3 = a^n b^n a^{n/2}$ which satisfies the lemma since $n = 2^{k-2}$. Now we show that if the formula is true for β_i $i \geq 3$ then it is true for β_{i+1}. So we find an SCS of β_i and α_{i+1} to get β_{i+1}.

$$\beta_i = a^n b^n a^{2^{k-i}} \bigoplus_{j=1}^{2^{i-3}-1} b^{\mathcal{F}(i,j)} a^{2^{k-i}}$$

$$\alpha_{i+1} = (b^{2^{k-(i+1)}} a^{2^{k-(i+1)}})^{2^{(i+1)-3}}$$

We claim that an LCS of β_i and α_{i+1} is $\gamma_i = (b^{2^{k-i-1}} a^{2^{k-i}})^{2^{i-3}}$. Firstly, γ_i is a subsequence of β_i since they have an equal number (2^{i-3}) of blocks of the form $b^x a^y$ for some x, y, and in each block, β_i has at least as many (2^{k-i}) b's and an equal number (2^{k-i}) of a's. Also, γ_i is a subsequence of α_{i+1} since, expressed differently,

$$\alpha_{i+1} = (b^{2^{k-(i+1)}} a^{2^{k-(i+1)}} b^{2^{k-(i+1)}} a^{2^{k-(i+1)}})^{2^{i-3}}$$

which is clearly a supersequence of $(b^{2^{k-(i+1)}} a^{2^{k-(i+1)}} a^{2^{k-(i+1)}})^{2^{i-3}} = \gamma_i$. There can be no longer CS of β_i and α_{i+1} since this would require at least one more block of the form $b^x a^y$, or at least one block containing more a's or more b's. More blocks are not possible since β_i has no more. More a's in any block is not possible since β_i has no more a's in each $b^x a^y$ block. More b's in any block is not possible since α_{i+1} has no more b's in each $b^x a^y$ block.

We can now construct a SCS of β_i and α_{i+1}.

$$SCS(\beta_i, \alpha_{i+1}) = a^n b^n a^{2^{k-i-1}} b^{2^{k-i-1}} a^{2^{k-i-1}} \bigoplus_{j=1}^{2^{i-2}-2} b^{\mathcal{F}(i,j)/2} a^{2^{k-i-1}}$$

$$= a^n b^n a^{2^{k-(i+1)}} \bigoplus_{j=1}^{2^{i-2}-1} b^{\mathcal{F}(i+1,j)} a^{2^{k-(i+1)}}$$

$$= a^n b^n a^{2^{k-i-1}} \bigoplus_{j=1}^{2^{(i+1)-3}-1} b^{\mathcal{F}(i+1,j)} a^{2^{k-(i+1)}}$$

$$= \beta_{i+1}$$

It follows that the supersequence generated by the Greedy algorithm is

$$\beta_k = a^n b^n a \bigoplus_{j=1}^{2^{k-3}-1} b^{\mathcal{F}(k,j)} a$$

The length of the prefix $a^n b^n$ is clearly $2n$. There are $2^{k-3} = n/2$ individual a's. The number of b's in the remaining part of the string is

$$\sum_{j=1}^{2^{k-3}-1} \max(2^p : 2^p | g) = \sum_{l=1}^{k-3} 2^{k-3-l} \cdot 2^{l-1}$$

$$= (k-3) \cdot 2^{k-4}$$

$$= \frac{(k-3)n}{4} \qquad \text{since } n = 2^{k-2}$$

The total length of the supersequence is then $5n/2 + (k-3)(n/4) = n(k+7)/4$. Since the length of the shortest common supersequence $(ab)^n$ is $2n$, the Greedy algorithm cannot guarantee to be within a factor better than $(n(k+7)/4)/2n = (k+7)/8$ of the optimal. $\square$

We can generalise the strategy in the proof of this theorem to alphabets of size greater than 2. For an alphabet $\{x_1, ..., x_a\}$ of size a, let $n = a^{k-a}$, and

$$\alpha_i = x_i^n \quad \text{for } 1 \leq i \leq a,$$

$$\alpha_i = (\bigoplus_{j=0}^{a-1} x_{a-j}^{a^{k-i}})^{a^{i-a-1}}, \quad \text{for } a+1 \leq i \leq k.$$

The length of the supersequence generated by applying rules analogous to those above is $na + n(a-1)/a + n(k-a-1)(1-(2a-1)/a^2)$ and the length of the shortest common supersequence is clearly na. Therefore the error in the approximation is $\geq \frac{(a-1)^2}{a^3} k$, which is maximised for $a = 3$, giving a lower bound of $(4k+17)/27$.

7 Conclusion and open problems

Although exact characterisation of the worst-case performance of the Tournament and Greedy algorithms for SCS remains open, close bounds have been found in the case of unbounded alphabet size in both cases, and for fixed alphabet size also in the case of the Greedy algorithm. These bounds show that, at least in the worst case, both algorithms perform very badly. Furthermore, in the case of fixed alphabet size, neither algorithm has a worst-case performance to compare with that of a trivial SCS algorithm that returns a fixed sequence.

The following problems remain open:

1) Is there a polynomial-time approximation algorithm for general SCS with worst-case performance guarantee better than $O(k)$?

2) Is there a polynomial-time approximation algorithm for SCS over an alphabet of size a with worst-case performance guarantee better than a?

As far as 1) is concerned, the strongest negative result is due to Jiang and Li [9] who established that if there is such an algorithm with ratio $O(\log \log k)$ then NP is contained in $DTIME(2^{polylogn})$.

In connection with 2), Jiang and Li [9] conjectured that the SCS problem on a binary alphabet is in the class MAX SNP-hard [13]. If this were the case, then in the light of recent results of Arora et al [2], there could not be a polynomial-time approximation scheme for SCS over a binary alphabet (if P $\neq$ NP).

References

1. A. Apostolico and C. Guerra. The longest common subsequence problem revisited. *Algorithmica*, 2:315–336, 1987.
2. S. Arora, C. Lund, R. Motwani, M. Sudan, and M. Szegedy. Proof verification and intractability of approximation problems. Unpublished manuscript, 1992.

3. J.H. Bradford and T.A. Jenkyns. On the inadequacy of tournament algorithms for the n-SCS problem. *Information Processing Letters*, 38:169–171, 1991.

4. D.S. Hirschberg. Algorithms for the longest common subsequence problem. *Journal of the A.C.M.*, 24:664–675, 1977.

5. W.J. Hsu and M.W. Du. Computing a longest common subsequence for a set of strings. *BIT*, 24:45–59, 1984.

6. J.W. Hunt and T.G. Szymanski. A fast algorithm for computing longest common subsequences. *Communications of the A.C.M.*, 20:350–353, 1977.

7. R.W. Irving and C.B. Fraser. Two algorithms for the longest common subsequence of three (or more) strings. In *Proceedings of the Third Annual Symposium on Combinatorial Pattern Matching*, pages 214–229. Springer-Verlag LNCS Vol. 644, 1992.

8. S.Y. Itoga. The string merging problem. *BIT*, 21:20–30, 1981.

9. T. Jiang and M. Li. On the approximation of shortest common supersequences and longest common subsequences. Submitted for publication, 1992.

10. D. Maier. The complexity of some problems on subsequences and supersequences. *Journal of the A.C.M.*, 25:322–336, 1978.

11. E.W. Myers. An O(ND) difference algorithm and its variations. *Algorithmica*, 1:251–266, 1986.

12. N. Nakatsu, Y. Kambayashi, and S. Yajima. A longest common subsequence algorithm suitable for similar text strings. *Acta Informatica*, 18:171–179, 1982.

13. C.H. Papadimitriou and M. Yannakakis. Optimization, approximation, and complexity classes. *Journal of Computer and System Sciences*, 43:425–440, 1991.

14. V.G. Timkovskii. Complexity of common subsequence and supersequence problems and related problems. *English Translation from Kibernetika*, 5:1–13, 1989.

15. E. Ukkonen. Algorithms for approximate string matching. *Information and Control*, 64:100–118, 1985.

16. R.A. Wagner and M.J. Fischer. The string-to-string correction problem. *Journal of the A.C.M.*, 21:168–173, 1974.

17. S. Wu, U. Manber, G. Myers, and W. Miller. An O(NP) sequence comparison algorithm. *Information Processing Letters*, 35:317–323, 1990.

An Algorithm for Locating Non-Overlapping Regions of Maximum Alignment Score

Sampath K. Kannan[†]   Eugene. W. Myers[‡]

Department of Computer Science
University of Arizona
Tucson, AZ 85721
email: {kannan|gene}@cs.arizona.edu

Abstract. In this paper we present an $O(N^2 \log^2 N)$ algorithm for finding the two non-overlapping substrings of a given string of length N which have the highest-scoring alignment between them. This significantly improves the previously best known bound of $O(N^3)$ for the worst-case complexity of this problem. One of the central ideas in the design of this algorithm is that of partitioning a matrix into pieces in such a way that all submatrices of interest for this problem can be put together as the union of very few of these pieces. Other ideas include the use of candidate-lists, an application of the ideas of Apostolico et al. [1] to our problem domain, and divide and conquer techniques.

1 Introduction

Let $A = a_1 a_2 \cdots a_N$ be a sequence of length N, and let $A[p..q]$ denote the substring $a_p a_{p+1} \cdots a_q$ of A. The problem we consider is that of finding the score of the best alignment between two substrings $A[p..q]$ and $A[r..s]$ under the the generalized Levenshtein model of alignment [2,7] which permits substitutions, insertions, and deletions of arbitrary score. This problem is a formalization of the problem, encountered by molecular biologists, of automatically detecting repeated regions in DNA and protein sequences. This problem has recently been considered by Miller [4]. When there is no restriction that the regions be non-overlapping, he points out that the problem can be solved in $O(N^2)$ time by a straight-forward modification of the algorithm of Smith and Waterman [6] that finds the highest-scoring local alignment between two sequences. Miller then goes on to consider the restriction that the regions be non-overlapping and presents a worst-case $O(N^3)$ algorithm which runs in $O(N^2)$ in practice. His method involves the use of the candidate-list paradigm which we review briefly in the next section since we use some of these ideas in our algorithm. Miller calls non-overlapping regions *twins*, and as his result indicates, this

† Supported in part by NSF grant CCR-9108969.
‡ Supported in part by NLM grant LM-04960 and NSF grant CCR-9002351.

constraint appears to make the problem much harder. For the rest of the paper we consider only the problem of finding the best scoring twins.

There is a related problem where the goal is to find non-overlapping regions which are exact repeats. This problem has been dealt with before and turns out to be of significantly lower complexity than the problem of finding the best scoring twins. When the exactly repeating non-overlapping regions are required to be contiguous, i.e. when the goal is to find the longest substring of A of the form ww, Main and Lorentz [3] provide an $O(N\log N)$ algorithm. If gaps are allowed between exactly repeating substrings, the problem becomes even simpler and can be handled in $O(N)$ time, by first creating a suffix tree and then computing the largest and smallest indices (of suffixes) which go through each internal node in post-order. The path to the deepest internal node whose smallest and largest index suffixes are sufficiently far apart, gives us the desired repeated substring.

The rest of the paper is organized as follows. In Section 2 we define some concepts and review the results that are used in the construction of our algorithms. In Section 3 we present a relatively simple algorithm for the problem of finding twins which runs in time $O(N^{2.5}\log^{0.5}N)$. This algorithm already incorporates some of the ideas used in the more complex algorithm and so serves as a useful preliminary exercise. In Section 4 we present an algorithm which achieves a running time of $O(N^2\log^2 N)$ which is within a polylog factor of being optimal. In Section 5 we consider the problem of finding the best twins under the condition that the best twins are of size no more than $O(N^{1-\varepsilon})$ for arbitrarily small values of ε. For this problem we present an algorithm which runs in time $O(N^2)$. In Section 6 we describe open problems mainly concerned with improving the space complexity of our algorithm.

2 Preliminaries

Throughout the paper we wish to think about the problem in terms of finding paths in a weighted *edit graph* [5] and performing the computation over the associated *dynamic programming matrix* [7]. The edit graph for sequence A versus itself consists of a lower triangular matrix of vertices (i, j) for $0 \le j \le i \le N$ with up to three edges directed into (i, j): a *substitution* edge from $(i-1, j-1)$ weighted $\delta(a_i, a_j)$, an *insertion* edge from $(i-1, j)$ weighted $\delta(a_i, \varepsilon)$, and a *deletion* edge from $(i, j-1)$ weighted $\delta(\varepsilon, a_j)$. Edges from nonexistent vertices and substitution edges on the main diagonal are not present. The scoring scheme δ may be chosen arbitrarily but in most application contexts is such that edge weights are negatively biased and only the substitution of similar symbols are given positive score. As illustrated in Figure 1, any path from vertex (p, r) to (q, s) models an alignment between $A[p+1..q]$ and $A[r+1..s]$ and the weight of the path is the score of the alignment. The correspondence is isomorphic, and so it suffices to think in terms of finding high-scoring paths in the edit graph. Limiting the graph to the lower-triangular part simply eliminates local alignments that cannot be twins, because any path crossing the main diagonal aligns overlapping regions.

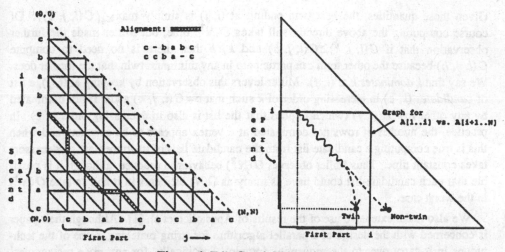

Figure 1: Edit Graph Illustrations.

The Smith-Waterman algorithm for local alignments applied to the edit graph for A reduces to evaluating the following fundamental recurrence for $C(i, j)$, the cost of the best path to (i, j) from some predecessor in the graph, in lexicographical order of i and j:

$$\text{For } j \leq i: \quad C(i, j) = \max \{ 0, \ C(i-1, j-1) + \delta(a_i, a_j) \text{ if } i > j > 0,$$
$$C(i-1, j) + \delta(a_i, \varepsilon) \qquad \text{if } i > 0,$$
$$C(i, j-1) + \delta(\varepsilon, a_j) \qquad \text{if } j > 0 \ \}$$

The terms qualified by an *if*-clause are present only if the condition is true. The score of the best substring alignment is given by $\max_{j \leq i} \{ C(i, j) \}$. Because the edit graph on A involves just the lower triangular part of the underlying dynamic programming matrix, the trivial answer of aligning A with itself is precluded.

However, while the above finds the best scoring path in the edit graph, it does not necessarily align non-overlapping substrings. Figure 1 illustrates that if a path starts at (i, j) and ends at (x, y) then it models a twin only if $y \leq i$, i.e. it ends in a column whose index is not greater than that of the row it starts in. Alternatively, a path is a twin if there exists an i such that the path lies entirely in the rectangular subgraph delimited by row i and column i, in which case we say the twins are *separated by* i. This observation leads to the obvious $O(N^3)$ algorithm for the twins problem: For each $i \in [1, N-1]$, run the Smith-Waterman algorithm for $A[1..i]$ versus $A[i+1..N]$, and record the best answer over all possible separators i.

Miller [4] obtained an algorithm for finding twins that is more efficient in practice by computing $C(i, j, k)$, the best path to (i, j) from row k for each value of $k \leq i$.

$$\text{For } j \leq k \leq i: \quad C(i, j, k) = \max \{ C(i-1, j-1, k) + \delta(a_i, a_j) \text{ if } i > j > 0 \text{ and } i > k,$$
$$C(i-1, j, k) + \delta(a_i, \varepsilon) \qquad \text{if } i > 0 \text{ and } i > k,$$
$$C(i, j-1, k) + \delta(\varepsilon, a_j) \qquad \text{if } j > 0 \text{ and } i \geq k,$$
$$0 \qquad\qquad\qquad\qquad \text{if } i = k \ \}$$

Given these quantities, the best twin ending at (i, j) is simply $\max_{j \leq k}\{C(i, j, k)\}$. Of course computing the above directly still takes $O(N^3)$ time, but Miller made the further observation that if $C(i, j, k) \geq C(i, j, h)$ and $k \geq h$ then there is no need to compute $C(i, j, h)$ because the other term can participate in any minimum twin that the former does. We say that k *dominates* h at (i, j). Miller levers this observation by keeping at (i, j), a list of *candidates* (k, c) in increasing order of k such that $c = C(i, j, k)$ and k is not dominated by any other row at (i, j) (which implies that the list is also in decreasing order of c). In practice, the number of rows not dominated at a vertex appears to be constant, and when this is true computing a candidate list from the candidate lists of its immediate predecessors takes constant time. Thus Miller observes $O(N^2)$ behavior in practice, although it is possible that each candidate-list could have as many as $\Omega(N)$ elements in it and so take $O(N^3)$ in the worst case.

We also make extensive use of the results of Apostolico et al. [1]. Although their paper is concerned with the design of a parallel algorithm for string matching, some of the techniques in it carry over to the sequential domain. Specifically, for any $m \times n$ rectangular[†] subgraph E of an edit graph where $m \leq n$, Apostolico et al. [1] show that the problem of computing the shortest distances between every one of the $n+m+1$ vertices on the left or top boundary of E to every one of the $n+m+1$ vertices on the right or bottom boundary of E can be done in $O((m+n)^2 \log n)$ time. We will refer to the resulting $(n+m+1) \times (n+m+1)$ table of distances between pairs of boundary vertices of E by $DIST_E$. Another result from this paper will also be relevant to us. This is an 'incremental' version of the previous result and states that given a square, $m \times m$ edit graph E that is decomposed into 4 $m/2 \times m/2$ subgraphs A, B, C, and D, by bisecting vertical and horizontal lines, $DIST_E$ can be computed from $DIST_A$, $DIST_B$, $DIST_C$, and $DIST_D$ in $O(m^2)$ time.

3 A Simple $O(N^{2.5} \log^{0.5} N)$ Algorithm

The candidate-list technique of Miller does not give us a better than $O(N^3)$ algorithm because each list can get as large as $\Omega(N)$. In this section we present an algorithm which achieves a better worst-case running time by eliminating the need for candidate-lists with more than b elements in them, where we will choose b as a function of N later. Consider partitioning the interval $[0, N]$ into N/b *panels*, $[0, b]$, $[b, 2b]$, $\cdots$, $[N-b, N]$, with the *partition indices* $b, 2b, 3b, \cdots, N-b$. Consider the following first phase:

1. For each partition index $i = b, 2b, \cdots, N-b$ run the Smith-Waterman algorithm on $A[1..i]$ versus $A[i+1..N]$.

The step above detects every twin that is separated by one of the partition indices. Thus the only twins not captured are those where the first part ends and the second part begins within one of the panels. The second phase of our algorithm processes each panel in search of such *panel twins*. The outer loop of the second phase is:

Here m and n refer to the length of the sides and not the number of vertices in them, which are $m+1$ and $n+1$, respectively.

2. For each panel $[j, i] = [0, b], [b, 2b], \cdots, [N-b, N]$ do the following 4 steps:

Consider a panel twin of $[j, i]$. Figure 2 shows the path of such a twin which must begin at a vertex between rows j and i, and which must end at a vertex between columns j and i. Thus it suffices to compute for all vertices (x, y) between columns j and i, the best paths originating from vertices between rows y and $\min(i, x)$. That is, $\max_{y \le z \le \min(i, x)} \{C(x, y, z)\}$ is the cost of the best panel twin ending at (x, y). Moreover, the maximum involves at most $b+1$ candidates, one from each row between j and i. To begin the processing of the panel, consider:

2.1 For each vertex on boundaries A and B shown in Figure 2, compute the best paths to them that originate in rows j through i using Miller's recurrence.

Rigorously stated this step computes $C(x, y, z)$ for $x \in [j, i]$, $y \in [0, j]$, and $z \in [j, x]$ in lexicographical order, and retains the computed quantities at vertices (x, y) such that $x = i$ or $y = j$ in candidate *arrays* indexed by z. Now the crux of the problem is to compute the best paths originating in rows j through i to the vertices on boundary C shown in Figure 2. Using Miller's recurrence and computing all the necessary intermediate quantities in the rectangular subgraph, $E(i, j)$, whose upper right corner is vertex (i, j), would require $O(N^2 b)$ time which is too costly. This is circumvented by the next two steps as follows:

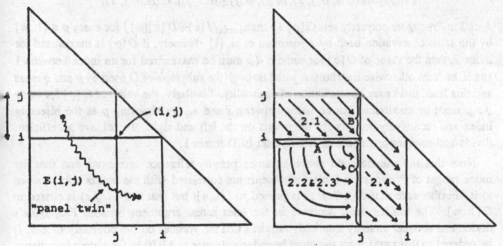

Figure 2: Panel Processing for the Simple Algorithm.

2.2 Compute $M = DIST_{E(i, j)}$ with the algorithm of Apostolico et al.

The table M of distances between pairs of points bounding $E(i, j)$ is used to efficiently "propagate" the candidate arrays on boundary A to boundary C. The basic observation is that:

$$C(x, j, z) = \max_{0 \le y \le j} \{C(i, y, z) + M[(i, y)][(x, j)]\} \text{ for all } x \in [i, N] \text{ and } z \in [j, i].$$

That is, a best path from row z to (x, j) must pass through row i, and so is decomposable into (1) a best path to some vertex (i, y) on this row, followed by (2) a best path from (i, y) to (x, j). With this preliminary, the next step is:

2.3 For each vertex on boundary C shown in Figure 2, compute the best paths originating in rows j through i by efficiently propagating the candidate lists from boundary A by consulting table M.

If the maximum embodying propagation were computed directly for each boundary C vertex, we would again take too much time. Fortunately, for a fixed originating row z, it is possible to compute $C(x, j, z)$ for all (x, j) on boundary C in $O(\log N)$ amortized time per vertex as follows:

$Propagate\ (I[1..n], D[1..n][1..m], O[1..m])$
    if $m > 0$ **then**
    { $p \leftarrow \lfloor(m+1)/2\rfloor$
        $O[p] \leftarrow \max_{1 \leq x \leq n}\{I[x]+D[x][p]\}$
        Determine $\bar{x}$ maximizing the above.
        $Propagate\ (I[1..\bar{x}], D[1..\bar{x}][1..p-1], O[1..p-1])$
        $Propagate\ (I[\bar{x}..n], D[\bar{x}..n][p+1..m], O[p+1..m])$
    }

    for $z \in [j, i]$ **do**
        $Propagate\ (C(i, 0..j, z), M[(i, 0..j)][(N..i, j)], C(N..i, j, z))$

A call to *Propagate* correctly sets $O[p] \leftarrow \max_{1 \leq x \leq n}\{I[x]+D[x][p]\}$ for every $p \in [1, m]$ by the same observation used by Apostolico et al. [1]. Namely, if $O[p]$ is maximized for index $\bar{x}$, then the value of $O[q]$ for some $q < p$ must be maximized for an index between 1 and $\bar{x}$, because otherwise the shortest paths through the subgraph of D used by p and q cross and this leads to an easy contradiction of optimality. Similarly, the value of $O[q]$ for some $q > p$ must be maximized for an index between $\bar{x}$ and n. By choosing p as the bisecting index and recursively solving for the points on the left and right, we achieve an efficient divide-and-conquer procedure as proven later in Theorem 1.

Note that the pseudo-code above assumes pass-by-reference semantics and that the index ranges of *Propagate*'s formal arguments are correlated with the matrix slices passed to it. Further note that $C(i, 0..j, z)$ is passed to $I[1..n]$, but that $C(N..i, j,z)$ is passed to $O[1..m]$. The indices must decrease in the later actual argument because *Propagate*'s divide and conquer strategy implicitly requires that the vertices on the boundary of $E(i, j)$ be ordered so that paths from the input boundaries (upper and left) to the output boundaries (lower and right) cross when their start and finish points are inverted with respect to this order. We use as the increasing input boundary order: up the left side and then across the top to the right, and as the increasing output boundary order: across the bottom to the right and then up the right side. For example, for $E(i, j)$ the increasing input boundary order is $(N, 0), (N-1, 0), \cdots, (i+1, 0), (i, 0), (i, 1), \cdots, (i, j-1), (i, j)$, and the increasing output boundary order is $(N, 0), (N, 1), \cdots, (N, j-1), (N, j), (N-1, j), \cdots, (i+1, j), (i, j)$. Our simple algorithm only requires propagation from the top part of the input boundary to the left part of the output boundary, so the call to *Propagate* only passes the appropriate sections of the matrices and distance table M, in the appropriate order. Later in Section 4, we will use *Propagate* for propagation across the entirety of each boundary and the issue of boundary vertex orders will again be important.

2.4 Given the candidate arrays for vertices on column j, compute the candidate arrays for every vertex between column j and i, and record the best scoring panel-twin ending at each.

This last step is easily accomplished using Miller's recurrence. Rigorously stated we compute $C(x, y, z)$ for all vertices (x, y) between columns j and i, and all $z \in [y, min(i, x)]$. Simultaneously, we determine the cost, $max_{y \leq z \leq min(i, x)}\{C(x, y, z)\}$, of the best panel twin ending at each (x, y).

Note that Steps 2.2 and 2.3 are not needed for panels $[0, b]$ and $[N-b, N]$ because the region $E(i, j)$ degenerates to a line. During the execution of Phase 1 and each execution of Step 2.4, the algorithm keeps a record of the best scoring twin so far. We state this as a final phase:

3. Output the score of the best twin found over all the stages.

Although the problem is to output just the score of the best twin, note that the algorithm actually computes the best scoring twin to every vertex in the edit graph of A, and so can produce more than one twin if desired.

Theorem 1: *The algorithm above computes twins in $O(N^{2.5} \log^{0.5} N)$ time and $O(N^2)$ space.*

Proof: Phase 1 of the algorithm takes $O(N^3/b)$ time since it involves solving N/b problems each taking time $O(N^2)$. In Phase 2 we perform the minor steps N/b times. Steps 2.1 and 2.4 are similar involving the computation of candidate arrays in rectangles whose sizes are less than $N \times b$. Since the candidate arrays have up to b elements in them this takes $O(Nb^2)$ for each pair of rectangles or a total of $O(N^2 b)$ time over all panels. Computing a distance table M in Step 2.2 takes $O(N^2 \log N)$ time for a total of $O(N^3 \log N/b)$ time over all panels.

It remains to analyze the complexity of Step 2.3. If $T(n, m)$ is the time for a call to *Propagate*, then it satisfies the recurrence: $T(n, m) \leq T(n_1, (m-1)/2) + T(n_2, (m-1)/2) + O(n)$ for any n_1 and n_2 such that $n_1 + n_2 = n+1$ and boundary condition $T(n, 1) = O(n)$. An easy induction shows that $T(n, m)$ is $O(n \log m + m)$. Thus each invocation of *Propagate* in Step 2.3 takes $O(N \log N)$ time for a total time in the step of $O(Nb \log N)$. Over the entire algorithm the time spent is then $O(N^2 \log N)$.

The overall complexity of the algorithm is determined by choosing b to equalize the $O(N^3 \log N/b)$ time spent in Step 2.2 and the $O(N^2 b)$ time spent in Steps 2.1 and 2.4. This is achieved when $b = \sqrt{N \log N}$ and gives a running time of $O(N^{2.5} \log^{0.5} N)$. For the space complexity, note that at worst one *DIST* table needs to be maintained at any given point in the algorithm. $\square$

4 An Improved Algorithm

In essence the algorithm above is designed around computing panel twins where the size of panels is roughly $N^{1/2}$. One immediately hopes that an $O(N^{7/3} polylog N)$ algorithm is possible using panels of size $N^{1/3}$, and if so, then one can get a progression of decreasing times by choosing panels of size $N^{1/K}$, ultimately yielding an $O(N^2 polylog N)$ algorithm for $K = \log N$. We are indeed able to pull off such a progression but doing so requires several refinements. First, we have to abandon computing distance tables from scratch for each panel problem. Instead, in a preprocessing step we produce a mesh of carefully chosen distance tables that permit us to subsequently propagate candidate lists through critical subgraphs in $O(N\log^2 N)$ time. Second, is that for a basic block size of $N^{1/K}$, we must proceed in K phases, where in the Jth phase the panels are of size $N^{J/K}$. In the Jth phase, we find the twins that are in a given $N^{J/K}$ panel but not in any $N^{(J-1)/K}$ sub-panel.

With this preamble we will begin the description of the algorithm. We describe the algorithm in the two subsections that follow and analyze its complexity in the third subsection. Throughout we will assume that $b = N^{1/K}$ is the *basic block size* for some $K \geq 2$ that will be chosen in the final subsection when the competing complexity terms are fully understood. For simplicity we assume that b is a power of 2 and consequently that N is a power of 2 as well.

4.1 Preprocessing and Propagation Technique

The preprocessing step consists of computing distance tables for a collection of subproblems. Term an edit graph vertex (i, j) *critical* if it meets the conditions: (1) i and j are multiples of b and (2) $0 < j < i < N$. For each critical (i, j) we will compute and associate a number of distance tables. Let $D(p, i, j)$ denote the distance table $DIST_{E(p, i, j)}$ for the square subgraph, $E(p, i, j)$ consisting of vertices (x, y) such that $x \in [i, i+p]$ and $y \in [j-p, j]$. In terms of the edit graph, $E(p, i, j)$ is a $p \times p$ square whose upper-right corner is the vertex (i, j). We say $E(p, i, j)$ is *cornered at* (i, j). The distance tables to be associated with a given critical vertex correspond to a doubling progression of square subgraphs cornered at the vertex. Specifically, the list of distance tables built for critical vertex (i, j) is: $D(b, i, j)$, $D(2b, i, j)$, $D(4b, i, j)$, $\cdots$, $D(2^x b, i, j)$ where $2^x b$ is the largest power of 2 times the block size that divides both i and j. In addition, let $maxp(i, j) = 2^x b$ be the size of (i, j)'s largest distance table. Note that because vertices for which $i = j$ are not critical, the largest subgraph for which a distance table is built has $N/4 + 1$ vertices on each side, i.e., $maxp(i, j) \leq N/4$. With this prelude our preprocessing algorithm is then simply:

for $p = b$, $2b$, $4b$, $\cdots$, $N/4$ do
  for $i = p$, $2p$, $3p$, $\cdots$, $N-p$ do
   for $j = p$, $2p$, $3p$, $\cdots$, $i-p$ do
   { $maxp(i, j) \leftarrow p$
    if $p = b$ then
      Compute $D(p, i, j)$ *de novo* using the algorithm in [1]
    else
      Compute $D(p, i, j)$ by fusing the 4 tables $D(p/2, i, j)$, $D(p/2, i+p/2, j)$
      $D(p/2, i, j-p/2)$, and $D(p/2, i+p/2, j-p/2)$ using the algorithm in [1]
   }

In a given iteration of the outer loop, observe that $O((N/p)^2)$ tables are built, and since p doubles with each iteration, $O((N/b)^2)$ tables are built over the entire algorithm. For the iteration where $p = b$, each distance table takes $O(b^2 \log b)$ time to build and occupies

$O(b^2)$ space, for upper bounds of $O(N^2 \log N)$ time and $O(N^2)$ space for the iteration. Since fusing 4 $p/2 \times p/2$ tables only takes time $O(p^2)$, the time spent in every iteration other than the first is bounded by $O(N^2)$ time and space. There are $O(\log N)$ iterations of the outer loop, for a grand total of $O(N^2 \log N)$ time and space for the preprocessing step.

The *mesh* of $O((N/b)^2)$ distance tables computed in the preprocessing step, have been chosen so that rectangular subgraphs of the edit graph relevant to our algorithm can be partitioned into $O(N)$ *mesh squares* for which tables have already been computed. For a critical vertex (i, j), as in Section 3 let $E(i, j)$, the *critical subgraph cornered at* (i, j), be the rectangular subgraph consisting of all vertices (x, y) such that $x \in [i, N]$ and $y \in [0, j]$. Suppose 2^x is the largest power of 2 dividing i/b, and 2^z is the largest power of 2 dividing j/b. Let $mp = maxp(i, j) = 2^{min(x, z)} b$. The subgraph $E(i, j)$ can be partitioned into mesh squares in the following way:

Case 1: $x < z$. In this case $E(i, j)$ is partitioned into (a) the row of $mp \times mp$ mesh squares $E(mp, i, mp)$, $E(mp, i, 2mp)$, $E(mp, i, 3mp)$, $\cdots$, $E(mp, i, j)$ along the top boundary of $E(i, j)$, and (b) the partition of $E(i+mp, j)$ obtained by recursively applying this procedure (unless $i+mp=N$ in which case one is done). Note that $maxp(i+mp, j)$ must be $2mp$ or greater in this case, and so the recursive partitioning of $E(i+mp, j)$ will involve squares of at least twice the size.

Case 2: $x > z$. The situation is symmetric and $E(i, j)$ is partitioned into (a) the recursive partitioning of $E(i, j-mp)$ if $j > mp$, and (b) the column of $mp \times mp$ mesh squares $E(mp, i, j)$, $E(mp, i+mp, j)$, $E(mp, i+2mp, j)$, $\cdots$, $E(mp, N-mp, j)$ along the right boundary of $E(i, j)$. Again note that $maxp(i, j-mp)$ must be $2mp$ or greater in this case, and so the recursive partitioning of $E(i, j-mp)$ will involve squares of at least twice the size.

Case 3: $x = z$. Here one "peels" off both a column and row of $mp \times mp$ squares along the top and right boundaries of $E(i, j)$, and then recursively partitions $E(i+mp, j-mp)$ if it is non-empty. Once again $maxp(i+mp, j-mp)$ must be $2mp$ or greater.

Note that in all cases the partition is into subgraphs cornered at critical vertices and so the distance tables for all the squares constituting the partition of $E(i, j)$ have been computed in the preprocessing step as part of the mesh. Note that the non-recursive part of the partitioning process employs at most $O(N/mp)$ $mp \times mp$ squares, with a total perimeter of $O(N)$ over all squares. Since mp at least doubles with each level of recursion, it follows that the $E(i, j)$ is partitioned into at most $O(N/mp)$ $mp \times mp$ squares, $O(N/2mp)$ $2mp \times 2mp$ squares, $O(N/4mp)$ $4mp \times 4mp$ squares, $\cdots$, and $O(1)$ $N/4 \times N/4$ squares. Therefore, $E(i, j)$ is in general partitioned into at most $O(N/maxp(i, j))$ mesh squares, whose total perimeter sum is at most $O(N \log N)$. This characteristic of the partitioning is important since a step in the ensuing algorithm is directly proportional to it.

Having now described how to partition the subgraphs $E(i, j)$, we next show how to propagate candidate list information from the left and top boundaries of $E(i, j)$ to its right and bottom boundaries. For the ensuing algorithm it will only be necessary, as in the simple algorithm of Section 3, to propagate candidates from the top boundary to the right boundary, but in the recursive process below we need to solve the more general left and top boundary to right and bottom boundary propagation problem. Recall from Section 3, the subprocedure *Propagate* (I [1..*n*], D [1..*n*][1..*m*], O [1..*m*]) that propagates a vector I through a distance table D to produce the vector O. Propagation through E (i, j) is accomplished by propagating candidates through the mesh partition using *Propagate* to propagate candidates through each mesh square.

```
Mesh_propagate (L [i..N], T [0..j], R [i..N], B [0..j])
 {  vector X [mp+1], Y [N+1]
    mp ← maxp (i, j)
    if j ≠ mp and mp = maxp (i, j−mp) then
    {  R [i..i+mp] ← L [i..i+mp]
       for k ← mp, 2mp, 3mp, · · · , j do
       {  X [0..mp] ← R [i..i+mp]
          Propagate (X [mp.. 1] · T [k−mp..k], D (mp, i, k), Y [k−mp..k−1] · R [i+mp..i])
          Y [k] ← R [i+mp]
       }
       if i < N−mp then Mesh_propagate (L [i+mp..N], Y [0..j], R [i+mp..N], B)
    }
    else
    {  if j > mp then Mesh_propagate (L, T [0..j−mp], Y [i..N], B [0..j−mp])
       B [j−mp..j] ← T [j−mp..j]
       for k ← i, i+mp, i+2mp, · · · , N−mp do
       {  X [0..mp] ← B [j−mp..j]
          Propagate (Y [k+mp..k] · X [1..mp], D (mp, k, j), B [j−mp..j−1] · R [k+mp..k])
          B [j] ← R [k+mp]
       }
    }
 }
```

The procedure *Mesh_propagate* above, propagates the values across E (i, j) where the formal parameters are as follows. Vector L [*i..N*] contains the input values on the left boundary vertices ($i..N$, 0) (denoting the sequence: (i, 0), ($i+1$, 0), $\cdots$, (N, 0)). Vector T [0..*j*] contains the input values on the top boundary vertices (i, 0..*j*), R [*i..N*] receives the propagated output values on the right boundary vertices ($i..N$, j), and B [0..*j*] receives the output values along the bottom boundary (N, 0..*j*). Note that the input and output vectors redundantly cover the upper-left and lower-right corner vertices, i.e., L [i] ≡ T [0] and R [N] ≡ B [j]. *Mesh_propagate* propagates the input vectors through the mesh using the recursive decomposition above. For example, if case 1 is true for (i, j) then the procedure begins by propagating T and L [*i..i+mp*] through the row of $mp \times mp$ mesh squares to a temporary vector Y [0..*j*] along the lower output boundary and R [*i..i+mp*] along the right boundary. Another temporary vector X is used to chain together the propagations through each mesh square via calls to *Propagate*. Note that as in Section 3, great care is taken to feed to *Propagate* the relevant portions of input and output vectors in the relevant order, so as to observe the ordering requirement for distance table boundary vertices. Once propagation through the row of mesh squares is accomplished, the task is completed by recursively mesh propagating $L[i+mp..N]$ and Y to B and $R[i+mp..N]$. Case 2 is also similarly reflected in the pseudo-code of *Mesh_propagate* and Case 3 is effectively handled by first applying Case 1 and then noting that the recursive call will immediately apply Case 2.

Recall that a call to *Propagate* with an $n \times n$ problem takes $O(n \log n)$ time. Because the mesh partition for $E(i, j)$ contains at most $O(N/mp)$ $mp \times mp$ mesh squares, it then follows that at most $O(N \log mp) = O(N \log N)$ time is spent propagating through mesh squares of this size. Moreover, square size doubles with each recursion, so there are a maximum of $O(\log(N/maxp(i, j))) = O(\log N)$ square sizes in the partition of $E(i, j)$. Thus the total time spent in *Mesh_propagate* is bounded above by $O(N \log^2 N)$.

4.2 The K-phase Algorithm

Recall that the basic block size b is $N^{1/K}$ where K is yet to be chosen. For $J \in [1, K]$. Let the N/b^J intervals $[0, b^J], [b^J, 2b^J], [2b^J, 3b^J], \cdots, [N-b^J, N]$ constitute the set of J-*panels*. A J-*panel twin* is a panel twin for some J-panel but not for any of the $(J-1)$-panels within it. Our K-phase algorithm proceeds through phases $J = K, K-1, \cdots, 1$, where in phase J all J-panel twins are found. Note that the simple algorithm of Section 3 is a specialization of this approach with $K=2$: Step 1 found the 2-panel twins of the sole 2-panel $[0, N]$, and Step 2 found the 1-panel twins. Note that the general algorithm succeeds in finding every twin as a twin must be a J-panel twin for some $J \in [1, K]$.

Finding the K-panel twins is an easy generalization of Step 1 of the algorithm of Section 3. Namely, for each partition index $i = b^{K-1}, 2b^{K-1}, \cdots, N-b^{K-1}$ run the Smith-Waterman algorithm on $A[1..i]$ versus $A[i+1..N]$. For phase $J < K$, the algorithm mimics steps 2.1 through 2.4 of the simple algorithm with two significant modifications. First, step 2.2 is no longer required as the mesh is precomputed earlier, and step 2.3 uses *Mesh_propagate* instead of the simpler *Propagate*. The second modification is in the nature of the C-values used in the phase. Since the goal of phase J is to find J-panel twins and no $(J-1)$-panel twins, one need only keep track of the best path that starts in a given $(J-1)$-panel. That is to know if a path is a J-panel twin of a given J-panel, it suffices to only know which of its b $(J-1)$-subpanels the path starts in. To this end, we introduce the quantity $C^J(i, j, k)$, the best path to (i, j) that begins in the kth $(J-1)$-panel of $[0, N]$, i.e., starts at some vertex between rows kb^{J-1} and $(k+1)b^{J-1}$. Note that for $J = 1$, the definition of C^J coincides with that of C. The recurrence of Section 2 is easily modified to describe the computation of C^J below.

For $j \le i$ and $k \in [\lceil j/b^{J-1} \rceil, \lfloor i/b^{J-1} \rfloor]$:

$$C^J(i, j, k) = \max \left\{ \begin{array}{ll} C^J(i-1, j-1, k) + \delta(a_i, a_j) & \text{if } i > j > 0 \text{ and } i > k, \\ C^J(i-1, j, k) + \delta(a_i, \varepsilon) & \text{if } i > 0 \text{ and } i > k, \\ C^J(i, j-1, k) + \delta(\varepsilon, a_j) & \text{if } j > 0 \text{ and } i \ge k, \\ 0 & \text{if } kb^{J-1} \le i \le (k+1)b^{J-1} \end{array} \right\}$$

With the introduction of C^J the entire computation of phase $J < K$ can now be described succinctly as follows.

Phase J:
For each J-panel $[j, i] \in [0, b^J], [b^J, 2b^J], [2b^J, 3b^J], \cdots, [N-b^J, N]$:

Step J.1:
For $x \in [j, i], y \in [0, j]$, and $z \in [j/b^{J-1}, \lfloor x/b^{J-1} \rfloor]$ compute $C^J(x, y, z)$.

Step J.2:
For $z \in [j/b^{J-1}, i/b^{J-1}]$
$Mesh\_propagate(<-\infty, -\infty \cdots >, C^J(i, 0..j, z), C^J(i..N, j, z), <-\infty, -\infty \cdots >)$

Step J.3:
For $y \in [j, i], x \in [y, N]$, and $z \in [\lceil y/b^{J-1} \rceil, \lfloor min(i, x)/b^{J-1} \rfloor]$
compute $C^J(x, y, z)$ and keep track of the maximum J panel twin found.

Each quantity $C^J(x, y, z)$ is called a panel-twin candidate and note that no more than $b+1$ such candidates are computed at any vertex (x, y) in any phase $J < K$. During the execution of the phases the best twin of any type is recorded and this twin is reported upon completion of all the phases.

4.3 Running Time Analysis

Theorem: *The above algorithm computes the optimal pair of twins in time $O(N^2 \log^2 N)$*

Proof: We will consider the total running time for each activity of the algorithm. Firstly, we have already argued that the preprocessing stage takes time $O(N^2 \log N)$. Phase K of the algorithm invokes the $O(N^2)$ Smith-Waterman algorithm b times for a total of $O(bN^2)$ time. Now consider any other phase $J < K$. Steps $J.1$ and $J.3$ compute $O(b)$ panel-twin candidates at each vertex in rectangles of size less than $N \times b^J$, for a total of $O(b^{J+1}N)$ time. Each mesh propagation in Step $J.2$ takes $O(N\log^2 N)$ as analyzed in Section 4.1 and is repeated for each of the $O(b)$ panel-twin candidates at the top boundary for a total of $O(bN\log^2 N)$ for the step. The total number of panels processed in phase J is N/b^J, giving a total time for phase J of $O(N^2(b + \log^2 N/b^{J-1}))$. Summing over all phases, gives a grand total for the algorithm of $O(KbN^2 + N^2 \log^2 N)$ time as the $\log^2 N/b^{J-1}$ term telescopes. Choosing $K = \log N$, makes $b = N^{1/K} = 2$ and gives us the bound $O(N^2 \log^2 N)$. Note that the dominant term comes from the cost of propagation through the mesh. $\quad\square$

5 Finding Short Twins Efficiently

In this section we describe a very simple algorithm to find twins efficiently if we know that the length of the twins is no more than $N^{1-\epsilon}$ for arbitrarily small ϵ. Let $L = N^{1-\epsilon}$ and consider the following recursive approach. Explicitly solve the problem of finding twins between $A[1..N/2]$ and $A[N/2+1..N]$. Recursively solve the subproblems of finding twins within $A[1..N/2+L]$ and $A[N/2-L..N]$ until the length of the string in the subproblem is no more than $3L$. For subproblems of this size solve them by using the algorithm of the previous section in time $O(N^{2-\epsilon})$. Letting $T(n)$ be the time to find optimal twins in a string of length n, we get the following recurrence:

$$T(n) = O(n^2) + 2T(L + n/2)$$

with initial condition, $T(3L) = O(N^{2-\epsilon})$. Solving this recurrence reveals that $T(N) = O(N^2)$.

This algorithm can be simplified somewhat if it is known that the twins sought are of size no more than $L = N^{2/3-\epsilon}$ in which case the algorithm of Section 3 is sufficient to handle the base case problems while still giving a time bound of $O(N^2)$. Finally if it is known that the twins are not of size more than $O(\sqrt{N})$ then the base cases can simply be handled by the brute-force cubic algorithm while still giving an overall running time of $O(N^2)$.

6 Open Problems

The algorithm presented in this paper is perhaps close to optimal in time complexity, but there is the vexing factor of $\log^2 N$ as opposed to $\log N$. Perhaps some modification of the algorithm would make this improvement. One important concern is also the space complexity of the algorithm. Can this be brought down from $O(N^2 \log N)$ to $O(N^2)$ or even less?

References

1. A. Apostolico, M.J. Atallah, L.L. Larmore, and S. McFaddin, "Efficient Parallel Algorithms for String Editing and Related Problems," *SIAM J. Comput.* **19** (1990), 968-988.

2. V.I. Levenshtein, "Binary codes of correcting deletions, insertions and reversals," *Soviet Phys. Dokl.* **10** (1966), 707.

3. M.G. Main and R.J. Lorentz, "An $O(n \log n)$ algorithm for finding all repetitions in a string," *J. of Algorithms* **5** (1984), 422-432.

4. W. Miller, "An Algorithm for Locating a Repeated Region", *manuscript*.

5. E.W. Myers, "An O(ND) difference algorithm and its variants," *Algorithmica* **1** (1986), 251-266.

6. T.F. Smith and M.S. Waterman, "Identification of common molecular sequences," *J. Mol. Biol.* **147** (1981), 195-197.

7. R.A. Wagner and M.J. Fischer, "The String-to-String Correction Problem," *J. of ACM* **21** (1974), 168-173.

Exact and approximation algorithms for the inversion distance between two chromosomes

John Kececioglu[1] and David Sankoff[2]

[1] Computer Science Department, University of California at Davis, Davis, CA 95616
[2] Centre de recherches mathématiques, Université de Montréal, C.P. 6128, succ. A, Montréal, Québec H3C 3J7

Abstract. Motivated by the problem in computational biology of reconstructing the series of chromosome inversions by which one organism evolved from another, we consider the problem of computing the shortest series of reversals that transform one permutation to another. The permutations describe the order of genes on corresponding chromosomes, and a *reversal* takes an arbitrary substring of elements and reverses their order.

For this problem we develop two algorithms: a greedy approximation algorithm that finds a solution provably close to optimal in $O(n^2)$ time and $O(n)$ space for an n element permutation, and a branch and bound exact algorithm that finds an optimal solution in $O(mL(n,n))$ time and $O(n^2)$ space, where m is the size of the branch and bound search tree and $L(n,n)$ is the time to solve a linear program of n variables and n constraints. The greedy algorithm is the first to come within a constant factor of the optimum, and guarantees a solution that uses no more than twice the minimum number of reversals. The lower and upper bounds of the branch and bound algorithm are a novel application of maximum weight matchings, shortest paths, and linear programming.

In a series of experiments we study the performance of an implementation. For random permutations we find that the average difference between the upper and lower bounds is less than 3 reversals for $n \leq 50$. Due to the tightness of these bounds we can solve to optimality random permutations on 30 elements in a few minutes of computer time.

1 Introduction

Much research has been devoted to efficient algorithms for the edit distance between two strings, that is, the minimum number of insertions, deletions, and substitutions to transform one string into another. Motivation comes in large part from computational biology: at the level of individual characters, genetic sequences mutate by these operations, so edit distance is a useful measure of evolutionary distance.

At the chromosome level, however, genetic sequences mutate by more global operations such as the reversal of a substring (*inversion*), the deletion and subsequent reinsertion of a substring far from its original site (*transposition*), duplication of a substring, and the exchange of prefixes or suffixes of two sequences in the same organism (*translocation*). An inversion, which takes takes a substring of unrestricted size and replaces it by its reverse in one operation, has the effect of reversing the order of the genes contained within the substring. It is perhaps the most common of these operations [17, pages 174–175] especially in organisms with one chromosome.

To take an example, the only major difference between the gene orders of two of the most well-known bactera, *Escherichia coli* and *Salmonella typhimurium*, is an inversion of a long substring of the chromosomal sequence [13]. In plants, Palmer, Osorio and Thompson [14] modeled the evolution of part of the pea chloroplast genome, which is also a single chromosome, in terms of five successive overlapping inversions. And in the fruit fly, inversions are a far more frequent reflection of differences both between and within species than either translocation or transposition [4, page 155]. The importance of inversion in these examples suggests that algorithmic study of gene rearrangement by inversions is a worthwhile first step in the study of evolutionary distance at the level of the chromosome.

In the mathematical problem that we consider, we are given the order of n genes in two related single-chromosome organisms or two related organelles, which we represent by permutations $\sigma = (\sigma_1 \; \sigma_2 \; \cdots \; \sigma_n)$ and $\tau = (\tau_1 \; \tau_2 \; \cdots \; \tau_n)$.[3] In this notation σ_i denotes $\sigma(i)$. Such gene orders often come from genetic *maps* that are the distillation of the work of many experimental geneticists. In current practice, the positions of the genes are increasingly found by sequence comparison or DNA hybridization, as opposed to the mapping experiments of traditional genetics.

We model an inversion by the reversal of an interval of elements. Formally, a *reversal* of interval $[i, j]$ is the permutation[4]

$$\rho = \begin{pmatrix} i & i+1 & \cdots & j \\ j & j-1 & \cdots & i \end{pmatrix}.$$

Applying ρ to σ by the composition[5] $\sigma \cdot \rho$ has the effect of reversing the order of genes $\sigma_i, \sigma_{i+1}, \ldots, \sigma_j$. Our problem is the following.

Definition 1. The *reversal distance problem* on permutations is, given permutations σ and τ, find a series of reversals $\rho_1, \rho_2, \ldots, \rho_d$ such that

$$\sigma \cdot \rho_1 \cdot \rho_2 \cdots \rho_d = \tau,$$

and d is minimum. $\qquad\qquad\square$

We call d the *reversal distance*[6] between σ and τ. Like edit distance, it satisfies the axioms of a metric. Reversal distance measures the amount of evolution that must have taken place at the chromosome level, assuming evolution proceeded by inversions.

---

[3] Genes in one organism may be missing in the other. We assume however that such genes can be removed from the analysis, and that gene insertions and deletions can be analyzed separately.

[4] This notation is shorthand for $\rho(i) = j$, $\rho(i+1) = j-1$, and so on. Outside interval $[i, j]$, ρ leaves elements unchanged.

[5] The *composition* of permutations σ and ρ indicated by $\sigma \cdot \rho$ is a permutation π where $\pi(i) = \sigma(\rho(i))$.

[6] We sometimes refer to d informally as the *inversion distance*.

Notice that the reversal distance between σ and τ is equal to the reversal distance between $\tau^{-1} \cdot \sigma$ and the identity permutation $\imath$, where τ^{-1} denotes the inverse of τ.[7] Hence we can take as input the permutation $\pi = \tau^{-1}\sigma$ and compute its distance from $\imath$. We call this formulation of the problem, *sorting by reversals*. Note also that any algorithm for the reversal distance between two strings that does not exploit a bounded-size alphabet must, as a special case, solve the reversal distance problem on permutations.

Little is known about reversal distance: even its computational complexity is open. The only reference to an algorithm appears to be Watterson, Ewens, Hall, and Morgan [20], which gives the first definition of the problem and a heuristic for computing reversal distance, which we describe in Section 2.

¿From the perspective of edit distance, Wagner [19] and Tichy [18] consider variations on the problem of editing one string into another where the operations include swapping of adjacent characters, or copying of substrings. More recently Schöniger and Waterman [16] present a heuristic for computing edit distance when non-overlapping inversions are allowed.

¿From the perspective of sorting, Gates and Papadimitriou [7] is interesting. They consider the problem of sorting a permutation by *prefix reversals*,[8] which are reversals of the form $[1, i]$, and derive bounds on the diameter of the problem. The *diameter* of the prefix reversal problem, which we denote by $d_{\text{prefix}}(n)$, is the maximum of the minimum number of prefix reversals over all permutations on n elements. Gates and Papadimitriou show that $d_{\text{prefix}}(n) \leq \frac{5}{3}n + \frac{5}{3}$, and that for infinitely many n, $d_{\text{prefix}}(n) \geq \frac{17}{16}n$. Under the requirement that each element is reversed an even number of times, which may be appropriate if elements have an orientation, they show $\frac{3}{2}n - 1 \leq d_{\text{prefix}}(n) \leq 2n + 3$. In other work, Aigner and West [1] consider the diameter of sorting when the operation is reinsertion of the first element, and Amato, Blum, Irani, and Rubenfeld [2] consider a variation inspired by the problem of reversing trains on a track.

¿From the perspective of group theory, Even and Goldreich [5] and Jerrum [9] are interesting. Even and Goldreich show that, given a set of generators of a permutation group G and a permutation π, determining the shortest product of generators that equals π is NP-hard. Their reduction implies that the problem remains NP-hard even if every generator is its own inverse, as is the case in our problem. Jerrum established that the problem is PSPACE-complete, and remains so when restricted to two generators. In our problem the generator set is fixed. Thus, while these complexity results give a feel for the problem, they do not imply the intractability of sorting by reversals. Nevertheless, we believe sorting by reversals is NP-complete.

In the next section we present an approximation algorithm for sorting by reversals and show it never exceeds the minimum number by more than a factor of 2. Section 3 develops an exact algorithm using the branch and bound technique. The lower bound uses a relaxation to maximum weight matchings and linear programming. Section 4 presents results from experiments with an implementation of these algorithms on biological data and random permutations.

---

[7] The *identity* permutation $\imath$ is $(1\ 2\ \cdots\ n)$, and the *inverse* of permutation π is the permutation π^{-1} satisfying $\pi^{-1} \cdot \pi = \imath$.

[8] This is also known as the *pancake flipping problem*.

2 An approximation algorithm

Perhaps the most natural algorithm for sorting by reversals, suggested by Watterson, Ewens, Hall and Morgan [20], is to bring element 1 into place, then element 2, and so on up to element n. Formally, at step i perform reversal $[i, \pi_i^{-1}]$ if $\pi_i \neq i$,. Once step $n - 1$ is completed element n must be in position n, so this sorts any n-element permutation in at most $n - 1$ reversals.

While it is conjectured that there are permutations for every n that require $n - 1$ reversals [8], which if true means this algorithm is worst-case optimal, for specific instances the algorithm can perform arbitrarily poorly. Consider for example the permutation $(n \ 1 \ 2 \ \cdots \ n-1)$. Bringing 1 into place, then 2, and so on, uses $n - 1$ reversals, yet the permutation can be sorted in two steps: reverse $[1, n]$, then $[1, n-1]$. Thus this natural algorithm can produce a solution $\frac{1}{2}(n - 1)$ times longer than the shortest, for arbitrarily large n. Using the idea of a breakpoint, also introduced in Watterson et al. [20], we show there is a simple algorithm guaranteed to use no more than twice the minimum number of reversals. To the best of our knowledge, this is the first constant-factor approximation for sorting by reversals. In order to describe the algorithm, we first define some terminology.

A *breakpoint* of a permutation π is a pair of adjacent positions $(i, i+1)$ such that $|\pi_{i+1} - \pi_i| \neq 1$. In other words, $(i, i+1)$ forms a breakpoint if values π_i and π_{i+1} are consecutively increasing or decreasing. To handle the boundaries we let π_0 have the value 0, π_{n+1} have the value $n + 1$, and allow i to range from 0 to n in the definition. Thus $(0, 1)$ is a breakpoint if $\pi_1 \neq 1$, and $(n, n+1)$ is a breakpoint if $\pi_n \neq n$. Notice that the identity permutation has no breakpoints, any other permutation has some breakpoint, and the number of breakpoints is at most $n + 1$.

When $|\pi_{i+1} - \pi_i| = 1$, we say values π_{i+1} and π_i are *adjacent*, and write $\pi_{i+1} \sim \pi_i$.

A *strip* of π is an interval $[i, j]$ such that $(i - 1, i)$ and $(j, j + 1)$ are breakpoints, and no breakpoint lies between them. In other words, a strip is a maximal run of increasing or decreasing elements.

A reversal ρ affects the breakpoints of π only at the endpoints of ρ. (In the interior, ρ only makes an increasing pair (π_i, π_{i+1}) decreasing, and vice versa.) Let us write $\Phi(\pi)$ for the number of breakpoints in π, and for a given reversal ρ, let

$$\Delta\Phi(\pi) \; = \; \Phi(\pi) - \Phi(\pi \cdot \rho).$$

Since a reversal $[i, j]$ changes the adjacency of only two points, namely $(i-1, i)$ and $(j, j+1)$, the only values $\Delta\Phi(\pi)$ can have are between -2 and 2. Since a solution must decrease the number of breakpoints from $\Phi(\pi)$ to zero, a *greedy strategy* is to choose a reversal that achieves the greatest decrease, or maximizes $\Delta\Phi(\pi)$. As any $\pi \neq \iota$ has a reversal with $\Delta\Phi(\pi) \geq 0$, we can always achieve a decrease of 2, 1, or 0.

Figure 1 specifies our greedy algorithm. With the rule of favoring reversals that leave decreasing strips, we are able to show it exceeds the minimum by at most a factor of 2.

In the following, a strip $[i, j]$ is *decreasing* iff $\pi_i, \pi_{i+1}, \ldots, \pi_j$ is decreasing. We consider a strip of one element to be decreasing, except for π_0 and π_{n+1}, which are always increasing. Thus the identity permutation forms one increasing strip extending from 0 to $n + 1$.

```
algorithm GREEDY(π)
    i := 0
    while π contains a breakpoint do
        i := i + 1
        Let ρᵢ be a reversal that removes the most breakpoints of π,
            resolving ties among those that remove one breakpoint
            in favor of reversals that leave a decreasing strip.
        π := π · ρᵢ
    end
    return i, (ρ₁, ρ₂, ..., ρᵢ)
end
```

Fig. 1. The greedy algorithm.

Lemma 2. *Every permutation with a decreasing strip has a reversal that removes a breakpoint.*

Proof. Consider the decreasing strip of π whose last element, π_i, is smallest. Element $\pi_i - 1$ must be in an increasing strip (else π_i is not smallest) that lies either to the left or to the right of the strip containing π_i, as illustrated in Figure 2. In either case, the indicated reversal removes at least one breakpoint. □

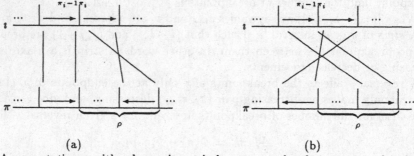

(a)   (b)

Fig. 2. A permutation π with a decreasing strip has a reversal ρ that removes a breakpoint. Element π_i is the smallest that lies in a decreasing strip.

Lemma 3. *Let π be a permutation with a decreasing strip. If every reversal that removes a breakpoint of π leaves a permutation with no decreasing strips, π has a reversal that removes two breakpoints.*

Proof. Again consider the decreasing strip of π containing the smallest element π_i. Case (b) of Figure 2 cannot occur, since ρ is a reversal that removes a breakpoint and leaves a decreasing strip. Thus the increasing strip containing $\pi_i - 1$ must be to the left of the strip containing π_i, as in case (a). Call the reversal of case (a), ρ_i.

Consider the decreasing strip of π whose first element, π_j, is largest. Element $\pi_j +$ 1 must be in an increasing strip (else π_j is not largest) that is to the right of the strip containing π_j. Otherwise, a reversal analogous to case (b) removes a breakpoint and leaves a decreasing strip. Call the reversal for π_j that is analogous to ρ_i, ρ_j.

Notice that π_j must lie in interval ρ_i and $\pi_j + 1$ must lie outside, else ρ_i leaves a decreasing strip. Similarly π_i must lie in ρ_j and $\pi_i - 1$ outside, else ρ_j leaves a decreasing strip. The situation is as shown in Figure 3. Intervals ρ_i and ρ_j overlap.

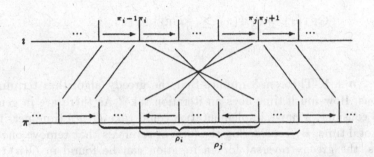

Fig. 3. If every reversal that removes a breakpoint of π leaves a permutation with no decreasing strips, ρ_i and ρ_j must overlap. Elements π_i and π_j are the smallest and largest in decreasing strips.

We now argue that not only do ρ_i and ρ_j overlap, they must be the same interval. For suppose $\rho_i - \rho_j$ is not empty. If it contains a decreasing strip, reversal ρ_j leaves a decreasing strip, and if it contains an increasing strip, reversal ρ_i leaves a decreasing strip. Similarly interval $\rho_j - \rho_i$ must be empty, which implies interval $\rho_i = \rho_j$.

Since reversal ρ_i removes the breakpoint on its left and reversal ρ_j removes the breakpoint on its right, and as these breakpoints are distinct, reversal $\rho = \rho_i = \rho_j$ removes two breakpoints. □

Lemma 4. *The greedy algorithm sorts a permutation π with a decreasing strip in at most $\Phi(\pi) - 1$ reversals.*

Proof. The proof is by Lemmas 2 and 3, and induction on $\Phi(\pi)$. Details are given in [10]. □

Theorem 5. *The greedy algorithm sorts every permutation π in at most $\Phi(\pi)$ reversals.*

Proof. If π has a decreasing strip, by Lemma 4, GREEDY sorts it within $\Phi(\pi)$ reversals. If π has no decreasing strip, any reversal chosen by GREEDY transforms π to a permutation π' with a decreasing strip such that $\Phi(\pi') \leq \Phi(\pi)$. By Lemma 4, GREEDY sorts π' in at most $\Phi(\pi') - 1$ reversals, which sorts π in at most $\Phi(\pi)$ reversals. □

An immediate consequence of Theorem 5 is the following. An *ε-approximation algorithm* for an optimization problem is an algorithm that runs in polynomial time and delivers a solution whose value is within a factor ϵ of optimal.

Corollary 6. *The greedy algorithm is a 2-approximation algorithm for sorting by reversals.*

Proof. Write $\text{OPT}(\pi)$ for the minimum number of reversals to sort a permutation π, and $\text{GREEDY}(\pi)$ for the number taken by the greedy algorithm. Since a solution must remove all breakpoints and any reversal can remove at most two,

$$\text{OPT}(\pi) \geq \frac{1}{2}\Phi(\pi) \geq \frac{1}{2}\text{GREEDY}(\pi).$$

□

Since $\Phi(\pi) \leq n + 1$, Theorem 5 implies that the greedy algorithm terminates in $O(n)$ iterations. How much time does an iteration take? As there are in general $\binom{n}{2}$ reversals to consider, a naive implementation could take $O(n^2)$ time per iteration, or $O(n^3)$ total time. By considering the form of reversals that remove one and two breakpoints, the greedy reversal for an iteration can be found in $O(n)$ time, which yields an $O(n^2)$ time algorithm. Details are given in [10]. We suspect that the algorithm also has an $O(n \log n)$ time implementation, though as we will see in Section 4, the approximation algorithm is far from being the dominant step in practice.

3 An exact algorithm

In the last section we obtained an algorithm that comes close to the optimum by applying a greedy strategy: of all reversals, select one that removes the most breakpoints. To obtain an algorithm that reaches the optimum we use a *branch and bound strategy*: consider all reversals, and eliminate those that cannot lead to an optimal solution.

Figure 4 shows the form of our branch and bound algorithm. We maintain three global variables: *Bound*, a dynamic upper bound on the solution value; *Best*, an array of reversals that sort the permutation in *Bound* steps; and *Current*, the series of reversals currently under consideration. At the start we initialize *Bound* and *Best* to values obtained from an upper bound algorithm. The algorithm we use is essentially GREEDY with a fixed-depth look-ahead, and is described in Section 3.3.[9]

After obtaining an upper bound we explore a tree of subproblems depth-first. Each invocation of SEARCH corresponds to a node of the tree and is labelled with π, a permutation to be sorted, and *Depth*, the number of edges from the root to the node. Array *Current* is maintained as a stack by SEARCH, and holds the reversals on the path from the root to the current node. We chose a depth-first strategy for

---

[9] We use $x_1, \ldots, x_n := e_1, \ldots, e_n$ as shorthand for the parallel assignments $x_i := e_i$. Function UPPERBOUND, like GREEDY, returns two values: an integer, followed by a list of reversals.

```
        global Bound, Current[1..n], Best[1..n]

        algorithm BRANCHANDBOUND(π)
            Bound, Best := UPPERBOUND(π)
            SEARCH(π, 0)
            return Bound, Best
        end

        algorithm SEARCH(π, Depth)
            if π is the identity permutation then
                if Depth < Bound then Bound, Best := Depth, Current
            else
                for each reversal ρ in order of decreasing ΔΦ(π) do
                    if LOWERBOUND(π · ρ) + Depth + 1 < Bound then
                        Current[Depth + 1] := ρ
                        SEARCH(π · ρ, Depth + 1)
                end
        end
```

Fig. 4. The branch and bound algorithm.

traversing the tree as this uses a polynomial amount of space even when the tree is of exponential size. Space, not time, is often the limiting resource.

Examining all reversals yields a very large tree: with $\binom{n}{2}$ children per node and a height of $n - 1$, there are $O(n^{2n}/2^n)$ nodes. The means we have to reduce the size of the tree is by ordering children and computing lower bounds. The algorithm of Figure 4 orders children by decreasing $\Delta\Phi$ on the assumption that the optimal solution uses reversals of greatest $\Delta\Phi$. We now explain how the lower bound is computed.

3.1 A lower bound from matchings

As stated in the proof of Corollary 6, a simple lower bound on $\mathrm{OPT}(\pi)$ is $\lceil \Phi(\pi)/2 \rceil$. While this is sufficient to prove an approximation factor of 2, it is extremely weak. It assumes every breakpoint of π can be eliminated by a reversal that removes two breakpoints, which can rarely be achieved. To obtain a better bound we ask, for a given permutation, how many breakpoints can possibly be eliminated by reversals that remove two breakpoints?

A pair of breakpoints $p = (i, i+1)$ and $q = (j, j+1)$ with values $(\pi_i, \pi_{i+1}) = (x, y)$ and $(\pi_j, \pi_{j+1}) = (x', y')$ can be eliminated in one reversal iff $x \sim x'$ and $y \sim y'$. This holds whether p is to the left or to the right of q. The only requirement is that x and y occur in the same order as x' and y'.

Notice that such a reversal also effects other pairs of breakpoints. A pair with values (x, y) and (y', x'), which cannot be eliminated immediately because adjacent values are not in the same order, can be eliminated in one step if preceded by a reversal that contains exactly one breakpoint of the pair. Such a reversal transforms the pair to the preceding case.

In general, determining when a collection of $2m$ breakpoints can be eliminated by a sequence of m reversals appears difficult. (In Section 5 we conjecture it is NP-complete.) To obtain a lower bound that can be efficiently computed, we ignore dynamic information about the order and interaction of reversals. The static information we retain is simply the adjacency of values between breakpoints, which can be represented by a graph.

Figure 5 shows the construction. Each breakpoint of π is mapped to a vertex of $G(\pi)$. We place an edge between breakpoints p and q iff either of the above cases apply. Effectively, if two breakpoints may be eliminated by one reversal, possibly after a sequence of other reversals that eliminate two breakpoints, they share an edge. Note that the order of the two values at a breakpoint is not important for the construction.

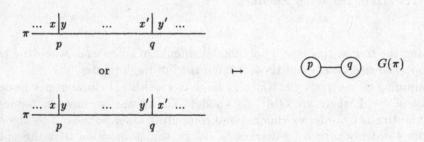

Fig. 5. Breakpoints $\{x, y\}$ and $\{x', y'\}$ share an edge in $G(\pi)$ iff $x \sim x'$ and $y \sim y'$.

Since each edge models a reversal and performing the reversal removes both vertices, a series of reversals of π that each eliminate two breakpoints corresponds to a set of vertex-disjoint edges in $G(\pi)$. A set of vertex-disjoint edges is a *matching*. The key property of $G(\pi)$ is that the most reversals we can possibly perform on π that each remove a pair of breakpoints, without performing any intervening reversals that remove less than two breakpoints, is the size of a maximum cardinality matching of $G(\pi)$.

Let m be the number of vertices in a maximum cardinality matching of $G(\pi)$, or in other words twice the number of edges in the matching. How many reversals must be performed to remove the remaining $\Phi(\pi) - m$ breakpoints of π? The best we can do is to expend a reversal that removes one breakpoint to set up a reversal that removes two breakpoints. Notice that we cannot remove one breakpoint, then two, then two again, since a reversal that removes one breakpoint can affect only one additional breakpoint, so the third reversal must have been available from the start. This is impossible since the matching is of maximum cardinality. In short the best we can do is to remove three breakpoints in two reversals. This gives a lower bound of

$$\left\lceil \tfrac{1}{2}m + \tfrac{2}{3}\left(\varPhi(\pi) - m\right)\right\rceil, \tag{1}$$

which has the extreme value $\left\lceil \tfrac{2}{3}\varPhi(\pi)\right\rceil$.

We can construct $G(\pi)$ from π in $O(n)$ time. Certainly the $O(n)$ breakpoints of π can be determined in $O(n)$ time. Every breakpoint has a constant number of edges, since the only values adjacent to x are $x+1$ and $x-1$, and with the help of π^{-1} we can determine all edges in $O(n)$ time as well. Finally, a maximum cardinality matching of a graph with V vertices and E edges can be computed in $O(E\sqrt{V})$ time [11]. As we have said, V and E are $O(n)$. Thus we can evaluate the lower bound in $O(n^{3/2})$ time.

3.2 A family of lower bounds

One can improve the bound further by considering 3-tuples of breakpoints, 4-tuples of breakpoints, and so on.

Let us call a reversal that eliminates k breakpoints a k-move. Thus a 2-move is a reversal that eliminates 2 breakpoints, and a (-2)-move is a reversal that creates 2 breakpoints.

In general for $k \geq 3$ we define a k-move as follows. Over all permutations consider all series of reversals that eliminate k breakpoints. A k-move for $k \geq 3$ is a shortest series that eliminates k breakpoints given that no 2-, 3-, up to $(k-1)$-moves are available.

The following two lemmas characterize the structure of a k-move. Proofs are given in [10].

Lemma 7. *For $k \geq 3$, a k-move is a series of $k-1$ reversals that decomposes into either*

    (i) *a 1-move followed by a $(k-1)$-move, or*
    (ii) *a 0-move followed by an i-move and a j-move where $i + j = k$.*

                          □

Lemma 8. *For $k \geq 2$, the sets of values at the breakpoints eliminated by a k-move have the form*

$$\{x_1, x_2'\}\ \{x_2, x_3'\}\ \cdots\ \{x_{k-1}, x_k'\}\ \{x_k, x_1'\} \tag{2}$$

where $x_i \sim x_i'$ for $1 \leq i \leq k$.   □

We now describe how to construct a graph $H(\pi)$ from permutation π that allows us to efficiently identify sets of breakpoints of form (2) for small k. In the construction, breakpoints of π are mapped to vertices of $H(\pi)$, and pairs of breakpoints that share an adjacency, such as $\{x, a\}\ \{x', b\}$ where $x \sim x'$, are mapped to edges as shown in Figure 6.

Edges of $H(\pi)$ are directed, but not in the standard sense. An edge touching v and w contributes to either the in- or out-degree of v and w. In a directed graph, an edge (v, w) that contributes to the out-degree of v necessarily contributes to the in-degree of w, and vice versa. However in what we call a *bidirected graph* there are two more possibilities: (v, w) may contribute to the in-degree of both v and w, or to the out-degree of both v and w.

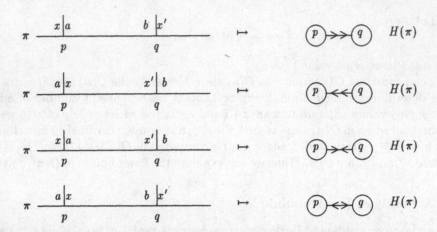

Fig. 6. Construction of graph $H(\pi)$. Values x and x' are adjacent, and induce an edge that contributes to the in- or out-degree of p and the in- or out-degree of q, depending on whether the value is on the left- or right-side of the breakpoint.

This gives rise to the four types of edges of Figure 6. We indicate the direction of an edge with double-ended arrows. When drawing an edge incident to v, we place an arrowhead pointing into v at the end touching v if the edge contributes to v's in-degree. Otherwise we direct the arrowhead out of v.

The utility of this construction is in the correspondence between cycles in the graph and k-moves on the permutation, as summarized in the following lemma. A *k-cycle* in a bidirected graph is a series of edges

$$(v_1, v_2)\,(v_2, v_3)\,\cdots\,(v_k, v_1)$$

such that the v_i are distinct, and every v_i has in- and out-degree 1.

Lemma 9. *The sets of breakpoints of π whose values have the form of k-moves are in one-to-one correspondence with the sets of vertices of $H(\pi)$ that form k-cycles.*

$\square$

We now have the tools to generalize the lower bound of Section 3.1. In outline, we construct a hypergraph $G^{(k)}(\pi)$ whose vertices correspond to breakpoints, but whose edges are sets of up to k vertices that correspond to k'-moves for $k' \leq k$. A series of moves on π maps to a matching of $G^{(k)}$, where a matching of a hypergraph is a collection of vertex-disjoint edges. Choosing a k-move corresponds to performing a series of $k - 1$ reversals. We weight edges by the number of reversals they represent and seek as before a maximum weight matching. However computing a maximum weight matching is in general NP-complete for hypergraphs [6]. We express the matching problem as an integer programming problem and relax the integrality constraint to obtain a linear programming problem. This gives a somewhat weaker lower bound that is computable in polynomial time.

This approach is summarized in the following theorem. The k-*girth* of a graph is the length of a shortest cycle of more than k edges. If the graph does not contain such a cycle, we define its k-girth to be $n + 1$, where n is the number of vertices.

Theorem 10. *Let (V, E) be the graph $G^{(k)}(\pi)$, g be the k-girth of $H(\pi)$, and $\mathcal{L}_k(\pi)$ be the solution value of the linear program*

$$\text{minimize } \frac{g-1}{g}\Phi(\pi) - \sum_{e \in E} \frac{g - |e|}{g} x_e$$
$$\text{subject to } 0 \le x_e \le 1, \text{ for all } e \in E,$$
$$\sum_{e|v \in e} x_e \le 1, \text{ for all } v \in V.$$

Then $\text{OPT}(\pi) \ge \lceil \mathcal{L}_k(\pi) \rceil$. □

Notice that the lower bound of Theorem 10 has the extreme value $\lceil \frac{g-1}{g}\Phi(\pi) \rceil$. This can be as large as $\Phi(\pi)$ even for small k, which meets the upper bound of Theorem 5.

How much time does it take to evaluate the lower bound $\mathcal{L}_k$? There are three tasks: constructing H and computing its k-girth, constructing $G^{(k)}$ and its associated linear program, and solving the linear program.

Constructing H takes time $O(n)$. There are $O(n)$ breakpoints, and each breakpoint has at most four in-edges and at most four out-edges, which can be identified in $O(1)$ time per edge using π^{-1}.

We can compute the k-girth of H in $O(4^k n^2)$ time as follows. A shortest cycle of more than k edges that contains a fixed vertex v is a path P of k edges from v to some vertex w, followed by a shortest path from w to v that does not visit any other vertices on P. Paths in a bidirected graph such as H alternate in- and out-edges: if we enter a vertex by an in-edge, we must leave by an out-edge, and vice versa. As every vertex of H has in- and out-degree at most four, there are at most 4^k paths of length k from a fixed vertex v. For each path P, we can mark its vertices and compute a shortest path from its end w back to v, taking care not to include marked vertices. This shortest path can be found by a breadth-first search from w in $O(n)$ time. Repeating for all start vertices v, all paths P, and recording the minimum over all cycles found takes time $O(4^k n^2)$.

Similarly, we can construct the edges of $G^{(k)}$ in $O(4^k n)$ time by enumerating the cycles of H of k or fewer edges in a depth-first search.[10] Space for all edges is $O(k4^k n)$.

The resulting linear program has $O(4^k n)$ variables and $O(n)$ constraints. Writing $L(a, b)$ for the time to solve a linear program of a variables and b constraints, the linear programming problem takes $O(L(4^k n, n))$ time. This dominates the time to compute the lower bound. Thus for arbitrary but fixed k, $\mathcal{L}_k$ can be computed in $O(L(n, n))$ time.

---

[10] Note that determining the edges of $G^{(k)}$ by examining all k-sets would take $O(n^k)$ time.

3.3 A family of upper bounds

As well as a lower bound on the solution value our exact algorithm requires an upper bound. The simplest approach is to use the approximation algorithm of Section 2, but for large n this gives too weak a bound to prune away much of the search tree.

Consider however a series of k reversals that removes the most breakpoints among series of that length. The greedy strategy of the approximation algorithm is really based on the observation that once we are k reversals away from sorting a permutation, such a series is optimal. GREEDY corresponds to the case $k = 1$.

One can find such a series by looking ahead k reversals, and this search can be made tractable by again employing branch and bound. The basic form of the computation is identical to BRANCHANDBOUND except that the recursion is stopped at depth k. The two requirements are a lower bound on the number of breakpoints that can be eliminated in a series of k reversals, and a method for computing an upper bound on the number eliminated by an optimal extension of a partial series.

We can compute the lower bound by running GREEDY. Computing an upper bound is a little more difficult but can be tackled by the methods of the previous section, as summarized in the following theorem.

Theorem 11. *Let* (V, E) *be the graph* $G^{(k+1)}(\pi)$ *and* $\mathcal{U}_k(\pi)$ *be the solution value of the linear program*

$$
\begin{aligned}
\text{maximize } & k + \sum_{e \in E} x_e \\
\text{subject to } & 0 \leq x_e \leq 1, && \text{for all } e \in E, \\
& \sum_{e \mid v \in e} x_e \leq 1, && \text{for all } v \in V, \\
& \sum_{e \in E} (|e| - 1) x_e \leq k.
\end{aligned}
$$

Then $\lfloor \mathcal{U}_k(\pi) \rfloor$ *is an upper bound on the number of breakpoints of* π *that can be eliminated in* k *reversals.* □

For fixed k the time and space to compute upper bound $\mathcal{U}_k$ is the same as for lower bound $\mathcal{L}_k$, which is $O(L(n, n))$ time and $O(n)$ space.

Given that we can find a series of k reversals that removes the most breakpoints, how should we piece together a solution from these series? One extreme is to perform only the first reversal of the series, arrive at a new permutation, and again look ahead k reversals. The other extreme is to execute all k of the series. We call the number of reversals that are performed from a series the *follow-through* of the algorithm. In our experiments, we chose a follow-through equal to the look-ahead.

To summarize, our exact algorithm constructs a solution in three stages, the first two of which are interleaved. The first stage runs the greedy algorithm to bound the number of breakpoints that can be eliminated within the look-ahead. The second stage improves this greedy solution by branch and bound to a fixed depth to obtain a series that is optimal within the look-ahead. Successive locally optimal series are concatenated to obtain a solution that upper bounds the global problem. The final stage improves this solution to a global optimum by a full branch and bound computation, now that a good upper bound is known.

This bootstrapping approach has proven to be quite effective, as is discussed in the next section.

4 Computational results

To examine the effectiveness of these ideas we tested a full implementation of the exact and approximation algorithms on biological and simulated data. The implementation comprises approximately 9,500 lines of C, of which roughly 2,500 lines are a sparse linear programming package. We tested the implementation on three types of data: all permutations on a fixed number of elements, published gene order data from the biology literature, and random permutations. The full paper also gives results from experiments on permutations generated by scrambling the identity with k random reversals [10].

In the first set of experiments we ran the implementation to optimality on all permutations of up to 8 elements. The exact distribution of reversal distance derived from these tests is given in Table 1. On these experiments we also measured the worst case performance ratio of the approximation algorithm. On permutations of up to 8 elements the maximum ratio is $\frac{8}{5}$, which is achieved on permutation (4 7 2 6 8 5 3 1).

Table 1. The number of permutations on n elements at distance d from the identity.

d	2	3	4	5	6	7	8
						n	
0	1	1	1	1	1	1	1
1	1	3	6	10	15	21	28
2		2	15	51	127	263	483
3			2	56	390	1,562	4,635
4				2	185	2,543	16,445
5					2	648	16,615
6						2	2,111
7							2

As an illustration of the algorithm on biological data we give the example of Figure 7. This permutation gives the order of the 36 genes that are common to the linearized mitochondrial genomes of mammals [3] and the flatworm *Ascaris suum* [21]. While we have been able to solve other permutations arising from mitochondria data in a small amount of time,[11] this 36 element permutation has proven extremely difficult to solve to optimality. The near-optimal solution of the figure was found after

---

[11] Computing the inversion distance between the mitochondrial genomes of mammals and the yeast *Schizosaccharomyces pombe* took 3.5 minutes (29 genes and 19 inversions), and between mammals and the fly *Drosophila yakuba* took 1.7 minutes (37 genes and 16 inversions). We make no claim for the biological significance of the particular solutions found, though the inversion distance tends to reflect evolutionary divergence.

```
12  [31 34 28 26 17 29 4 9 36 18 35 19 1 16 14 32 33 22 15 11 27 5 20 13]  30 23 10 6 3 24 21 8 25 2 7
12 13 20 5 27 11 15 22 33 32 14 16 1 19  [35 18]  36 9 4 29 17 26 28 34 31 30 23 10 6 3 24 21 8 25 2 7
12 13 20  [5 27 11 15 22 33 32 14 16 1 19 18 35 36 9]  4 29 17 26 28 34 31 30 23 10 6 3 24 21 8 25 2 7
12 13 20  [9 36 35 18 19 1 16 14 32 33 22 15 11 27 5 4 29 17 26 28 34 31 30 23 10 6 3 24 21]  8 25 2 7
12 13 20 21 24  [3 6 10 23 30 31 34 28 26 17 29 4 5 27 11 15 22 33 32 14 16 1 19 18 35 36 9 8 25]  2 7
[12 13 20 21 24 25 8 9 36 35 18 19 1 16 14 32 33 22 15]  11 27 5 4 29 17 26 28 34 31 30 23 10 6 3 2 7
15  [22 33 32 14 16 1 19 18 35 36 9 8 25 24]  21 20 13 12 11 27 5 4 29 17 26 28 34 31 30 23 10 6 3 2 7
15 24 25  [8 9 36 35 18 19 1]  16 14 32 33 22 21 20 13 12 11 27 5 4 29 17 26 28 34 31 30 23 10 6 3 2 7
1 19 18  [35 36 9 8 25 24 15 16 14 32 33 22 21 20 13 12 11 27 5 4 29 17 26 28]  34 31 30 23 10 6 3 2 7
1 19 18  [28 26 17]  29 4 5 27 11 12 13 20 21 22 33 32 14 16 15 24 25 8 9 36 35 34 31 30 23 10 6 3 2 7
1  [19 18 17 26 28 29 4 5 27 11 12 13]  20 21 22 33 32 14 16 15 24 25 8 9 36 35 34 31 30 23 10 6 3 2 7
1  [13 12 11 27 5 4 29 28 26 17 18 19 20 21 22 33 32]  14 16 15 24 25 8 9 36 35 34 31 30 23 10 6 3 2 7
1  [32 33 22 21 20 19 18 17 26 28 29 4 5 27 11 12 13 14 16 15 24 25 8 9 36 35 34 31 30 23 10 6 3 2]  7
1 2 3 6  [10 23 30 31 34 35 36]  9 8 25 24 15 16 14 13 12 11 27 5 4 29 28 26 17 18 19 20 21 22 33 32 7
1 2 3 6  [36 35 34 31 30 23 10]  9 8 25 24 15 16 14 13 12 11 27 5 4 29 28 26 17 18 19 20 21 22 33 32 7
1 2 3  [6 7 32 33 22 21 20 19 18 17 26 28 29 4]  5 27 11 12 13 14 16 15 24 25 8 9 10 23 30 31 34 35 36
1 2 3 4 29 28 26  [17 18 19 20 21 22 33 32 7 6 5 27 11 12 13 14]  16 15 24 25 8 9 10 23 30 31 34 35 36
1 2 3 4 29 28 26 14 13 12 11  [27 5 6 7 32 33 22 21 20 19 18 17 16 15 24 25 8 9 10]  23 30 31 34 35 36
1 2 3 4 29 28 26  [14 13 12 11 10 9 8 25 24]  15 16 17 18 19 20 21 22 33 32 7 6 5 27 23 30 31 34 35 36
1 2 3 4 29 28 26  [24 25 8 9 10 11 12 13 14 15 16 17 18 19 20 21 22 33 32 7 6 5 27]  23 30 31 34 35 36
1 2 3 4  [29 28 26 27 5 6 7 32 33 22 21 20 19 18 17 16 15 14 13 12 11 10 9 8 25 24 23]  30 31 34 35 36
1 2 3 4  [23 24 25 8 9 10 11 12 13 14 15 16 17 18 19 20 21 22 33 32 7 6 5]  27 26 28 29 30 31 34 35 36
1 2 3 4 5 6 7  [32 33 22 21 20 19 18 17 16 15 14 13 12 11 10 9 8]  25 24 23 27 26 28 29 30 31 34 35 36
1 2 3 4 5 6 7 8 9 10 11 12 13 14 15 16 17 18 19 20 21 22  [33 32 25 24 23 27 26 28 29 30 31]  34 35 36
1 2 3 4 5 6 7 8 9 10 11 12 13 14 15 16 17 18 19 20 21 22  [31 30 29 28 26 27 23 24 25]  32 33 34 35 36
1 2 3 4 5 6 7 8 9 10 11 12 13 14 15 16 17 18 19 20 21 22  [25 24 23]  27 26 28 29 30 31 32 33 34 35 36
1 2 3 4 5 6 7 8 9 10 11 12 13 14 15 16 17 18 19 20 21 22 23 24 25  [27 26]  28 29 30 31 32 33 34 35 36
```

Fig. 7. Near-optimal solution for gene orders from the mitchondrial genomes of mammals and the flatworm *Ascaris suum*. This solution of 27 reversals is within 2 reversals of optimal.

24 seconds of computation on a 33 Mhz Silicon Graphics Iris 4D/300GTX. The lower and upper bounds from this run, of 25 and 27 reversals, were found using a lower bound family of 6 and an upper bound look ahead of 5 reversals.[12] The search tree during look-ahead had a maximum size of 2,234 nodes, where this counts all nodes at which a linear programming problem was solved (including nodes that were pruned by the lower bound). ¿From the difference between the upper and lower bounds we know that this solution is within 2 reversals of optimal. A family of 10 and a look-ahead of 8, which required a search of 408,653 nodes and terminated after 7.5 hours of computation, failed to improve the bounds by even one reversal.

The limit of what we can reliably solve to optimality is around 30 elements. Table 2 gives the running time and search tree size to solve a sample of 10 random permutations on 30 elements to optimality. Though the lower bound family and upper bound look-ahead were the same across these runs, the execution time and search tree size varied by three orders of magnitude. Since the average running time

---

[12] As described in Section 3.3, the follow-through for the upper bound was equal to the look-ahead in all experiments.

and tree size are skewed by these outliers we also give median values, which are more representative of the sample.

Table 2. Running time and search tree size for exact solution of random permutations on 30 elements.

Running time				Tree size							
				Upper bound algorithm				Exact algorithm			
min	max	med	avg	min	max	med	avg	min	max	med	avg
0:05	43:48	1:29	6:53	228	372,563	11,419	45,393	0	8,291	29	843

Lower bound family is 8, upper bound look-ahead is 6 reversals, and sample size is 10 permutations. Running time is the total time for the exact algorithm in minutes, which includes the time to compute the upper bound. Tree size for the upper bound algorithm is the maximum size of the search tree during look-ahead. The average reversal distance of the sample is 20.6.

That we could solve these instances to optimality is due to the tightness of the bounds. Over the 10 permutations the maximum difference between the upper and lower bound was 2 reversals, and in each case the exact solution value was equal to the lower bound. Except for the one permutation on which the exact algorithm examined over 8,000 nodes, a common picture emerged from these runs. Optimal solution involves the upper bound algorithm exploring a rather large tree to find a solution that is within one or two reversals of optimal; with this solution in hand the exact algorithm hugs the left chain of its search tree to make up the difference to the lower bound, which is exact.

The effect of varying the family and look-ahead is shown in Table 3. On random permutations of 50 elements the average lower bound reached a maximum of roughly 36 reversals at family 6, and the average upper bound did not improved much on 38 reversals beyond look ahead 5.[13] Consequently these values of family 6 and look-ahead 5 were used in the remaining experiments.

The search tree for these runs was limited to 50,000 nodes. Once this limit was exceeded, the best series known within the look-ahead was used to form the upper bound. Column T gives the median tree size for the sample, where the tree size of a run is the maximum size of the search tree during look-ahead. For a look-ahead of 6 or more the majority of runs had a tree size meeting the limit. Column C gives the median number of cycle sets on k or few breakpoints, which is equivalent to the number of variables in the linear program for the lower bound at family k. The number of constraints on these variables is essentially the number of breakpoints in the permutation.

In the third set of experiments we studied the quality of the approximation algorithm and the upper bound algorithm on random permutations. Results are given in Table 4, where *dev* denotes the standard deviation. Note that the maximum

---

[13] Notice that the average upper bound actually increased at look-ahead 8.

Table 3. Lower bound L, upper bound U, number of cycle sets C, and tree size T at various families and look-aheads for random permutations on 50 elements.

	L	U	C	T
	avg	avg	med	med
1	26.0	43.7	0	917
2	32.1	40.8	4	2,907
3	34.5	39.5	14	4,312
4	35.5	39.1	40	24,242
5	35.7	38.4	119	36,676
6	35.9	38.2	353	50,000
7	35.9	38.0	910	50,000
8	35.9	38.9	2,464	50,000

Maximum tree size is 50,000 nodes and sample size is 10 permutations. Row k gives the lower bound for family k, the upper bound for look-ahead k, the number of cycle sets of at most k breakpoints for the lower bound, and the maximum tree size for the upper bound during look-ahead.

difference between the upper and lower bound, which is what limits the range of optimal solution, was at most 2 reversals for n up to 30, while the average difference between the bounds was around 2.5 reversals for n up to 50. This suggests that while we can find optimal solutions for at most around 30 elements, we can find near-optimal solutions of acceptable quality for up to 50 elements.

The average performance ratio of the approximation algorithm for the sample was around $\frac{5}{4}$, while the worst ratio was less than $\frac{4}{3}$.

5   Conclusion

Algorithmic study of the reversal distance between permutations is in its earliest stages. We have presented two algorithms: a greedy approximation algorithm and a branch and bound exact algorithm. By analyzing the $\Delta\Phi$-sequence of the greedy algorithm we were able to show it achieves a worst-case approximation factor of 2, and by applying matchings, shortest paths, and linear programming we derived a class of nontrivial lower bounds for our exact algorithm. Experiments with the exact algorithm indicate that we can solve to optimality random permutations of up to 30 elements usually within a few minutes of computer time.

Acknowledgements

Guillaume Leduc, in analyzing signed reversals, contributed to the line of proof that $\Phi(\pi)$ bounds $\mathrm{OPT}(\pi)$. Holger Golan began the study of the distribution of $\mathrm{OPT}(\pi)$ for small n, and with his conjectures on the diameter of the symmetric group generated by reversals, stimulated the search for better lower bounds. Michel Berkelaar

Table 4. Lower bound L, upper bound U, approximation A, and tree size T for random permutations on n elements.

n	L avg	L dev	U avg	U dev	$U-L$ avg	$U-L$ max	A avg	A dev	A/L avg	A/L max	T med
10	6.0	1.1	6.0	1.1	0.0	0	6.1	1.0	1.02	1.20	0
20	12.6	1.2	13.2	1.5	0.6	2	14.8	1.5	1.18	1.31	517
30	20.8	0.9	21.8	1.6	1.0	2	24.9	1.5	1.20	1.25	4,615
40	28.5	1.1	30.3	1.6	1.8	4	33.8	2.0	1.19	1.24	23,712
50	35.9	1.4	38.4	1.8	2.5	5	43.7	2.7	1.22	1.31	36,676
60	43.6	1.1	46.7	1.5	3.1	5	52.6	1.6	1.21	1.26	50,000
70	51.7	1.3	56.7	2.4	5.0	8	63.4	3.0	1.23	1.30	50,000
80	58.9	1.0	64.5	2.0	5.6	8	72.3	2.3	1.23	1.26	50,000
90	67.6	1.4	74.1	2.1	6.5	8	83.2	1.8	1.23	1.25	50,000
100	74.2	1.1	82.4	2.6	8.2	10	91.9	2.7	1.24	1.27	50,000

Lower bound family is 6, upper bound look-ahead is 5 reversals, maximum tree size is 50,000 nodes, and sample size is 10 permutations. Approximation A is the number of reversals from the greedy algorithm. Tree size T is the maximum tree size for the upper bound during look-ahead.

generously provided the sparse linear programming code used in our experiments. The first author also wishes to thank Dan Gusfield, Paul Stelling, and Lucas Hui for many helpful discussions.

Research of JK was supported by a postdoctoral fellowship from the Program in Mathematics and Molecular Biology of the University of California at Berkeley under NSF Grant DMS–8720208, and by a fellowship from the Centre de recherches mathématiques of the Université de Montréal. Research of DS was supported by grants from the Natural Sciences and Engineering Research Council of Canada, and the Fonds pour la formation de chercheurs et l'aide à la recherche (Québec). DS is a fellow of the Canadian Institute for Advanced Research. Electronic mail addresses: kece@cs.ucdavis.edu and sankoff@ere.umontreal.ca.

A full version of this paper will be appearing in *Algorithmica* [10].

References

1. Aigner, Martin and Douglas B. West. Sorting by insertion of leading elements. *Journal of Combinatorial Theory* (Series A) 45, 306–309, 1987.
2. Amato, Nancy, Manuel Blum, Sandra Irani, and Ronitt Rubinfeld. Reversing trains: a turn of the century sorting problem. *Journal of Algorithms* 10, 413–428, 1989.
3. Bibb, J.J., R.A. van Etten, C.T. Wright, M.W. Walberg, and D.A. Clayton. *Cell* 26, 167–180, 1981.
4. Dobzhansky, Theodosius. *Genetics of the Evolutionary Process*. Columbia Univeristy Press, 1970.
5. Even, S. and O. Goldreich. The minimum-length generator sequence problem is NP-hard. *Journal of Algorithms* 2, 311–313, 1981.

6. Garey, Michael R. and David S. Johnson. *Computers and Intractability: A Guide to The Theory of NP-Completeness.* W.H. Freeman, New York, 1979.

7. Gates, William H. and Christos H. Papadimitriou. Bounds for sorting by prefix reversal. *Discrete Mathematics* 27, 47–57, 1979.

8. Golan, Holger. Personal communication, 1991.

9. Jerrum, Mark R. The complexity of finding minimum-length generator sequences. *Theoretical Computer Science* 36, 265–289, 1985.

10. Kececioglu, John and David Sankoff. Exact and approximation algorithms for the inversion distance between two chromosomes. To appear in *Algorithmica.*

11. Micali, S. and V. Vazirani. An $O(\sqrt{|V|} \cdot |E|)$ algorithm for finding maximum matchings in general graphs. In *Proceedings of the 21st Symposium on Foundations of Computer Science*, 17–27, 1980.

12. Nadeau, J.H. and B.A. Taylor. Lengths of chromosomal segments conserved since divergence of man and mouse. *Proceedings of the National Academy of Sciences USA* 81, 814, 1984.

13. O'Brien, S.J., editor. *Genetic Maps.* Cold Spring Harbor Laboratory, 1987.

14. Palmer, J.D., B. Osorio, and W.F. Thompson. Evolutionary significance of inversions in legume chloroplast DNAs. *Current Genetics* 14, 65–74, 1988.

15. Sankoff, David, Guillame Leduc, Natalie Antoine, Bruno Paquin, B. Franz Lang, and Robert Cedergren. Gene order comparisons for phylogenetic inference: evolution of the mitochondrial genome. *Proceedings of the National Academy of Sciences USA* 89, 6575–6579, 1992.

16. Schöniger, Michael and Michael S. Waterman. A local algorithm for DNA sequence alignment with inversions. *Bulletin of Mathematical Biology* 54, 521–536, 1992.

17. Sessions, Stanley K. Chromosomes: molecular cytogenetics. In *Molecular Systematics*, David M. Hillis and Craig Moritz editors, Sinauer, Sunderland, Massachusetts, 156–204, 1990.

18. Tichy, Walter F. The string-to-string correction problem with block moves. *ACM Transactions on Computer Systems* 2:4, 309–321, 1984.

19. Wagner, Robert A. On the complexity of the extended string-to-string correction problem. In *Time Warps, String Edits, and Macromolecules: The Theory and Practice of Sequence Comparison*, David Sankoff and Joseph B. Kruskal, editors, Addison-Wesley, Reading Massachusetts, 215–235, 1983.

20. Watterson, G.A., W.J. Ewens, T.E. Hall, and A. Morgan. The chromosome inversion problem. *Journal of Theoretical Biology* 99, 1–7, 1982.

21. Wolstenholme, D.R., J.L. MacFarlane, R. Okimoto, D.O. Clary, and J.A. Wahleithner. *Proceedings of the National Academy of Sciences USA* 84, 1324–1328, 1987.

The maximum weight trace problem in multiple sequence alignment

John Kececioglu

Computer Science Department, University of California at Davis, Davis, California 95616

Abstract. We define a new problem in multiple sequence alignment, called *maximum weight trace*. The problem formalizes in a natural way the common practice of merging pairwise alignments to form multiple sequence alignments, and contains a version of the minimum sum of pairs alignment problem as a special case.

Informally, the input is a set of pairs of matched characters from the sequences; each pair has an associated weight. The output is a subset of the pairs of maximum total weight that satisfies the following property: there is a multiple alignment that places each pair of characters selected by the subset together in the same column. A set of pairs with this property is called a *trace*. Intuitively a trace of maximum weight specifies a multiple alignment that agrees as much as possible with the character matches of the input.

We develop a branch and bound algorithm for maximum weight trace. Though the problem is NP-complete, an implementation of the algorithm shows we can solve instances on as many as 6 sequences of length 250 in a few minutes. These are among the largest instances that have been solved to optimality to date for any formulation of multiple sequence alignment.

1   Introduction

Multiple sequence alignment is among the outstanding problems of computational biology for which a satisfactory solution is unknown.

Briefly, multiple alignment may be defined as the following general problem. Given a set of sequences S_1, S_2, ..., S_k and a scoring function d, find a *multiple sequence alignment* $\mathcal{A} = (a_{ij})_{1 \leq i \leq k}$ that minimizes

$$\sum_j d(a_{1j} a_{2j} \cdots a_{kj}).$$

The entries a_{ij} of matrix $\mathcal{A}$ are either symbols from the sequence alphabet Σ or the *null character* ε, which is the identity under concatenation. The only constraint on $\mathcal{A}$ is that, concatenating characters $a_{i1} a_{i2} \cdots$ in any row i of the matrix, we must obtain the ith sequence S_i. Function d assigns a score to a column based on the characters that appear in it. In words, we seek an alignment $\mathcal{A}$ that minimizes the sum of the scores of its columns.

All the standard multiple sequence alignment problems may be cast in this form. For example in the *shortest common supersequence problem*, d counts the number of distinct symbols from Σ that appear in a column. The score of a multiple alignment is then the length of the supersequence it encodes.

In the *longest common subsequence problem*, d assigns the value -1 to a column consisting of k copies of a symbol from Σ, and assigns any other column the value 0. Minimizing the alignment score then maximizes the length of the common subsequence.

In the *minimum sum of pairs alignment problem* of Carrillo and Lipman [2],

$$d(a_{1j}\, a_{2j} \cdots a_{kj}) = \sum_{p<q} \delta(a_{pj}, a_{qj}),$$

where δ is a scoring function for pairs of symbols, including ε, satisfying $\delta(\varepsilon, \varepsilon) = 0$. In this problem the score of a multiple alignment is the sum of the scores of its pairwise alignments.

Even the problem of *multiple alignment under a fixed evolutionary tree*, in which we have a tree T with leaves labelled $S_1, ..., S_k$ and we seek ancestral sequences to label the internal nodes of T so as to minimize the sum of the edit distances along its edges, can be reduced to this general form, as shown by Sankoff [14].

As can be expected for a problem of such generality, multiple sequence alignment, as formulated above, is NP-complete. This follows from the NP-completeness of the longest common subsequence problem for a set of sequences [12], which we have noted is a special case of general multiple alignment.

On the other hand, by the early 1970s it was known that multiple sequence alignment could be solved by dynamic programming for k sequences of length n using $O(2^k n^k)$ evaluations of the scoring function d and $O(n^k)$ space. Thus for a fixed number of sequences, the problem is polynomial-time solvable.

This is hardly a practical solution, however. For a reasonable sized problem such as five sequences of length 100, the space requirements for dynamic programming are already more than ten gigabytes, even if the user could afford to wait for completion of the algorithm.

Consequently biologists have opted for suboptimal methods, such as the following. For every pair of sequences an optimal pairwise alignment is computed; then a set of pairwise alignments is selected that connect the sequences, and the pairwise alignments are merged to form a multiple alignment. Such an approach is fast, usually $O(k^2 n^2)$ time to compute the pairwise alignments and $O(k^2 \log k + kn)$ time to select a subset and merge them, and there is a well-developed theory for alignment of two sequences.

In the method of Feng and Doolittle [4] for instance, a maximum weight spanning tree of pairwise alignments is selected, where the weight of a pairwise alignment is its similarity score. This exploits the fact that any tree T of pairwise alignments can be merged into a multiple alignment that agrees with the alignments in T.

The main drawback of this and similar methods is that less than k of all $\binom{k}{2}$ pairwise alignments are used. Indeed, it is not hard to construct instances in which the merged alignment agrees with *only* the $k-1$ alignments of the tree, and none of the remaining alignments, even when there is a multiple sequence alignment that agrees with more than $\binom{k}{2} - k$ of the pairwise alignments [11].

This leads to the question we address in this paper, namely,

> *Given a set of pairwise alignments, how can we find a multiple alignment that is as close as possible to all pairwise alignments in the set?*

2 The maximum weight trace problem

To answer our question we must define "as close as possible."

The set of pairwise alignments may be represented by a graph, whose vertices are the characters[1] of the sequences, and whose edges are the pairs of characters matched in the alignments. The interpretation of vertices as characters in a sequence can be maintained by a vertex relation that captures the ordering of characters within a sequence.

Definition 1. An *alignment graph* $(V, E, \prec)$ for a set S of sequences is a graph (V, E) whose vertices V correspond to the characters of the sequences in S, together with a relation $\prec$ on V. Relation $v \prec w$ holds if and only if character v immediately precedes character w in a sequence of S. $\qquad\square$

Consider a path of edges in an alignment graph. Each edge on the path is a pair of aligned characters; transitively the path is a set of characters to be aligned in one column. The connected components[2] then of a set of edges correspond to the columns of a multiple alignment. These columns form a valid alignment if they can be ordered so as to respect $\prec$ character by character.

We can represent this ordering of columns by shrinking each component to a supervertex, and directing an edge from one supervertex to another if a character in one component immediately precedes a character in the other. We express this formally by extending $\prec$ to a relation $\prec^*$ on sets of vertices. For subsets X and Y of V,

$$X \prec^* Y \text{ if and only if } (\exists x \in X)\,(\exists y \in Y) \text{ such that } x \prec y.$$

Relation $\prec^*$, unlike $\prec$, may not be a partial order. Figure 1 gives an example. The point is that a multiple alignment exists that agrees with a set E of edges precisely when the components of E have a linear ordering under $\prec^*$.

Definition 2. A multiple sequence *trace* of an alignment graph $(V, E, \prec)$ is a subset $T \subseteq E$ of the edges such that $\prec^*$ on the connected components of T is acyclic. $\quad\square$

A trace of maximum cardinality agrees with as many of the matches in an alignment graph as possible. We gain a little more flexibility by weighting edges and seeking a trace of maximum total weight.

Definition 3. The *maximum weight trace problem* is, given alignment graph $(V, E, \prec)$ and edge weight function w, find a trace $T \subseteq E$ maximizing $\sum_{e \in T} w(e)$.
$\qquad\square$

Maximum weight trace was originally defined in [11] as a way of forming a multiple alignment from pairwise overlaps to determine a consensus sequence for DNA sequence reconstruction. We make a few remarks on the problem.

---

[1] Throughout the paper, a *character* of a sequence S is a position in S together with the symbol at that position.

[2] A *connected component* of a graph (V, E) is a maximal set $C \subseteq V$ such that every pair of vertices in C is joined by a path in E. *Maximal* means C is not contained in a larger set with this property.

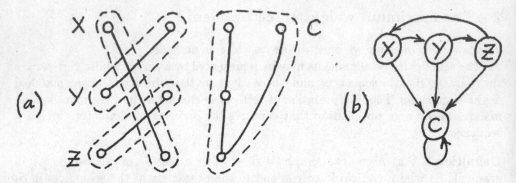

Fig. 1. (a) An alignment graph on three sequences. We use the convention of drawing the characters in a sequence horizontally left to right. (b) Relation $\prec^*$ on its connected components.

First, notice that multiple sequence trace, when restricted to two sequences, coincides with the definition of trace in sequence comparison [15, page 12]. It generalizes the definition and conforms to established terminology.

Second, when the alignment graph is a complete graph, maximum weight trace contains the minimum sum of pairs alignment problem [2], for symbol-independent insertion and deletion costs, as a special case. To see this, note that we can treat the minimum sum of pairs problem as a maximization problem by negating the cost function, and then by adding twice the insertion or deletion cost to all substitution scores we can transform the objective function so that the set of solutions is unchanged, while making all insertion and deletion scores zero. Weighting the edges of the complete alignment graph by the resulting substitution scores reduces this version of the minimum sum of pairs problem to maximum weight trace. When the alignment graph is not complete, we can score the multiple alignments considered by our maximum weight trace algorithm with the minimum sum of pairs objective function by performing the same transformation and simply taking the transitive closure of the input alignment graph, as will become clear in Section 4.1.

Third, we note that the alignment graph in the definitions may be a hypergraph, in which edges can be triples or quadruples of vertices, rather than simply pairs. The connected components for a set of edges are then given by the partition of vertices induced by successive union of intersecting edges. As before a trace is a collection of edges such that $\prec^*$ on its connected components is acyclic. This accomodates a very general notion of trace.

One advantage of the trace formulation is that a combinatorial structure is imposed on the problem from the start. When the edges of the graph come from an optimal alignment for each pair of sequences, this structure can often be exploited to speed up the computation. We call a graph whose edges between pairs of sequences form pairwise traces, a *pairwise alignment graph*.

In this paper we design an exact algorithm for the class of pairwise alignment graphs. Much of what we develop can be extended to more general inputs, and as we shall see in the next section, even for pairwise alignment graphs the problem remains hard.

3 The complexity of maximum weight trace

Theorem 4. *Maximum weight trace is NP-complete.*

Proof. Certainly maximum weight trace is in NP, as a trace can be guessed, tested for acyclicity, and its weight determined, all in polynomial time. The NP-hardness of maximum weight trace can be shown by a simple reduction from the *feedback edge set problem* [5]. In feedback edge set we are given a directed graph (V, E) together with an integer m, and we ask whether there is a subset $F \subseteq E$ called a *feedback edge set* with at most m edges such that $(V, E - F)$ is acyclic.

Given an instance (V, E) of feedback edge set, we construct an instance $(V', E', \prec)$ of maximum weight trace as follows. For every vertex $v \in V$ we create a vertex sequence S_v, and for every edge $(v, w) \in E$ we create an edge sequence S_{vw}. Sequence S_v contains one character, while sequence S_{vw} contains two. Edges in E' connect the first character of S_{vw} to S_v, and the second character of S_{vw} to S_w. The effect is that the connected components of (V', E') are star graphs centered on vertex sequences, which are ordered under $\prec^*$ by the edge sequences.

It is straightforward to show that, giving the edges of E' unit weight, (V, E) has a feedback edge set of at most m edges if and only if $(V', E', \prec)$ has a trace of weight at least $|E| - m$. □

We remark that a consequence of this reduction is that maximum weight trace is NP-complete even if every sequence has only two characters and the edges between every pair of sequences form a pairwise trace. In other words, maximum weight trace is NP-complete for pairwise alignment graphs over sequences of bounded length.

4 A branch and bound algorithm

As with the general alignment problem of the introduction, maximum weight trace can be solved by dynamic programming, which in turn can be expressed as a shortest or longest path problem. The standard breadth-first search solution of longest path on an acyclic graph is the starting point of our branch and bound algorithm. In order to describe the algorithm, we quickly review dynamic programming and its expression as a path problem.

Recall that a multiple alignment given by a maximum weight trace ends in a column containing a subset of the characters $S_1[n] \cdots S_k[n]$, where for convenience all sequences have length n. Remove this last column from the alignment. The remaining columns correspond to a maximum weight trace over the prefix of the sequences obtained by deleting the characters of the final column.

This gives a recurrence for the weight of an optimal trace. Writing $D(x_1 \cdots x_k)$ for the weight of an optimal trace over prefixes $S_1[1, x_1], \ldots, S_k[1, x_k]$, the recurrence has the form[3]

$$D(x_1 \cdots x_k) = \max_{(b_1 \cdots b_k)} \left\{ D\big((x_1-b_1) \cdots (x_k-b_k)\big) + d\big((S_1[x_1])^{b_1} \cdots (S_k[x_k])^{b_k}\big) \right\},$$

where

- $(b_1 \cdots b_k)$ is a binary vector from the set $\{0,1\}^k - \{0\}^k$,
- a^1 denotes the character a, and
- a^0 denotes the null character ε.

Function d gives the score of a column, which is the total weight of the edges in the subgraph induced by the column. Value $D(n \cdots n)$ is the weight of an optimal trace of the alignment graph.

One can evaluate D by filling in a k-dimensional table in order of lexicographically increasing coordinates. Each entry involves a maximum over $2^k - 1$ terms, and evaluating d for a term takes $O(k^2)$ time. As there are $O(n^k)$ entries, evaluating the dynamic program takes $O(2^k n^k k^2)$ time and $O(n^k)$ space.

In turn one can view the dynamic program as finding a longest path through an acyclic graph. Each vertex of the graph is associated with a subproblem $(x_1 \cdots x_k)$. Edges correspond to columns, and are weighted by the score of the column. An edge is directed from vertex $(x_1 \cdots x_k)$ to $(y_1 \cdots y_k)$ if $y_i - x_i \in \{0,1\}$ for each coordinate, and the vertices are distinct. A maximum weight trace alignment corresponds to a longest path from the source vertex $(0 \cdots 0)$ to the sink vertex $(n \cdots n)$.

This longest path problem may be solved by a breadth-first search from the source. The search maintains a queue of vertices with the property that the length of a longest path is known from the source to the vertex at the head of the queue. The generic step removes vertex v from the head of the queue, examines all edges (v, w) leaving v, compares the length of the best known path to w with the length of the longest path to v followed by (v, w), and adds w to the queue in lexicographic order if it is not already queued.

We can recast this breadth-first search in the framework of a *branch and bound algorithm*. We view examining the edges (v, w) leaving v as a *branch step*, in which a set of subproblems, namely the vertices w, are generated.[4] The dynamic program generates $2^k - 1$ subproblems by considering all possible columns with which to extend an alignment. Section 4.1 describes how to consider a smaller set of subproblems and still guarantee optimality.

We can add a *bound step* before placing w on the rear of the queue. Suppose we have a lower bound L on the length of a longest path from the source to the sink, for instance the weight of a multiple sequence trace computed by a heuristic. If we can quickly determine an upper bound $U(w)$ on the length of the longest path from w to the sink, or equivalently, an upper bound on the weight of an optimal trace over a suffix of the sequences, we may be able to avoid adding subproblem w to the queue. Writing $\ell(v)$ for the length of the longest path from the source to v, and $d(v, w)$ for

---

[3] To avoid cluttering the equation we have left out the boundary conditions. The conditions are that no term with a coordinate less than zero is evaluated and that $D(0 \cdots 0) = 0$.

[4] Instead of the computation tree of a standard branch and bound algorithm, we have a computation dag.

the weight of the column represented by edge (v, w), we can avoid adding w to the queue if

$$\ell(v) + d(v, w) + U(w) \le L.$$

Section 4.2 describes how to compute bounds L and $U(w)$.

4.1 The branch step

Given vertex $v = (x_1 \cdots x_k)$, which edges (v, w) do we have to consider? Recall that v represents the problem $S_1[1, x_1], \ldots, S_k[1, x_k]$, and (v, w) represents a column over $\{S_i[x_i + 1]\}$.

Call set $\{S_i[x_i + 1]\}$ the *exposed characters* on the *frontier* $(x_1 \cdots x_k)$, as shown in Figure 2. The alignment subgraph exposed by the frontier consists of all edges that touch an exposed character at one end, and touch a character on or to the right of the frontier at the other end.

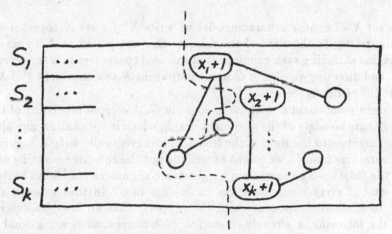

Fig. 2. The alignment subgraph exposed by frontier $(x_1 \cdots x_k)$.

Any column that extends frontier $(x_1 \cdots x_k)$ is a subset of the exposed characters. Our first observation is that we never have to consider subsets that are disconnected.

Observation 1. *It suffices to consider columns of characters that are connected in the alignment graph.*

Proof. Any column that contains two or more connected components can be divided into columns of one component without changing the weight of the alignment. □

Given that we can consider columns over each connected component of the exposed subgraph separately, which connected components do we have to consider?

Figure 3 shows a set of components on the frontier, together with edges of the exposed subgraph that leave the components. A maximum weight trace avoids cutting edges as much as possible, where an edge is *cut* if it is not included. Trying to avoid cutting exposed edges sets up a precedence among components.

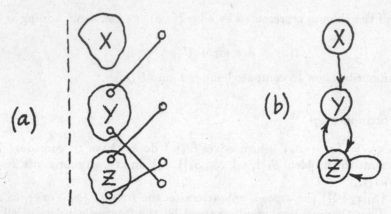

Fig. 3. (a) Components on the frontier, together with their exposed edges. (b) Avoiding cutting the exposed edges imposes an ordering on components.

For a set X of exposed characters, let us write X^* for set X together with the characters strictly to the right of the frontier that are joined to X by an exposed edge. Imagine shrinking each component of exposed characters to a supervertex in a graph G, and directing an edge in G from component X to component Y if $X \prec^* Y^*$, as in part (b) of the figure.

Consider a component C with no in-edges in G. A column consisting of the characters in C cuts no edges of the alignment graph, which is optimal. As any alignment over the characters to the right of the frontier must eventually output a column containing characters from C, we might as well output the column consisting of C, and advance the frontier. The maximum weight trace alignment can do no better.

But what if every component has an in-edge in G? In that case we can generalize C to a set of components with the property that no in-edges enter the set in G. In the following, a *strongly connected supercomponent* is a maximal set $\mathcal{C}$ of components such that, for every pair X, Y of components in $\mathcal{C}$, there is a path in G from X to Y and from Y to X. We use the term supercomponent to distinguish this set from a component, which is a connected set of exposed characters.

Observation 2. *It suffices to consider columns over the components in a strongly connected supercomponent of G with no incoming edges.*

Proof. Let $\mathcal{C}_1 \mathcal{C}_2 \cdots$ be the strongly connected supercomponents of G. Certainly considering all columns over all components in the $\mathcal{C}_i$ is correct.

Notice however that columns from supercomponents that are incomparable in G may be arbitrarily ordered without affecting the weight of an alignment. Thus considering columns from one supercomponent is also correct, as long as supercomponents are considered in topological order. Choosing a supercomponent with no incoming edges effectively chooses one of the many topological orders. ☐

Now that we know we can guarantee optimality by considering columns over each component X in one supercomponent, which columns over X do we have to consider?

If there are no exposed edges leaving component X, simply outputing the column consisting of X is correct. But if there are edges leaving the component, we are forced to cut an edge. Figure 4 shows the three kinds of edges that affect a component.

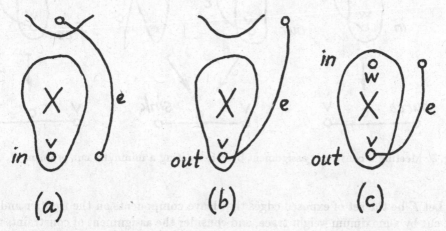

Fig. 4. The three types of edges that affect a component, and the constraints they induce.

Unless we are willing to look arbitrarily far past the frontier, we cannot tell in general which of the edges leaving a component will be cut in a maximum weight trace. If an edge of type (a) is not cut, we must output a column from component X that contains a character from the sequence to which e is incident; in the figure, this is character v. We represent this by placing an *in-constraint* on vertex v, which means we will consider a column that contains v.

Now consider an edge of type (b). If edge e is not cut, we must reserve character v so that it may be aligned with the other endpoint of e. This is represented by an *out-constraint* on v, which means we will consider a column that does not contain v.

Finally, an edge of type (c) is handled similarly to types (a) and (b). We place an in-constraint on vertex w and an out-constraint on vertex v.

The combination of edges not cut by an optimal trace assigns some set of in- and out-constraints to the characters of X. For a given constraint assignment, we can find the optimal column over X that meets the constraints by solving a minimum cut problem.[5]

Figure 5 shows the construction. Every character of the component has either the in-constraint, the out-constraint, or it has no constraint, in which case we say it is *free*. We construct a graph H that contains a vertex for every free character, and an additional source and sink vertex. An exposed edge between a free character and an in-contrained character is mapped to an edge of H touching the source, and an exposed edge between a free character and an out-constrained character is mapped to an edge touching the sink. Any parallel edges that arise are collapsed to a single edge that is given the sum of the weights of the parallel edges.

---

[5] A *minimum weight cut* of a graph is a partition of the vertices into two non-empty sets, such that the total weight of the edges that span the partition is minimized. A minimum cut can be found in polynomial time by a maximum flow computation [6].

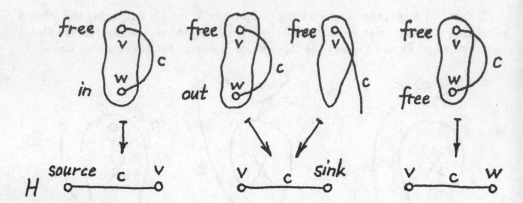

Fig. 5. Meeting a constraint assignment by constructing a minimum cut problem H.

Let F be the set of exposed edges that leave components on the frontier and are not cut by a maximum weight trace, and consider the assignment of constraints that F induces on component X. The effect of the construction is that the set of free characters that are in the source-half of a minimum weight source-sink cut of H, together with the in-constrained characters, form an optimal column over X that does not cut the edges of F. This gives our final observation.

Observation 3. *For each combination of constraints on the vertices in a component, it suffices to consider the column given by a minimum weight cut of H.*

Proof. Let E be the set of edges that leave components on the frontier, X be the component we are considering columns over, and C be a set of characters from X that are together in a column of a maximum weight trace alignment.

The maximum weight trace uses some subset F of the edges in E, and cuts the rest. Using F constrains some characters of X to be in or out of C, as in Figure 4. Solving the associated minimum cut problem of Figure 5 over H yields a column $\widetilde{C}$ that satisfies the same constraints.

Since $\widetilde{C}$ is given by a minimum cut, the weight of the edges cut by $\widetilde{C}$ is no more than the weight of the edges cut by C. Thus $\widetilde{C}$ is also optimal for the given component X and the given choice of edges F. □

Combining our observations, we get the following branch step.

1. Collect the connected components of exposed characters on the frontier for v.
2. Construct G and determine its strongly connected supercomponents.
3. Find a supercomponent C with no incoming edges in G.
4. For each connected component $C \in \mathcal{C}$, and for every combination of constraints on C, branch[6] on the column (v, w) given by a minimum cut of H.

---

[6] By *branch* we mean update the length of the best path to the new frontier w, and add w to the queue if not pruned by the bound step.

Theorem 5. *The set of columns considered by the branch step is sufficient to guarantee optimality.*

Proof. By Observations 1, 2, and 3. □

4.2 The bound step

The bound step requires a lower bound L on the weight of an optimal trace and an upper bound U on the weight of an optimal trace over a suffix of the sequences.

We can obtain L by the following *greedy heuristic.* Instead of computing a minimum cut for every assignment of constraints to the characters in a component, compute one cut for each component, in which every character is free. From all the components in the chosen supercomponent, select that column that has the minimum cut weight, and advance the frontier. This chooses the column that gives up the least immediate trace weight.

For upper bound U in a pairwise alignment graph, we can take the total weight of the edges in the alignment graph over the suffixes of the sequences. More elaborate upper bounds are also possible, such as computing a maximum weight trace over the suffixes for all triples of sequences, summing the weights of the optimal traces, and dividing by the number of sequences minus two, which is the number of times a pair of sequences is counted in triples.

5 Preliminary computational results

We have implemented a version of the branch and bound algorithm that exploits branching to reduce the search space, but at present does no bounding.

The current implementation is roughly 2,000 lines of C. The code computes minimum cuts using the push-relabel maximum flow algorithm of Goldberg and Tarjan [6]. Vertices of the longest path graph are stored and looked up by coordinate using the lexicographic splay tree of Sleator and Tarjan [16]. Since this effectively factors vertex coordinates in a trie, the space per vertex in practice is less than $O(k)$ for k sequences.

We give an example in Figure 6 of a maximum weight trace alignment of 6 tyrosine kinase protein sequences of length 273 to 285. The alignment graph consisted of one optimal alignment from each pair of sequences. Pairwise alignments were computed using a standard scoring scheme: the PAM 250 matrix of Margaret Dayhoff to score substitutions, which contains integer similarities in the range 0 to 25, with insertions and deletions receiving similarity 0 and a length-independent gap penalty of 8. Edges of the alignment graph were weighted by the substitution matrix. Notice that, as is common with protein sequences, there are many substitutions and few gaps in the alignment.

This problem was solved to optimality in 176.6 seconds (roughly 3 minutes) and 2.6 megabytes on a NeXT machine running at 33 megahertz. The branch and bound algorithm explored a subgraph of 119,046 vertices, out of a dynamic programming graph of 4.14×10^{14} vertices. This is less than .00000003% of the dynamic programming graph.

```
-----GLA-K-DAWEIPRESLRLEAKLGQGCFGEVWMGT-WND-TTRVAIKTLK-PGTMSPEA-FLQEAQVMKKLRHEK
-----GLA-K-DAWEIPRESLRLEVKLGQGCFGEVWMGT-WNG-TTKVAIKTLK-LGTMMPEA-FLQEAQIMKKLRHDK
TIY--GVSPNYDKWEMERTDITMKHKLGGGQYGEVYEGV-WKKYSLTVAVKTLKE-DTMEVEE-FLKEAAVMKEIKHPN
-VLNRAVP-K-DKWVLNHEDLVLGEQIGRGNFGEVFSGRLRAD-NTLVAVKSCRETLPPDIKAKFLQEAKILKQYSHPN
-VLTRAVL-K-DKWVLNHEDVLLGERIGRGNFGEVFSGRLRAD-NTPVAVKSCRETLPPELKAKFLQEARILKQCNHPN
-------S-S-YYWKMEASEVMLSTRIGSGSFGTVYKGK-WHG-DVAVKILKVVDPTPEQLQA-FRNEVAVLRKTRHVN

LVQLYAVVSE-EPIYIVIEYMSKGSLLDFLKGEMGKYLRLPQLVDMAAQIASGMAYVERMNYVHRDLRAANILVGENLV
LVPLYAVVSE-EPIYIVTEFMTKGSLLDFLKEGEGKFLKLPQLVDMAAQIADGMAYIERMNYIHRDLRAANILVGDNLV
LVQLLGVCTREPPFYIITEFMTYGNLLDYLRECNRQEVSAVVLLYMATQISSAMEYLEKKNFIHRDLAARNCLVGENHL
IVRLIGVCTQKQPIYIVMELVQGGDFLTFLRT-EGARLRMKTLLQMVGDAAAGMEYLESKCCIHRDLAARNCLVTEKNV
IVRLIGVCTQKQPIYIVMELVQGGDFLSFLRS-KGPRLKMKKLIKMMENAAAGMEYLESKHCIHRDLAARNCLVTEKNT
ILLFMGYMTK-DNLAIVTQWCEGSSLYKHLHV-QETKFQMFQLIDIARQTAQGMDYLHAKNIIHRDMKSNNIFLHEGLT

CKVADFGLARLIEDNEYTARQGAK-FPIKWTAPEA--ALY-GRFTIKSDVWSFGILLTELTTKGRVPYPGMVNRE-VLD
CKIADFGLARLIEDNEYTARQGAK-FPIKWTAPEA--ALY-GRFTIKSDVWSFGILLTELVTKGRVPYPGMVNRE-VLE
VKVADFGLSRLMTGDTYTAHAGAK-FPIKWTAPES--LAY-NKFSIKSDVWAFGVLLWEIATYGMSPYPGIDLSQ-VYE
LKISDFGMSREAADGIYAASGGLRQVPVKWTAPEA--LNY-GRYSSESDVWSFGILLWETFSLGASPYPNLSNQQ-TRE
LKISDFGMSRQEEDGVYASTGGMKQIPVKWTAPEA--LNY-GWYSSESDVWSFGILLWEAFSLGAVPYANLSNQQ-TRE
VKIGDFGLATVKSRWSGSQQVEQPTGSVLWMAPEVIRMQDDNPFSFQSDVYSYGIVLYELMA-GELPYAHINNRDQIIF

QVERGY---RMPCP-PECPESLHDLMCQCWRKDPEERPTFKYLQAQLLPACVLEVAE-------
QVERGY---RMPCP-QGCPESLHELMKLCWKKDPDERPTFEYIQS-FLEDYFTAAEPSG-----
LLEKDY---RMERP-EGCPEKVYELMRACWQWNPSDRPSFAEIH-----Q-AFETMFQESS-IS
FVEKGG---RLPCP-ELCPDAVFRLMEQCWAYEPGQRPSFSAIY-----Q-ELQSIRKRHR---
AIEQGV---RLEPP-EQCPEDVYRLMQRCWEYDPHRRPSFGAVH-----Q-DLIAIRKRHR---
MVGRGYASPDLSRLYKNCPKAIKRLVADCVKKVKEERPLFPQIL-----S-SIELLQHSLPKIN
```

Fig. 6. An optimal maximum weight trace alignment of six tyrosine kinase sequences. The alignment is wrapped to fit lines of a fixed width.

On this data set we could not add a seventh tyrosine kinase sequence and solve the problem to optimality within 32 megabytes of memory. Incorporating the bound step should extend the range of solution.

6   Conclusion

In summary, we have introduced a new problem in multiple sequence alignment, maximum weight trace, which we believe is a natural formulation of merging partial alignments to form multiple alignments. We have studied the problem from the point of view of exact solution, developed a branch and bound algorithm, and demonstrated that we can solve nontrivial instances to optimality in a reasonable amount of time and space. We close with some lines for future research.

Further research

With an applied problem there are two basic questions: Does the problem have a practical solution? and Is it a good model? One way to approach these questions is with a computational study. Practicality could be examined by generating a random sequence of length n, making k copies, editing each copy with $\epsilon \cdot n$ random errors, computing a minimum edit distance alignment between every pair of copies, and then measuring the mean and variance of the size of the subgraph explored by the branch and bound algorithm, as function of k, n, and ϵ. The modeling issue could then be addressed by measuring how close the optimal maximum weight trace alignment is to the "true alignment" of the generated sequences. These experiments, on a full implementation of the branch and bound algorithm, are forthcoming for the final journal paper.

While it is interesting to explore how large a problem can be solved to optimality, approximation algorithms will most likely be the choice in practice. Is there a polynomial-time approximation algorithm for maximum weight trace with a constant worst-case approximation factor? In particular, is the greedy heuristic of Section 4.2, or a simple variant, such an algorithm?

Finally, much of the effort of the branch and bound algorithm is in untangling edges where the sequences do not align well, yet we suspect users are less interested in such areas of the alignment. This suggests constructing an alignment graph from the set of all significant pairwise local similarities.[7] This would give a rigorous way to form multiple alignments from local similarities, which we envision as the main application of maximum weight trace.

Acknowledgements

We thank Dr. David Lipman for the protein sequences of Figure 6.

This research was supported by a postdoctoral fellowship from the Program in Mathematics and Molecular Biology of the University of California at Berkeley under NSF Grant DMS–8720208. Author's electronic mail address: kece@cs.ucdavis.edu.

References

1. Altschul, Stephen F. and David J. Lipman. Trees, stars, and multiple biological sequence alignment. *SIAM Journal on Applied Mathematics* 49:1, 197–209, 1989.
2. Carrillo, Humberto and David Lipman. The multiple sequence alignment problem in biology. *SIAM Journal on Applied Mathematics* 48, 1073–1082, 1988.
3. Chan, S.C., A.K.C. Wong and D.K.Y. Chiu. A survey of multiple sequence comparison methods. To appear in the *Bulletin of Mathematical Biology*, 1992.
4. Feng, Da-Fei and Russell F. Doolittle. Progressive sequence alignment as a prerequisite to correct phylogenetic trees. *Journal of Molecular Evolution* 25, 351–360, 1987.
5. Garey, Michael R. and David S. Johnson. *Computers and Intractability: A Guide to the Theory of NP-Completeness*. W.H. Freeman, New York, 1979.

---

[7] A *local similarity* is an alignment between a substring of one sequence and a substring of another. See [17] for a definition commonly used in practice.

6. Goldberg, Andrew V. and Robert E. Tarjan. A new approach to the maximum flow problem. *Journal of the Association for Computing Machinery* 35:4, 921–940, 1988.

7. Gotoh, Osamu. Consistency of optimal sequence alignments. *Bulletin of Mathematical Biology* 52:4, 509–525, 1990.

8. Gusfield, Dan. Efficient methods for multiple sequence alignment with guaranteed error bounds. *Bulletin of Mathematical Biology* 55:1, 141–154, 1993.

9. Hsu, W.J. and M.W. Du. Computing a longest common subsequence for a set of strings. *BIT* 24, 45–59, 1984.

10. Irving, Robert W. and Campbell B. Fraser. Two algorithms for the longest common subsequence of three (or more) strings. In Proceedings of the 3rd Symposium on *Combinatorial Pattern Matching*, 211–226, 1992.

11. Kececioglu, John. *Exact and Approximation Algorithms for DNA Sequence Reconstruction*. PhD dissertation, Technical Report 91-26, Department of Computer Science, The University of Arizona, Tucson, Arizona 85721, 1991.

12. Maier, David. The complexity of some problems on subsequences and supersequences. *Journal of the Association for Computing Machinery* 25:2, 322–336, 1978.

13. Pevzner, Pavel. Multiple alignment, communication cost, and graph matching. To appear in *SIAM Journal on Applied Mathematics*.

14. Sankoff, David. Minimal mutation trees of sequences. *SIAM Journal on Applied Mathematics* 28:1, 35–42, 1975.

15. Sankoff, David and Joseph B. Kruskal, editors. *Time Warps, String Edits, and Macromolecules: The Theory and Practice of Sequence Comparison*. Addison-Wesley, Reading, Massachusetts, 1983.

16. Sleator, Daniel D. and Robert E. Tarjan. Self-adjusting binary search trees. *Journal of the Association for Computing Machinery* 32:3, 652–686, 1985.

17. Smith, Temple F. and Michael S. Waterman. Identification of common molecular sequences. *Journal of Molecular Biology* 147, 195–197, 1981.

18. Vingron, Martin and Patrick Argos. A fast and sensitive multiple sequence alignment algorithm. *Computer Applications in the Biosciences* 5:2, 115–121, 1989.

19. Waterman, M.S. and R. Jones. Consensus methods for DNA and protein sequence alignment. *Methods in Enzymology* 188, 221–237, 1990.

An Algorithm for Approximate Tandem Repeats

Gad M. Landau* and Jeanette P. Schmidt**

Polytechnic University, 6 MetroTech, Brooklyn, NY 11201.

Abstract. A *perfect* tandem repeat within a string S is a substring $r = r_1, \ldots r_{2\ell}$ of S, for which $r_1 \ldots r_\ell = r_{\ell+1} \ldots r_{2\ell}$. An *approximate* tandem repeat is a substring $r = r_1, \ldots, r_{\ell'}, \ldots r_\ell$, for which $r_1, \ldots, r_{\ell'}$ and $r_{\ell'+1}, \ldots r_\ell$ are similar. In this paper we consider two criterions of similarity: the Hamming distance (k mismatches) and the edit distance (k differences). For a string S of length n and an integer k our algorithm reports all locally optimal approximate repeats, $r = \bar{u}\hat{u}$, for which the Hamming distance of $\bar{u}$ and $\hat{u}$ is at most k in $O(nk \log(n/k))$ time, or all those for which the edit distance of $\bar{u}$ and $\hat{u}$ is at most k, in $O(nk \log k \log n)$ time.

1 Introduction

The *perfect* tandem repeat problem is a well studied problem. Main and Lorentz [ML-84] present an $O(n \log n)$ algorithm, which reports all *perfect* tandem repeats and Apostolico [Ap-92] describes an optimal speed-up parallel algorithm for the problem. Motivations for the exact repeat problem can be found in research in formal languages (see a survey in [ML-85]).

Important motivations for the *approximate* tandem repeat problem are found in different areas. In molecular Biology, tandem repeats play an important role in both DNA and protein sequences. At the DNA level they act as "hot spots" that enable these regions to more rapidly conform to environmental changes. Such repeats are also frequent in bacterial proteins, where their function is less understood. The repeats in these applications are not exact. One can use different criterions to measure the similarity of the repeats. In this paper we consider two simple measures of similarity. While these measures are suitable for several of the above applications, they also lend themselves to the design of fast algorithms.

Given a string S and an integer k the algorithm finds all non empty substrings $r = \bar{u}\hat{u}$, for which:
(i) the Hamming distance of $\bar{u}$ and $\hat{u}$ is at most k; or
(ii) the edit distance of $\bar{u}$ and $\hat{u}$ is at most k.

The Hamming distance of u and v is defined as the number of substitutions necessary to get v from u, (u and v must be of same length). The edit distance, as defined by Levenshtein [L-66], is the minimum number of deletions in u, substitutions or insertions in v necessary to get v from u. In the case of the Hamming distance

* e-mail: landau@pucs2.poly.edu. Partially supported by the New York State Science and Technology Foundation Center for Advanced Technology.
** e-mail: jps@pucs4.poly.edu. Partially supported by NSF grant CCR-9110255 and the New York State Science and Technology Foundation Center for Advanced Technology.

our algorithm runs in $O(nk \log(n/k))$ time. For the case of the edit distance our algorithm runs in $O(nk \log k \log n)$ time.

The framework of the algorithm, which is the same in both cases, is described in Section 2. The procedure to be used by the algorithm when looking for the Hamming distance is described in Section 3, and the one for the edit distance is given in Section 4.

2 Algorithm's framework

The framework of our algorithm is similar to the algorithm for *perfect* repeats described by Main and Lorentz [ML-84, ML-85]. Given a string $S = s_1 \ldots s_n$, the algorithm has $O(\log n)$ iterations. We assume for simplicity of presentation that n is a power of 2. (No difficulty arises in the general case). In iteration i the string is divided into the $n/2^i$ substrings, $(s_1 \ldots s_{2^i}), \ldots (s_{\ell 2^i + 1} \ldots s_{(\ell+1)2^i}), \ldots (s_{n-2^i+1} \ldots s_n)$. For each substring, $s_{\ell 2^i + 1} \ldots s_{(\ell+1)2^i}$, separately, all approximate tandem repeats $r = \bar{u}\hat{u}$, which lie entirely within the substring and for which the middle characters $s_{\ell 2^i + 2^{i-1}}, s_{\ell 2^i + 2^{i-1}+1} \in r$, are found. Each such iteration has two steps. In the **first step** one finds all approximate tandem repeats for which $s_{\ell 2^i + 2^{i-1}}$ is in $\bar{u}$, and in the **second step** the approximate tandem repeats for which $s_{\ell 2^i + 2^{i-1}}$ is in $\hat{u}$ are identified. For the Hamming distance our algorithm will start with iteration $\log k$, i.e. with strings of length k. The repeats within these strings of length k will be found by a straightforward algorithm in $O(k^2)$ time per string.

In what follows we are going to describe the first step of any given iteration. Since iterations only differ in that they consider strings of different sizes, consider the iteration which finds an approximate tandem repeat $r = \bar{u}\hat{u}$ in the string $S = s_1 \ldots s_n$, for which $s_{n/2} \in \bar{u}$. As in [ML-84, ML-85] for each j, $n/2 < j \le n$, we determine the indexes ℓ, for which $s_{n/2+1} \ldots s_\ell, \ell \le j$ is similar to some prefix $s_{j+1} \ldots s_\pi$ of $s_{j+1} \ldots s_n$, and $s_{\ell+1} \ldots s_j$ is similar to some suffix $s_\sigma \ldots s_{n/2}$ of $s_1 \ldots s_{n/2}$.

We then have: $\quad \bar{u} = \begin{array}{cc} s_\sigma \ldots s_{n/2} & s_{n/2+1} \ldots s_\ell \\ \hat{u} = s_{\ell+1} \ldots s_j & s_{j+1} \ldots \quad s_\pi \end{array}$ is an *approximate* repeat if $\bar{u}$ is "sufficiently" similar to $\hat{u}$. To find all indices ℓ which might be candidates for such a repeat, (for a given j, $n/2 < j \le n$), we will determine for each $k_1 \le k$ the longest suffix $s_{\ell_{k_1}} \ldots s_j$ of $s_1 \ldots s_j$ whose Hamming (or edit) distance with some suffix of $s_1 \ldots s_{n/2}$ is no more than k_1. We also determine for each $k_2 \le k$ the *longest* prefix $s_{n/2+1} \ldots s_{\ell_{k_2}}$ of $s_{n/2+1} \ldots s_n$ whose Hamming (or edit) distance with some prefix of $s_{j+1} \ldots s_n$ is no more than k_2:

$$\left. \begin{array}{cc} \cdots & \cdots s_{n/2} \\ s_{\ell_{k_1}} \cdots s_{\ell+1} & \cdots s_j \end{array} \right\} \quad distance \; k_1, \qquad \left. \begin{array}{cc} s_{n/2+1} \cdots & s_\ell \cdots s_{\ell_{k_2}} \\ s_{j+1} & \cdots \quad \cdots \end{array} \right\} \quad distance \; k_2.$$

For any pair k_1, k_2 for which $k_1 + k_2 \le k$, and $\ell_{k_1} - 1 \le \ell \le \ell_{k_2}$ an *approximate repeat* is identified. Note that for the Hamming distance $|\bar{u}| = |\hat{u}| = j - n/2$.

3 Hamming distance

In this section we describe the use of the Hamming distance as the similarity measure. We also translate the alignments described in the previous section to alignment paths, (or in fact diagonals) in the dynamic programming matrix. Although this is not essential to follow the computation of the Hamming distance it does set the ground for a very easy understanding of the computations necessary when the measure of similarity is the edit distance. We describe the algorithm for identifying all approximate tandem repeats in S for which $s_{n/2}$ is in $\bar{u}$, (identifying those for which $s_{n/2}$ is in $\hat{u}$ requires simple straightforward modifications). The algorithm has two parts.

Part 1. We align the string S with itself and consider the dynamic programming matrix $D[0\ldots n, 0\ldots n]$. We divide the bottom half of D into two sub-matrices: $PR[n/2\ldots n; n/2\ldots n] = D[n/2\ldots n; n/2\ldots n]$, (to compute common prefixes), and $SU[n/2\ldots n; 0\ldots n/2] = D[n/2\ldots n; 0\ldots n/2]$, (to compute common suffixes). SU and PR share the middle column of D, which both use for the purpose of initialization, (see Figure 1 below). For any $j > n/2$ any sub-diagonal through $D[j, n/2]$ of length $j - n/2$ represents a tandem repeat $r = \bar{u}\hat{u}$, with $s_{n/2} \in \bar{u}$. (The algorithm then, as in [ML-84, ML-85], recursively continues with sub-matrices $D[0\ldots n/2, 0\ldots n/2]$ and $D[n/2\ldots n, n/2\ldots n]$.)

It remains to identify those diagonals which represent tandem repeats with less than k mismatches. Each diagonal of PR and SU is computed separately. In PR the computation starts from the left side of the matrix, with $PR[j+c, n/2+c]$ holding the value of the alignment of $s_{j+1}\ldots s_{j+c}$ with $s_{n/2+1}\ldots s_{n/2+c}$. In SU the computation starts from the right side of the matrix, with $SU[j - c, n/2 - c]$ holding the value of the alignment of $s_j \ldots s_{j-c+1}$ with $s_{n/2}\ldots s_{n/2-c+1}$. ($SU[j, n/2] = PR[j, n/2] = 0$.) The computation is similar for all diagonals.

Let $j > n/2$ and consider the diagonal through $D[j, n/2]$. It can be computed in the following straightforward manner:

```
Computing PR
    e = 0;
    For h = 0 to k do
        PR[j + e, n/2 + e] = h  (an auxiliary value when h = 0)
        While (j + e + 1 ≤ n) and (s_{j+e+1} = s_{n/2+e+1})   do
            e = e + 1;  PR[j + e, n/2 + e] = h
        od
        P*[j, h] = e;  e = e + 1;
    od

Computing SU
    e = 0;
    For h = 0 to k do
        SU[j + e, n/2 + e] = h  (an auxiliary value when h = 0)
        While (j - e > n/2) and (s_{j-e} = s_{n/2-e})   do
            e = e + 1;  SU[j - e, n/2 - e] = h
        od
        S*[j, h] = e;  e = e + 1;
    od
```

Successive values along the diagonals increase and differ by at most one and only the last row (respectively column) in each diagonal that contains a value h less or

The matrix D

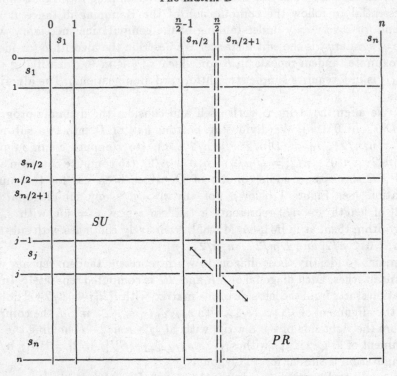

Fig. 1. Subdivision of the matrix D; shown are the matrices PR and SU in the computation for index j. The point $D[i, i]$ correpsonds to the intersection point of the vertical line labelled i and the horizontal line labelled i. In PR, for $c', c > 0$, the value of an alignment $s_{j+1} \ldots s_{j+c}$ and $s_{n/2+1} \ldots s_{n/2+c'}$ is stored in the bottom right corner of the square $s_{j+c}, s_{n/2+c'}$, (i.e. in $PR[j + c, n/2 + c']$); in SU, for $c', c \geq 0$, an alignment $s_j \ldots s_{j-c}$ and $s_{n/2} \ldots s_{n/2-c'}$ is stored in the upper left corner of the square $s_{j-c}, s_{n/2-c'}$, (i.e. in $SU[j - c - 1, n/2 - c' - 1]$).

equal to k is of interest. These values are recorded in the matrices P^* and S^*. The last row with value h on the diagonal that goes through $D[j, n/2]$ is $(j + P^*[j, h])$. Similarly $(j - S^*[j, h])$ is the last row, (upward from j), with value h on that diagonal. $P^*[j, h]$ (resp. $S^*[j, h]$) can be computed from $P^*[j, h - 1]$ (resp. $S^*[j, h - 1]$) in $O(1)$ time using suffix trees, as shown in [LV-88], in place of the above while loop. Therefore, for a given j, Part 1 of step 1 (resp. 2) is computed in $O(k)$ time. It remains to combine the pieces.

Part 2. We can now identify all repeats $r = \bar{u}\hat{u}$ for which $s_{n/2} \in \bar{u}$ and in which s_j $(j > n/2)$ is aligned with $s_{n/2}$. Define $j^* = j - n/2$, and note that all the indexes on this diagonal are of the form $[m, m - j^*]$. The repeats correspond to all the sub-diagonals of length j^* through $D[j, n/2]$, i.e. from any $[\ell, \ell - j^*]$ to $[\ell + j^*, \ell]$, $\ell > n/2$, for which $SU[\ell, \ell - j^*] + PR[\ell + j^*, \ell] \leq k$.

It is reasonable to report only those repeats $r = \bar{u}\hat{u}$ through $D[j, n/2]$ for which the Hamming distance between $\bar{u}$ and $\hat{u}$ is minimal among such repeats.

More generally we compute for each $h \leq k$ the value $f(h) = \min_{h' \leq k-h}\{h' \mid (P^*(j,h) + S^*(j,h') \geq j^*)\}$. If no such h' can be found $f(h)$ is undefined. These values can be easily computed by a merge like procedure from the data collected in Part 1 in $O(k)$ time for each point j, as illustrated below.

Computing optimal repeats
```
h' = k + 1;
for h = 0 to k
    while (h' > 0) and (P*(j, h) + S*(j, h' - 1) ≥ j*)  do
        h' = h' - 1;
    od
    if ((h' + h) ≤ k) then f(h) = h' else h' = h' - 1;
od
```

For all those h for which $h + f(h)$ is minimal we can now report all sub-diagonals of length j^* that start between $D[j - S^*(j, f(h)), n/2 - S^*(j, f(h))]$ and $D[j + P^*(j,h) - j^*, n/2 + P^*(j,h) - j^*]$. (A sub-diagonal from $D[j - c, n/2 - c]$ to $D[j + j^* - c, j - c]$ corresponds to a repeat $r = \bar{u}\hat{u}$, with $\hat{u} = s_{j-c+1} \ldots s_{j+j^*-c}$ and $\bar{u} = s_{n/2-c+1} \ldots s_{j-c}$.)

Strings of size k. (Sketch) For small strings of size $\ell \leq k$ we use a simple $O(\ell^2)$ algorithm. For a given $t \leq \ell/2$, it gives the number of mismatches corresponding to all repeats $r = \bar{u}\hat{u}$ of length $2t$ in $O(\ell)$ time.

Complexity. Our iterations follow the $O(n \log n)$ Main and Lorentz algorithm, but our algorithm starts with strings of size k, and has therefore $\log(n/k)$ iterations. For each substring of length ℓ, our algorithm takes $O(\ell k)$, and hence each iteration takes $O(nk)$ time. The initial iteration, (for strings of length k), takes $O((n/k)k^2) = O(nk)$ time. The total running time of our algorithm is therefore $O(nk \log(n/k))$.

4 Edit distance

The computation of the edit distance will proceed similarly to the computation of the Hamming distance. The nature of the repeats is somewhat different than before in that now a repeat $r = \bar{u}\hat{u}$ no longer has the property that $|\bar{u}| = |\hat{u}|$. We will still carry out the computation using the matrices D, (with its sub-matrices SU and PR) and consider all repeats for which the alignment between $\bar{u}$ and $\hat{u}$ goes through $D[j, n/2]$, $j > n/2$. The repeats in this case correspond to paths in the matrix which start at $D[j - r_1, \ell_1]$, $(j - r_1 \geq n/2, \ell_1 \leq n/2)$, and end at $D[\ell_2, j - r_1]$, for some $\ell_2 \geq j$. In the previous section (Hamming distance) one diagonal in D was computed for each j. Here, the computation for each j progresses along several diagonals. Therefore, we are obliged to compute a matrix D for each j.

Consider the computation of PR for j. A straightforward, $O(n^2)$, computation would be:

```
for a = j to n
    PR[a, n/2] = a - j
    for b = n/2 to n
```

$$PR[n/2, b] = b - n/2$$

for $a = j + 1$ to n

    for $b = n/2 + 1$ to n do

        $PR[a, b] = \min\{PR[a, b-1] + 1, PR[a-1, b] + 1, ((PR[a-1, b-1]$ if

        $s_a = s_b)$ and $(PR[a-1, b-1] + 1$ otherwise$)\}$

od

As before the computation in PR starts from the left side of the matrix, and the computation of SU from the right side. In PR for each $h \leq k$ we identify the longest prefix of $s_{n/2+1} \ldots s_n$ which matches some prefix of $s_{j+1} \ldots s_n$ with h differences. This will correspond to a path starting at $[j, n/2]$ which stretches as far right as possible. In SU, for each $h' \leq k$, we identify the longest suffix of $s_{n/2} \ldots s_j$ that matches some suffix of $s_1 \ldots s_{n/2}$ with h' differences. This will correspond to a path in SU that starts at $[j, n/2]$ and stretches as high as possible. (See Figure 1 for the direction of these paths). Two such path can be combined to an approximate repeat with at most k differences if $h + h' \leq k$.

Consider the computation of PR. We compute the edit distance matrix for the strings $s_{n/2+1} \ldots s_j$ and $s_{j+1} \ldots s_n$. The values along the diagonals in PR increase and successive values differ by at most one, it is hence sufficient to record, for each $h \leq k$, the last element on each diagonal whose value is h, as observed in [U-83]. In order to simplify the explanation we give the diagonals fixed numbers, which are independent of the reference point j. Define the diagonal through $[d, n/2]$ as diagonal d. Note that all diagonals in any PR have positive numbers now. Let $p_j^*(d, h)$ be the column number, (in PR for a given j), of the last element with value h on diagonal d. Since the smallest value on diagonals $j \pm x$ is $|x|$ it is sufficient to compute the values on the $2k + 1$ diagonals corresponding to $x \in [j + k, j - k]$. The list $p_j^*(j + h, h) \ldots p_j^*(j, h) \ldots p_j^*(j - h, h)$ is defined as *wave* h, and denoted by L_j^h, $(p_j^*(\ell, h) = L_j^h(\ell))$. Note that by slight abuse of notation, but without ambiguity, $L_j^h(\ell)$ will denote both the point on diagonal ℓ on wave h, as well the column number of that point. Also note that for a given j the information stored in PR, is now stored in k waves each of size (at most) $2k + 1$.

We define $P^*(j, h) = max_\ell \, L_j^h(\ell)$, the right most column reached on wave h. Since there might be more than one such point we define $\pi(j, h)$ as the *set* of corresponding end points on the wave.

In Section 4.1 we discuss how to compute the k waves (wave 0, wave 1, ... wave k). Section 4.2 describes an algorithm to find $P^*(j, h)$ and $\pi(j, h)$ for each wave.

Until the end of this subsection assume that $P^*(j, h)$ and $\pi(j, h)$ have already been computed for all the waves.

The computations for SU are very similar to the ones of PR. The diagonals are defined exactly as before, and $s^*(d, h)$ will be the number of rows upwards from j, of the last element with value h on diagonal d. $S^*(j, h) = max_d \, s^*(h, d)$, is the maximum number of rows upwards reached with at most h differences by a path that starts on $[j, n/2]$; the set of corresponding end points on the wave is denoted by $\sigma(j, h)$.

Tandem repeats $r = \bar{u}\hat{u}$ with $s_{n/2} \in \bar{u}$ aligned with $s_j \in \hat{u}$ with at most k differences correspond to pairs $S^*(j, h'), P^*(j, h)$, for which $h' + h \leq k$ and $S^*(j, h') + P^*(j, h) \geq j - n/2$.

As before to obtain the tandem repeats with the least number of differences we perform the same merge like procedure as described for the Hamming distance. In particular we compute for each $h \leq k$ the value $f(h) = \min_{h' \leq k-h}\{h' \mid (P^*(j,h) + S^*(j,h') \geq j - n/2$ and consider only those pairs $(h, f(h))$ for which $h + f(h)$ is minimal. The computation here is the same as in "Computing optimal repeats" in Section 3. For such optimal pairs (h, h'), the repeats are given by the sub-paths that start at any point in $\sigma(j, h')$ go through $[n/2, j]$ and end at any point in $\pi(j, h)$. Since these pairs are optimal, whenever the corresponding values $P^*(j,h) + S^*(j,h') > j - n/2$, these paths start with matches along the diagonal and end with matches along the diagonal. Let $(i, i_1) \in \sigma(j, h')$ and let $(i', i_1') \in \pi(j, h)$. Notice that by definition all points in $\sigma(j, h')$ have the same coordinate $i = j - S^*(j, h')$ and those in $\pi(j, h)$ have the same coordinate $i_1' = n/2 + P^*(j, h)$. In addition if $P^*(j, h) + S^*(j, h') = j - n/2 + c$, for $c > 0$ then $i_1' = i + c$ for all such points, and the repeats correspond to the sub-paths starting at $(i + r, i_1 + r)$ and ending at $(i' - c + r, i_1' - c + r)$, for $0 \leq r \leq c$.

4.1 Computing the waves

The problem of incrementally computing the waves for all indices j, was addressed by Myers in [M-86]. Myers, however, considers a different model, in which only insertions and deletions are allowed. A substitution in his model counts as two differences, (an insertion followed by a deletion). Using his model Myers computes the k waves of PR for a given j in $O(k)$ time, when given the k waves for $j + 1$. Under the Levenshtein model, where substitutions are counted as a single difference, Landau and Vishkin [LV-88] compute the k waves of PR for a given j in $O(k^2)$ time.

An important contribution of the current paper is an algorithm to compute the k waves for a given j, in time $O(1)$ per wave, when given the k waves for $j + 1$. This also answers an open question raised in Myers' paper. Our algorithm modifies Myers' algorithm to handle mismatches as well as insertions and deletions.

The main body of the algorithm is given below.

for $j = n$ downto $n/2 + 1$ do
    for $h = 0$ to k do
        Compute the linked list $L_j^h = p_j^*(h, j + h) \dots p_j^*(h, j - h)$.
    od
od

Theorem 1 below proves that L_j^h, which contains one point for each diagonal in $[j + h \dots j - h]$ in this order, is composed the concatenation of the following: (i) a prefix of L_{j+1}^{h-1}; possibly followed by (ii) a point p_{ℓ_1} between $L_{j+1}^{h-1}(\ell_1)$ and $L_{j+1}^h(\ell_1)$ on column ℓ_1; followed by (iii) a sublist of L_{j+1}^h; possibly followed by (iv) a point p_{ℓ_2} between $L_{j+1}^h(\ell_2)$ and $L_{j+1}^{h+1}(\ell_2)$ on column ℓ_2; followed by (v) a suffix of L_{j+1}^{h+1}. Each of the pieces may be empty.

To compute L_j^h in $O(1)$ we show that the new points can be computed in $O(1)$ time and that the cut points of the three linked lists L_{j+1}^{h-1}, L_{j+1}^h, and L_{j+1}^{h+1} can also be accessed in $O(1)$ time.

Remark 1. The computation of the point on diagonal d of L_j^h, $L_j^h(d)$ can be done in two steps. First we preliminary set $L_j^h(d) = max\ (L_j^{h-1}(d+1) + 1, L_j^{h-1}(d) + 1, L_j^{h-1}(d-1))$. In the second step, as long as the corresponding characters match, $L_j^h(d)$ is increased by one. [LV-88] describe how augmented suffix trees can be used to compute the second step in $O(1)$ time. The entire computation of $L_j^h(d)$ can hence be performed in $O(1)$ time. $L_j^h(d)$ is therefore *obtained* either from $L_j^{h-1}(d+1)$, (i.e. from the left), or from $L_j^{h-1}(d)$, (i.e. from the same diagonal), or from $L_j^{h-1}(d-1)$, (i.e. from above).

The next Observation is an immediate consequence of Remark 1.

Observation I. If for a given j, the three points of wave $h-1$ on diagonals $d+1$, d and $d-1$ are all less or equal (resp. all greater or equal) to the corresponding three points on those diagonals for some other index j' and wave $h'-1$, then the point on the middle diagonal d of wave h, (of j) is less or equal (resp. greater equal) than the point on diagonal d and wave h' (of j'). Formally if the coordinates in triplet $[L_j^{h-1}(d+1), L_j^{h-1}(d), L_j^{h-1}(d-1)]$ are all less or equal than (resp. all greater or equal than) those of triplet $[L_{j'}^{h'-1}(d+1), L_{j'}^{h'-1}(d), L_{j'}^{h'-1}(d-1)]$ then $L_j^h(d) \leq L_{j'}^{h'}(d)$, (resp. $L_j^h(d) \geq L_{j'}^{h'}(d)$)

The above of course immediately implies that if $[L_j^{h-1}(d+1), L_j^{h-1}(d), L_j^{h-1}(d-1)] = [L_{j'}^{h'-1}(d+1), L_{j'}^{h'-1}(d), L_{j'}^{h'-1}(d-1)]$ then $L_j^h(d) = L_{j'}^{h'}(d)$

Theorem 1. *Wave h in PR of j (L_j^h) is the concatenation of (up to) five pieces: (i) a prefix of L_{j+1}^{h-1}, (ii) an in between point p_1, (iii) a sublist of L_{j+1}^h, (iv) an in between point p_2, and (v) a suffix of L_{j+1}^{h+1}. Each piece may be empty; the in between point p_1, (if any) is between wave $h-1$ and wave h of $j+1$, while the in between point p_2, (if any) is between wave h and wave $h+1$ of $j+1$.*

Proof. The proof of Theorem 1 is essentially by induction on h, but we will first introduce several notations and establish the equivalence of certain properties.

Since the computation of the waves for j are done after those for $j+1$ have been completed, we shall refer to L_{j+1}^h as the old h wave and to L_j^h as the new h wave.

For a point p in L_j^h, let $key(p)$ be the old wave number of p, if p belonged to a wave of $j+1$, otherwise (p is an in between point) if p is in between the old wave d and the old wave $d+1$ we set $key(p) = d + 1/2$.

Let q_j be the value of a given point in the dynamic programming matrix for index j and let q_{j+1} be the value of the same point in the computation for index $j+1$. Clearly $q_j \leq q_{j+1} + 1$ and $q_{j+1} \leq q_j + 1$, and hence the *key* value of points in L_j^h are between $h-1$ and $h+1$.

To prove Theorem 1 it suffices therefore to prove that the following property holds for all h waves:

The first key property:

The *key* values of the points along L_j^h, from left to right, are non decreasing and strictly increasing around in between points. Formally let $L_j^h = p_{j+h} \cdots p_{j-h}$, then for $j + h \geq i > j - h$, $\lceil key(p_i) \rceil \leq key(p_{i-1})$.

We shall prove by induction on h that the *first key property* holds for all h waves.

The following *second key property*, which is implied by the *first key property*, (as shown below in Lemma 2), will help in the proof of Theorem 1. (A similar second key property was also used by Myers, [M-86]).

The second key property

If for index j, diagonal ℓ contains an in between point p_ℓ, for which $h < key(p_\ell) < h + 1$, then the old $h + 1$ point on diagonal ℓ, $L_{j+1}^{h+1}(\ell)$, was obtained from the left, (i.e. from $L_{j+1}^h(\ell + 1)$).

Recall that we identify the points on a given diagonal by their column number. An inequality $p > q$ is hence always interpreted as "the column number of p is higher than the column number of q".

Lemma 2. *If the first key property holds for all waves h then the second key property also holds for all waves h.*

Proof. The Lemma is also proven by induction on h.

Wave L_j^0 contains only one point, (which is on diagonal j). The old 1 point on diagonal j was obtained from the (only) old zero point on diagonal $j + 1$, i.e. from the left.

Let $L_j^h(\ell)$ be the in between point so that $h' < key(L_j^h(\ell)) < h' + 1$, for some h'. If $h \neq 0$ then $L_j^h(\ell)$ was obtained from a point p on the L_j^{h-1}.

If this point p is either right or left of $L_j^h(\ell)$ then $key(p) > h' - 1$, otherwise if p is also on diagonal ℓ then $key(p) \geq h'$.

We distinguish three cases:

1. $L_j^h(\ell)$ was obtained from the left, (i.e. from the point $L_j^{h-1}(\ell + 1)$), and $key(L_j^{h-1}(\ell + 1)) > h' - 1$. By *key property 1* the key values of the points in L_j^{h-1} on diagonals, $\ell + 1$ through $\ell - 1$ are:

$$\begin{cases} & \ell+1 & \ell & \ell-1 \\ L_j^{h-1}: & >h'-1 & \geq h' & \geq h' \\ L_j^h: & & <h'+1 & \end{cases}.$$

Since the old $h' + 1$ point was not reached on diagonal ℓ, it must have been obtained originally from the left. (In this case, both $L_j^{h-1}(\ell + 1)$ and $L_j^h(\ell)$ are in between points.)

2. $L_j^h(\ell)$ was obtained diagonally, (i.e. from the point $L_j^{h-1}(\ell)$), and $key(L_j^{h-1}(\ell)) \geq h'$. By *key property 1* the key value on diagonal $\ell - 1$ was also at least h'. Again the old $h' + 1$ point must have been obtained from the left.

3. $L_j^h(\ell)$ was obtained from the right, (i.e. from the point $z_{\ell-1} = L_j^{h-1}(\ell - 1)$), and $key(z_{\ell-1}) > h' - 1$. Note that the old $h' + 1$ point could not have been obtained from $L_{j+1}^{h'}(\ell)$ on its own diagonal, since in this case there could be no in between point strictly between $L_{j+1}^{h'}(\ell)$ and $L_{j+1}^{h'+1}(\ell)$. If $key(z_{\ell-1}) \geq h'$ then as before the old $h' + 1$ point must have been obtained from the left. We show that $key(z_{\ell-1}) = h' - 1/2$ leads to a contradiction. In this case $z_{\ell-1}$ is itself an in between point.

   a. $z_{\ell-1} > x_\ell^{h'}$, the old h' point on diagonal ℓ, since it reached our point whose key value is greater than h'.

b. On the other hand, by our inductive assumption the old h' point on $\ell - 1$, (which is clearly beyond $z_{\ell-1}$), was obtained from diagonal ℓ, i.e. from $x_\ell^{h'-1}$; this implies that $x_\ell^{h'-1} \geq z_{\ell-1}$, which contradicts a.

□

Assume now that the *first* and *second key properties* hold for wave h; we shall prove that the *first key property*, denoted henceforth by $P1$ also holds for wave $h+1$. To achieve this it suffices to examine all adjacent pairs of diagonals in wave $h + 1$ and to prove the required inequality on their key values.

We will examine the pairs in wave h based on their *key* values.

Case 0. If $L_j^h(\ell + 1)$ and $L_j^h(\ell)$ have the same *key* value h' then by $P1$ we have:
$$\begin{cases} \quad\ell+1 \quad \ell \quad \ell-1 \\ L_j^h\!: h' \quad\;\; h' \;\geq h' \end{cases}, \text{ (provided that } \ell \text{ is not the rightmost diagonal). This implies}$$
that $key(L_j^{h+1}(\ell)) \geq h'+1$ and $key(L_j^{h+1}(\ell+1)) \leq h'+1$. If ℓ is the rightmost diagonal in L_j^h then $\ell = j - h$ is the rightmost diagonal of wave $h+1$ for $j+1$. Since diagonal $j - h$ does not have an old h point we necessarily have $\begin{cases} \quad\ell+1 \quad \ell \quad\;\; \ell-1 \\ L_j^h\!: h+1 \quad h+1 \;\; \bullet \end{cases}$, (i.e.

$h' = h + 1$). The new $h + 1$ wave and the old $h + 2$ wave have the same point on diagonal ℓ as well as $\ell - 1$. In other words $key(L_j^{h+1}(\ell)) = h + 2$. Since all *key* values on wave $h + 1$ are at most $h + 2$, the *first key property* holds for $(\ell + 1, \ell)$ in wave $h + 1$.

It remains to verify that $P1$ holds for all pairs of diagonals $\ell + 1$, ℓ which have different key values in the new h wave, and those pairs for which either $\ell + 1$ or ℓ is an extreme diagonal in the new $h + 1$ wave.

Case 1. Let diagonal ℓ be the first diagonal that contains a point p from the old $h - 1$ wave. Either there is no such diagonal, or it is the leftmost (the $j + h$) diagonal in the new h wave. Since $j + h$ is also the leftmost diagonal of the old $h - 1$ wave, $L_{j+1}^h(j+h+1)$ got its value from p, hence $L_j^{h+1}(j+h+1) = L_{j+1}^h(j+h+1)$, which translates to $key(L_j^{h+1}(j+h+1)) = h$. Since all *key* values in L_j^{h+1} are greater or equal to h, $P1$ is satisfied for the pair $(\ell + 1, \ell)$ in wave $h + 1$.

Case 2. Suppose that diagonal ℓ contains the in between point p that is between point $o_{h'}$ from the old h' wave and point $o_{h'+1}$ from the old $h' + 1$ wave, ($h' \in \{h - 1, h\}$). $o_{h'+1} = L_{j+1}^{h'+1}(\ell)$, was obtained from the left by our inductive assumption, (i.e. from $L_{j+1}^{h'}(\ell+1)$); hence ℓ is not the leftmost diagonal. Provided that ℓ is not the rightmost diagonal we have the following *key* values: $\begin{cases} \quad\ell+1 \quad \ell \qquad\; \ell-1 \\ L_j^h\!: \leq h' \quad h'+1/2 \;\geq h'+1 \end{cases}.$

We distinguish between two scenarios:

1. $\begin{cases} \quad\ell+1 \quad \ell \qquad \ell-1 \\ L_j^h\!: h' \quad\;\; h'+1/2 \;\geq h'+1 \end{cases};$ Since $L_{j+1}^{h'+1}(\ell)$ was obtained from $L_{j+1}^{h'}(\ell+1)$ we

must have $L_{j+1}^{h'}(\ell+1) \geq p$, even though $key(p) = h' + 1/2$. Diagonal ℓ therefore "loses" to $\ell + 1$ and $key(L_j^{h+1}(\ell + 1)) \leq h' + 1$. Since all 3 diagonals have *key* values at least h', $key(L_j^{h+1}(\ell)) \geq h'+1$, which establishes $P1$. If ℓ is the rightmost diagonal then, similar to **Case 0**, we necessarily have $h' = h$, which gives: $\begin{cases} \ell+1 & \ell \\ L_j^h: h & h+1/2 \end{cases}$. Diagonal $\ell-1$ contains no old h value, hence $key(L_j^{h+1}(\ell)) \geq h+1$. The key value on diagonal $\ell+1$ is similar to before at most $h+1$.

2. $\begin{cases} \ell+1 & \ell \\ L_j^h: <h' & h'+1/2 \end{cases}$; diagonal $\ell-1$ is not needed in this case. We still have that $L_{j+1}^{h'}(\ell+1)$ is greater or equal than the column of our point p, with *key* value $h'+1/2$. p can therefore not give $\ell+1$ a point with *key* value greater than h'. The other relevant *key* values for $\ell+1$ are: $\begin{cases} \ell+2 & \ell+1 \\ .L_j^h: \leq h'-1 & <h' \end{cases}$, hence $key(L_j^{h+1}(\ell+1)) \leq h'$. $key(L_j^{h+1}(\ell))$ is clearly greater or equal than h', since already $key(L_j^h(\ell) > h')$. Again, we showed that $P1$ holds.

Case 3. Let diagonal ℓ be the first diagonal in L_j^h that contains a point with $key(p) = h'$, $(h' \in \{h, h+1\})$. (The case $h' = h-1$ corresponds to **Case 1**.) Provided that ℓ is not the rightmost diagonal we have: $\begin{cases} \ell+1 & \ell & \ell-1 \\ L_j^h: <h' & h' & \geq h' \end{cases}$. Even if ℓ is the leftmost diagonal in L_j^h, diagonal $\ell+1$ contains an old h' point $L_{j+1}^{h'}(\ell+1)$, since $h' \geq h$. We distinguish between two cases:

1. $L_{j+1}^{h'}(\ell+1) \geq L_{j+1}^{h'}(\ell)$. In this case $L_{j+1}^{h'}(\ell)$, the old h' point on ℓ, cannot give a better point than $L_{j+1}^{h'}(\ell+1)$ and $key(L_j^{h+1}(\ell+1)) \leq h'$. Clearly $key(L_j^{h+1}(\ell)) > h'$, since already $L_j^h(\ell)$ had *key* value h'.

2. $L_{j+1}^{h'}(\ell+1) < L_{j+1}^{h'}(\ell)$. In this case $L_{j+1}^{h'}(\ell+1)$ could not compete with $L_{j+1}^{h'}(\ell)$. If ℓ is not the rightmost diagonal for L_j^h then $L_j^h(\ell-1)$ has *key* value at least h', and we get that: $key(L_j^{h+1}(\ell)) \geq h'+1$. On the other hand $key(L_j^{h+1}(\ell+1))$ is clearly less or equal to $h'+1$, since all *key* values on the diagonal ℓ or left of ℓ are less or equal than h', and $P1$ holds. Finally if ℓ is the rightmost diagonal we necessarily have: $\begin{cases} \ell+1 & \ell \\ L_j^h: <h+1 & h+1 \end{cases}$, (i.e. $h' = h+1$). But in the current subcase we assumed that the old $h+1$ point on $\ell+1$ could not compete with the old $h+1$ point on ℓ, hence $key(L_j^{h+1}(\ell)) = h+2$. The key value on diagonal $\ell+1$ is as before at most $h+2$.

Hence $P1$ holds in this case.

Case 4. $\ell = j-h$ is the rightmost diagonal on the new h wave. Its key value is strictly greater than h, since the last diagonal of the old h wave is $j-h+1$. If its key value is $h+1$ then the key value of $L_j^{h+1}(\ell-1)$ is $h+2$ and $P1$ holds for the pair $\ell, \ell-1$, (since all *key* values on wave $h+1$ are at most $h+2$). If the key value of $L_j^h(j-h)$

is such that: $h < key(L_j^h(j - h)) < h + 1$, then we have $\begin{cases} & \ell+1 \quad \ell \\ L_j^h: & \leq h \quad < h+1 \end{cases}$ and the

new $h + 1$ point on ℓ has key value at most $h + 1$, i.e. $key(L_j^{h+1}(\ell)) \leq h + 1$. The key value of $L_j^{h+1}(\ell - 1)$, the rightmost point on the new $h + 1$ wave, is always greater than $h + 1$, and $P1$ holds in this case as well.

$\square$

We have shown that L_j^h can be pasted together from L_{j+1}^{h-1}, L_{j+1}^h and L_{j+1}^{h+1} and at most two additional in between points, which may occur between sublists. Note that all the in between points on L_j^h occur in the vicinity of diagonals ℓ whose key value is different than the key value of $\ell - 1$. There are hence only $O(1)$ points on L_j^{h-1} that have to be examined to compute L_j^h. $L_j^{h-1}(d)$ will also be cross linked to $L_j^h(d)$, to allow finding the breakpoint in $L_j^h(d)$ quickly. Notice that all but $O(1)$ of these cross pointers will be imported directly from the old waves, and only $O(1)$ pointers need to be changed. It follows that L_j^h, $(h < k)$ can be computed in $O(1)$ time from L_j^{h-1} and the computations for $j + 1$. The computation of L_j^k may take $O(k)$ time since the required L_{j+1}^{k+1} piece has not been constructed previously. In total the computation of the k waves for j takes $O(k)$ time, when given the k waves for $j + 1$, cross linked as described above.

4.2 Finding the maximum on each wave

In the previous section we described an algorithm that computes the waves in the dynamic programming matrix of PR for a given j. In wave h, $h \leq k$, the rightmost column c on each diagonal d, $j + h \leq d \leq j - h$, with value h is reported. In this section we compute $P^*[j, h]$ - the maximum value on L_j^h.

Computing $P^*[j, h]$ in $O(1)$ time, would provide a $O(nk)$ running time for the entire algorithm. Unfortunately, we were only able to compute $P^*[j, h]$ in $O(\log k)$ time. The reports are concatenated in a linked list starting with the report of diagonal $j+h$ and ending with report of diagonal $j-h$. In the previous section, we showed how to compute wave h by concatenating up to three sub waves from the computation of PR for index $j+1$, (a prefix of the wave $h-1$ a sub list of h and a suffix of $h+1$) and adding up to two new points. This was done in $O(1)$ time, by changing $O(1)$ pointers. Finding the maximum of the concatenated list can be reduced to the following more general problem. Given three unsorted linked lists L^1, L^2 and L^3 of $O(h)$ elements each and their respective maximum, as well as two pointers p_1^i, p_2^i to list L^i, $(i = 1..3)$, the problem is to construct the new list $L^* = p_1^1 \ldots p_2^1, p_1^2 \ldots p_2^2, p_1^3 \ldots p_2^3$ and to find its maximum. Since the problem of sorting the elements of a linked list L can easily be reduced to the above problem, there is little hope for a faster algorithm for this general problem. It is not clear however that the special circumstances we are dealing with could not be somehow exploited.

In the current paper we are using mergeable 2-3 heaps to compute $P^*[j, h]$ in $O(\log k)$ time, as for example described in [AHU-74], Sections 4.11-4.12.

Each list corresponds to the list of leaves of a 2-3 tree. The leaves of each subtree are hence reports of consecutive diagonals. The information is stored in each internal node is the maximum value of the leaves in its subtree.

To build a 2-3 heap on the linked list of the new wave h, the algorithm merges a prefix of the 2-3 heap of the old $h - 1$ wave with a sublist of the 2-3 heap of the old h wave and the suffix of the 2-3 heap of the old $h + 1$ wave. In addition at most 2 new leaves, (reports) need to be inserted.

The algorithm described in the previous section provides the pointers to each starting and end point of the sub waves. The 2-3 heap corresponding to the new h wave is constructed in $O(\log h)$ time. $P^*[j, h]$ is therefore obtained by accessing the root of the 2-3 heap.

Complexity. The total time to find all new maxima for index j is hence $O(k \log k)$.

4.3 Time complexity

The algorithm follows the $O(n \log n)$ Main and Lorentz algorithm. In the case of the edit distance we need $O(\ell k \log k)$ time for strings of size ℓ. The total running time of our algorithm is therefore, $O(nk \log k \log n)$.

5 Conclusions

We have shown how to find all locally optimal repeats whose Hamming distance is at most k in $O(nk \log n/k)$ time, and all locally optimal repeats whose edit distance is at most k in $O(nk \log k \log n)$ time. It would be interesting to find an algorithm for the edit distance, whose complexity matches the one for the Hamming distance.

Kannan and Myers have recently addressed the problem of finding the repeat with highest score, when given an arbitrary penalty matrix, [KM-93].

Acknowledgement. We are grateful to Gene Myers, for pointing out that it might be possible to improve an earlier version of this algorithm, by using an incremental approach like the one in [M-86].

References

[AHU-74] A.V. Aho, J.E. Hopcroft and J.D. Ullman, *The Design and Analysis of Computer Algorithms*, Addison-Wesley, Reading, MA, 1974.

[Ap-92] A. Apostolico (1992) "Fast Parallel Detection of Squares in Strings," *Algorithmica*, Vol. 8, pp. 285-319.

[KM-93] S. K. Kannan and E. W. Myers (1993) "An algorithm for locating non-overlapping regions of maximum alignment score," these proceedings.

[L-66]   V. I. Levenshtein (1966) "Binary Codes Capable of Correcting Deletions, Insertions and Reversals," *Soviet Phys. Dokl*, Vol. 10, pp. 707-710.

[LV-88] G.M. Landau and U. Vishkin (1988) "Fast string matching with k differences," *JCSS*, Vol. 37, No. 1, pp. 63-78.

[M-86]   E. Myers (1986) "Incremental Alignment Algorithms and Their Applications," Tech. Rep. 86-22, Dept. of Computer Science, U. of Arizona, Tucson, AZ 85721.

[ML-84] M.G. Main and R.J. Lorentz (1984) "An $O(n \log n)$ algorithm for finding all repetitions in a string," *J. of Algorithms*, Vol. 5, pp. 422-432.

[ML-85]   M.G. Main and R.J. Lorentz (1985) "Linear time recognition of square free strings," A. Apostolico and Z. Galil (editors), *Combinatorial Algorithms on Words*, NATO ASI Series, Series F: Computer and System Sciences, Vol. 12, Springer-Verlag, pp. 272-278.

[U-83]   E. Ukkonen (1983) "On approximate string matching," *Proc. Int. Conf. Found. Comp. Theor.*, Lecture Notes in Computer Science 158, Springer-Verlag, pp. 487-495.

Two Dimensional Pattern Matching in a Digitized Image

Gad M. Landau[1]*, Uzi Vishkin[2]**

[1] Dept. of Computer Science, Polytechnic University, 333 Jay Street, Brooklyn, NY 11201.
[2] Institute for Advanced Computer Studies, and Department of Electrical Engineering, University of Maryland, College Park, MD 20742 and Dept. of Computer Science, Tel Aviv University, Tel Aviv, Israel 69978;

Abstract. For motivation purpose, imagine the following *continuous pattern matching* problem. Given are two continuous pictures, each consisting of unicolor regions; one picture is called the *scene* and the other the *pattern*. The problem is to find all occurrences of the pattern in the scene.
As a step towards efficient algorithmic handling of the continuous pattern matching problem by computers, where discretized representations are involved, we consider in this paper a two-dimensional pattern matching problem where the pattern and the text are specified in terms of exemplar digitized images.
From the wider perspective of areas such as computer vision or image processing, our problem definitions identify an important gap in the fundamental theory of image formation and image processing - how to determine, even in the absence of noise, if a digitized image of a scene could contain an image of a given pattern?

1 Introduction

Images are continuous, but their computer representations are discrete (or digitized). This paper is about bridging this "digitization gap" where pattern matching and image formation problems are involved.

By way of motivation, consider the following **continuous pattern matching** problem. Consider two two-dimensional pictures, each consisting of unicolor regions; one picture is called the *scene* and the other the *pattern*; find all occurrences of the pattern in the scene; that is, specify all instances where the pattern can be laid over the scene, resulting in a perfect match between regions and colors. We suggest some preliminary ideas for handling such a problem in an efficient algorithmic way on computers, where representations are digitized.

Our introduction extends into sections 2 and 3 where definitions and additional rationale are given. In order not to have a too verbose introduction we delayed some less essential parts of it to section 7. Algorithms are presented in sections 4, 5 and 6. The main contribution is the algorithm of section 5.

* Partially supported by the New York State Science and Technology Foundation Center for Advanced Technology.
** Partially supported by NSF grants CCR-8906949 and CCR-9111348.

2 A Framework for Problem Definitions

Each problem we consider is actually a *pair* consisting of an origin problem and its digitized counterpart. Typically, the *origin problem* is to find a single *pattern* in a *scene*. Both the pattern and the scene are defined in a k-dimensional Euclidean space, where k is 1, 2, or 3 (same k for the pattern and the scene) and are referred to later as "continuous pattern" and "continuous scene", respectively. However, while the continuous pattern and scene will not be available to us, we will have the input for the *digitized problem*. In digitized problems the input will include: (1) a *text* string in one (or two, or higher) dimensions, representing the scene; (2) a *pattern* string, representing the continuous pattern. A text string is *sampled* from the continuous scene in some way, which will be explained later; similarly, a pattern string is sampled from the continuous pattern. It is possible to have more than one text string for representing the same scene. It is also possible to have more than one pattern string for representing the same continuous pattern. The motivation is that the pattern string (or strings) is actually an exemplar, or *incarnation*, of the continuous pattern; this incarnation represents a whole family of pattern strings that includes all possible *incarnations* of the continuous pattern in the string representation. (The same applies for the text string and the scene.) A string s qualifies as an incarnation of the continuous pattern if it passes the following *compatibility test* with respect to the input pattern string: *there exists a continuous string such that (a) the input pattern string can be sampled from it, and (b) the string s itself can also be sampled from it.* Our *problems* will be to find every location in the text where an incarnation of the continuous pattern may occur.

Meaning of results. We elaborate on the relationship between origin problems and their digitized counterparts. If the output of an algorithm for the digitized problem provides one or more occurrences, it would have the following meaning: *based on the input data, we do not have any evidence at hand that enables contradicting any of the declared occurrences.* This seems to be the best we can hope for, since our reasoning is bounded by the data we have.

Imperfect algorithms. Suppose that we have only inefficient algorithms for some digitized problem. It may make sense to settle for a lesser solution if it can be achieved by a considerably more efficient algorithm. Consider an algorithm for the same digitized problem that provides the following kind of output. For each text location, if the algorithm outputs that "the location does not have an occurrence" then this is correct. However, if it outputs that "the location has an occurrence" then it might be wrong. We call such an algorithm a *negative recognition (NR)* algorithm, since only its negative statements are always valid. Similarly, a *positive recognition (PR)* algorithm is one that provides the following kind of output. For each text location, if the algorithm outputs that "the location has an occurrence" then this is correct. However, if it outputs that "the location does not have an occurrence" then it might be wrong.

3   Problem Definitions

We first give the flavor of problems considered in this paper, by defining a representative problem.

3.1   The discrete one dimensional (D-1-D) pattern matching problem

We first present the origin problem and then its digitized counterpart.

The origin problem. Let $\bar{P}$ and $\bar{T}$ be two straight-line *intervals*, where the interval of $\bar{T}$ is longer than the one of $\bar{P}$. Each interval is partitioned into unicolor subintervals (*SIs*, in short). Adjacent SIs have different colors, and the set of possible colors is fixed. The **continuous one-dimensional (C-1-D) pattern matching problem** is: Find all occurrence of $\bar{P}$ in $\bar{T}$. For convenience, we will always assume that each interval or SI is "half open".

Recall that in a one-dimensional Euclidean space (i.e., a straight line) the half open interval (or SI) $[a, b)$ includes all reals x, $a \leq x < b$. We define one *unit* as our basic unit of length, and a *unit-interval* as an interval whose length is one unit.

Motivation for the digitized counterpart. The intervals and the SIs for the C-1-D problem may have any length, but a computer can store boundary information up to some finite precision only. We overcome this problem by sampling the intervals $\bar{T}$ and $\bar{P}$. Suppose $\lceil |\bar{P}| \rceil = m$ for some integer m (that is, the length of interval $\bar{P}$ is m units or slightly less). The interval $\bar{P}$ is sampled at m points, each sample reports the color of the interval in its point. Each successive pair of points is exactly one unit apart. Replace each point by its color to produce the string $P = p_1 p_2 \ldots p_{m-1} p_m$ (actually, we may have only $m - 1$ samples) and have the string P represent the interval $\bar{P}$. Note that the first point (that provides p_1) is sampled from the first half-open unit-interval in $\bar{P}$. Locations of the other points are determined by the first point. However, there is a continuum of possibilities for selecting the first point. This paper makes the simplifying *assumption* that the length of each half open SI of $\bar{P}$ is at least one unit. This guarantees that:

(1) For each SI at least one point will be sampled; and

(2) $\bar{m} \leq m - 1$, where $\bar{m}$ denotes the number of different strings P that may be sampled to represent interval $\bar{P}$. To see this, put the point for p_1 at the leftmost end of the interval $\bar{P}$. The samples taken for $p_1 \ldots p_m$ provide the first string. Now consider shifting all the (sampling) points simultaneously to the right until the first time one of the points (for some p_i) is moved into a new SI, resulting in a different p_i. This defines a new possible string for P. Repeat this process. Stop when the points have traversed a unit-interval each. During this process, p_1 does not change and each other p_i can change at most once. (The length of each SI is at least one unit and every p_i does not traverse more than one unit.) Therefore, $\bar{m}$, the number of different strings for P, is at most $m - 1$.

We call these $\bar{m}$ possible strings for P *incarnations of the continuous pattern* (*IP*, in short) and denote them by $IP_1 IP_2 \ldots IP_{\bar{m}}$. The IP_i's are stored in the following way: (i) IP_1 (ii) A list of the $\bar{m} - 1$ changes. IP_i can be constructed from IP_1 and the first $i - 1$ changes in the list. The difference between IP_i and IP_{i+1} is in one location. Note that if change number k, in the list, is that p_j changes from red to black, then p_j must have been red in $IP_1 \ldots IP_k$, and p_j must remain black in $IP_{k+1} \ldots IP_{\bar{m}}$.

Example :

$$
\begin{array}{ccccccccc}
p_1 & p_2 & p_3 & p_4 & p_5 & p_6 & p_7 & p_8 & p_9 \\
IP_1 = r & r & g & g & g & r & b & b & g \\
IP_2 = r & r & g & g & g & r & b & g & g \\
IP_3 = r & r & g & g & r & r & b & g & g \\
IP_4 = r & r & g & g & r & b & b & g & g \\
IP_5 = r & g & g & g & r & b & b & g & g
\end{array}
$$

The discrete one dimensional (D-1-D) pattern matching problem.

Input: (i) A 1-dimensional array $IP_1 = p_1 p_2 \dots p_m$ (the basic pattern).

(ii) A list of $\bar{m} - 1$ changes in the pattern.

(iii) A 1-dimensional array $T = t_1 t_2 \dots t_n$, $n \geq m$, (the text), where T represents some sample of points such that each successive two are one unit apart in $\bar{T}$. We assume that each character in IP_1 (and therefore in any other IP_i) and in T is taken from some alphabet Σ whose size is fixed.

Output: All locations in the text in which one of the IP_i's occurs.

The paper [LV-92a] presents a linear time algorithm for the discrete one dimensional (D-1-D) pattern matching problem (assuming that the colors are taken from a set whose size is constant). In other words the cost (in terms of asymptotic running time) of searching for $\bar{m}$ different incarnations is the same as searching for a single incarnation.

In this paper we present algorithms that solve single incarnation digitized problems.

3.2 The single incarnation digitized one-dimensional (SID-1-D) problem.

The origin problem is the C-1-D problem (as for the D-1-D problem). We also make the same assumption as in the D-1-D problem that the length of each unicolor region in the scene or in the continuous pattern is at least one unit. The input for the SID-1-D problem consists of: (i) a text string $T = t_1, \dots, t_n$, that was obtained by sampling (in unit intervals) from the continuous text (of the C-1-D problem); and (ii) a pattern string $P = p_1, \dots, p_m$, that was obtained by sampling (in unit intervals) from the continuous pattern. Our assumptions imply that each p_i and t_j is sampled from a unicolor interval whose length is at least one. This pattern string is the only incarnation of the continuous pattern that is available. Nothing else is known about the origin problem. The SID-1-D problem is to find all locations in the text in which an incarnation of the continuous pattern occurs.

3.3 The single incarnation digitized two-dimensional (SID-2-D) problem.

The continuous two-dimensional (C-2-D) pattern matching problem. Consider two continuous two-dimensional images, the *scene* ($\bar{T}$) of size $n \times n$ and the pattern ($\bar{P}$) of size $m \times m$. Each of these images is partitioned into unicolor region. Adjacent regions have different colors, and the set of possible colors is fixed (and finite). Find all occurrence of $\bar{P}$ in $\bar{T}$.

A *semi open square region* [a, b; c, d) of the continuous pattern contains all points x, y such that $a \leq x < b$ and $c \leq y < d$. Note that the sides of the above region are parallel to the X and Y axes. The region $[a, a+1; c, c+1)$ is defined as a semi open unit square (*unit square* in short).

In the present paper, we will elaborate on situations where the following assumption holds:

Assumption **A1**. Each point of a unicolor region belongs to a unit square which is fully contained in the same unicolor region.

Other assumptions are, of course, possible.

The input for the SID-2-D problem consists of: (1) a two-dimensional text string $T = [t_{i,j}]_{i=1,...,n;j=1,...n}$, that was obtained by sampling at grid intersection points, called pixels, from the $n \times n$ continuous scene; and (2) a two-dimensional pattern string $P = [p_{i,j}]_{i=1,...,m;j=1,...m}$, that was obtained by sampling at grid intersection points from the continuous pattern. The distance between two successive pixels on a row or on a column is one unit. This pattern string is the only incarnation of the continuous pattern that is available. Nothing else is known about the origin problem. The SID-2-D problem is to find all locations in the text in which an incarnation of the continuous pattern occurs.

4 The Single Incarnation Digitized One-Dimensional (SID-1-D) Problem and the Parallel-Seaming Technique

Consider an exact string matching algorithm that finds the occurrences of P in T. This will be a PR-algorithm. However, the definition of the SID-1-D problem implies that exact matches rarely occur. Actually, it can be argued that the number of possible incarnations is exponential in the number of color changes in P (that is, the number of characters p_i for which $p_i \neq p_{i+1}$). Therefore, such exact matching algorithms will likely fail to report some occurrences.

Section 4.1 gives an algorithm which is both NR and PR. Section 4.2 gives an alternative presentation of the same algorithm which has the following advantage: it can be generalized to more involved problems. Intuitively, if the one-dimensional version resembles balanced binary trees the two-dimensional version resembles quad-trees (see [S-90]). We call the alternative mode of presentation *parallel-seaming*.

4.1 An algorithm for the SID-1-D problem

In this section, we describe a simple algorithmic approach, which is both NR and PR.

The pattern string is a sampling of the continuous pattern at m points. Suppose that when sampled the continuous pattern covered a half open interval $[b, e)$ where $0 < b < 1$, $m \leq e < m + 1$ and characters $p_1, p_2, ..., p_m$ were sampled at points $1, 2, ..., m$, respectively. This assumption is without loss of generality. These sample points partition the continuous pattern (which is unknown to us) into the following $m + 1$ half open intervals: $[b, 1), [1, 2), ..., [m, e)$. Observe that the exact values of b and e, as well as any information about the continuous pattern beyond the sample is assumed not to be available. For our algorithm, we will consider instead the following set of intervals $I = \{I_0, I_1, ..., I_m\} = \{[0,1), [1,2), ..., [m, m+1)\}$, with respect to the continuous pattern.

For each substring $t_i, t_{i+1}, ..., t_{i+m}$, where $0 \leq i \leq n - m + 1$, we want to find out whether it represents a possible occurrence of the continuous pattern. (Observe that the above statement refers to two characters, t_0 and t_{n+1}, which are not part of the input string T; this is actually only a convenience since later we view each of these two characters as a possible match for any other character.) This means that the algorithm should determine whether, for every $0 \leq j \leq m$, t_{i+j} can belong to I_j.

Claim 4.1 *A possible occurrence of the pattern starts at t_i if and only if the following two conditions are satisfied for every $0 \leq j \leq m$:*
(i) (each text character equals one of the encompassing pixels) t_{i+j} can belong to I_j for each $0 \leq j \leq m$; specifically, check whether t_{i+j} equals either p_j or p_{j+1} for each $1 \leq j \leq m - 1$, and no checks are needed for t_i and t_{i+m}; and
(ii) (each character in the pattern occurs in at least one of the intervals it borders) for every p_j, at least one of the following two equalities is satisfied: $p_j = t_{i+j}$ and $p_j = t_{i+j-1}$.

Proof. It is trivial to see that if conditions (i) or (ii) do not hold then it is impossible to have a continuous pattern such that each of $t_i t_{i+1}...t_{i+m}$ and $p_1 p_2...p_m$ can be an incarnation of the same image.

Suppose that conditions (i) and (ii) hold. We prove the claim by constructing a continuous pattern so that each of $t_i t_{i+1}...t_{i+m}$ and $p_1 p_2...p_m$ is a possible incarnation. Specifically, we construct a continuous pattern such that $t_i t_{i+1}...t_{i+m}$ and $p_1 p_2...p_m$ are two possible incarnations of it. Consider an interval of length $m + 1$ of the form $[0, m+1)$. Each text character t_{i+j} gives its color to the half open interval $[j + 0.25, j + 0.75)$ and each pattern character p_j gives its color to the half open interval $[j - 0.25, j + 0.25)$. Conditions (i) and (ii) guarantee that each color extends to cover an interval whose total length is at least one. (If $t_i \neq p_1$ then the interval starts at 0.75 and if $t_{i+m} \neq p_m$ then the interval ends at $m+0.25$.) To see that $t_i t_{i+1}...t_{i+m}$ and $p_1 p_2...p_m$ are two possible incarnations of it, align pattern point p_j with integer point j, and text character t_{i+j} with the half integer point $j.5$. □

4.2 The generic parallel-seaming algorithm

A high-level description of the algorithm is followed by detailed description and complexity analysis. A discussion of the new method is provided at the end of the section.

High-level description of parallel seaming

We suggest the following parallel approach, that works in $\lceil \log(m+2) \rceil + 1$ iterations. Following Iteration 0, for each character t_i in the text we will have a list of all the intervals I_j that it can match. In other words these are all intervals which are compatible with character t_i (as in condition (i) above).

Following iteration k, $1 \leq k \leq \lceil \log(m+2) \rceil$, for each i, where i is an integer multiple of 2^k and $0 \leq i \leq n + 1$, we will have a list of all interval sequences of length 2^k, $I_j, I_{j+1}, ..., I_{j+2^k-1}$, that are *compatible* with the string $t_i, t_{i+1}, ..., t_{i+2^k-1}$.

Next, we discuss some implementation details of the algorithm. Following iteration k, $1 \leq k \leq \lceil \log(m+2) \rceil$, we will have the following detailed information: (i) a list of all interval sequences of length 2^k, $I_j, I_{j+1}, ..., I_{j+2^k-1}$, that can match $t_i, t_{i+1}, ..., t_{i+2^k-1}$ (called *full matches*); (ii) all interval sequences $I_0, I_1, ..., I_\ell$, for each $\ell \leq 2^k - 2$, that can match $t_{i+2^k-\ell-1}, t_{i+2^k-\ell}, ..., t_{i+2^k-1}$ (called *prefix matches*); and (iii) all interval sequences $I_{m-\ell}, I_{m-\ell+1}, ..., I_m$ that can match $t_i, t_{i+1}, ..., t_{i+\ell}$ (called *suffix matches*). Next, we overview iteration $k+1$ (assuming $k+1 \leq \lceil \log(m+2) \rceil$).

Let i be an integer multiple of 2^{k+1} and $0 \leq i \leq n+1$. Suppose that: (i) the interval sequence $I_j, I_{j+1}, ..., I_{j+2^k-1}$ matches $t_i, t_{i+1}, ..., t_{i+2^k-1}$; and (ii) the interval sequence $I_{j+2^k}, I_{j+2^k+1}, ..., I_{j+2^{k+1}-1}$ (which is called the *successor* of interval sequence $I_j, I_{j+1}, ..., I_{j+2^k-1}$) matches $t_{i+2^k}, t_{i+2^k+1}, ..., t_{i+2^{k+1}-1}$. In order to determine whether $I_j, I_{j+1}, ..., I_{j+2^{k+1}-1}$ matches $t_i, t_{i+1}, ..., t_{i+2^{k+1}-1}$, it remains to check whether condition (ii) in the above claim holds for the "seam" pattern character p_{j+2^k}; specifically, check whether p_{j+2^k} equals at least one among t_{i+2^k-1} and t_{i+2^k}. This check is called the "seam test". The above description takes care of extending a full match for k and a successor full match into a full match for $k+1$. Extending a prefix match for k and a successor full match into a prefix match for $k+1$, or a full match for k followed by a successor suffix match into a suffix match for $k+1$, is similar. In addition a prefix match for k at t_{i+2^k} will automatically become a prefix match for $k+1$ at t_i, and a suffix match for k at t_i will remain a suffix match for $k+1$ at t_i. Finally, at round $\lceil \log(m+2) \rceil$ we may match a suffix match and prefix match to get a match of $I_0, I_1, ..., I_m$, which means a match of a whole incarnation of the pattern.

Iteration k consists of taking every (full or prefix) match at location t_i that has a successive (full or suffix) match at location t_{i+2^k} and apply to this pair of matches the seam test.

Complexity. The number of operations is $O(nm)$ since each location of the text which participates in an iteration may have at most $2m$ intervals that it can match. The number of locations of the text that participate in an iteration is half the number of locations that participated in the previous iteration. Therefore, the total number of operations is $O(nm)$. (The running time of the algorithm is independent of the size of the alphabet; that is, the number of different colors.)

Why is the generic parallel-seaming algorithm interesting?
We expect this very simple algorithmic approach to be efficient in practice; since only "search directions" that potentially lead to an occurrence of at least one incarnation of the pattern will be explored: starting with actual matches to single characters of the text, the algorithm progresses by restricting them further to actual matches to substrings of the text of increasing size. This is clearly more efficient than, for instance, naively generating all possible incarnations of the pattern (whose number may be exponential), and searching for all of them. A preliminary investigation shows that the parallel seaming approach handles general enough extensions including, for instance, pattern matching subject to rotations and scaling, and can be implemented with considerable increase in time complexity; we guess that parallel-seaming will be relatively efficient though. We leave details and extensions to the interested reader. No fundamentally new ideas appear to be needed.

The book [BB-82] reviews a somewhat related idea due, independently, to Huffman [H-71] and Clowes [C-71], which has been implemented by Waltz. Informally, it has been applied for "labeling" problems in computer vision. (Note that our parallel-seaming algorithm is a labeling algorithm in the sense that it labels substrings of the text by all matching substrings of the pattern.) The book considers this idea as very useful for deriving efficient algorithms, where the structure of the graphs involved is rather degenerate; however, for more involved structures the idea turned out to be insufficient for producing efficient algorithms, since the "possible vertex labelings multiply enormously". There are two main differences between the Huffman-Clowes-Waltz work and ours. (1) *They work on different problems.* Their work fits into the matching-description-recognition stage of the contemporary approach to computer vision rather than on the image formation and early vision stage, and the issue of digitization, which is the main novelty of the present paper, is completely missing from it. (2) *They need more general algorithms, which are apparently much more time consuming.* Their labeling requires more general combinatorial structures, implying more elaborate computations for its algorithmic implementation. In the extreme, their structures can get as general as any graph. In contrast to this, our generic parallel seaming algorithm explores matches in a string-matching-like setting (for two, or three, dimensions this would mean two, or three, dimensional strings). Strings are "rigid" combinatorial objects that preserve distances and can be described as a very degenerate family of graphs. This should enable more efficient algorithmic implementations.

5 An Algorithm for the Single Incarnation Digitized Two-Dimensional (SID-2-D) Problem

5.1 An NR algorithm

In this section we present an approach that provides only an NR algorithm for the SID-2-D problem. (The reader may use Figure 1 as a guide through the current section.)

Claim 5.1 *A possible occurrence of the pattern starts at $t_{i,j}$ only if the following two conditions hold: (i) $t_{i+x,j+y}$ equals either $p_{x,y}$ or $p_{x+1,y}$ or $p_{x,y+1}$ or $p_{x+1,y+1}$, and (ii) every $p_{x,y}$ equals either $t_{i+x-1,j+y-1}$, or $t_{i+x-1,j+y}$, or $t_{i+x,j+y-1}$, or $t_{i+x,j+y}$, for every $1 \leq x, y \leq m$.*

Claim 5.1 can be readily proven.
The algorithm that the claim implies should be clear.

Claim 5.2 *The algorithm is NR but not PR.*

Proof. Claim 5.1 shows that the algorithm is NR. It remains to show that it is not PR. We actually provide two separate examples where the algorithm reports an occurrence of an incarnation of the pattern in the text, but it is impossible to construct a continuous image that satisfies the following three requirements: (1) the input pattern is one of its incarnation, (2) a subarray of the input text is one of its incarnations, and (3) it satisfies Assumption A1.

Here are the two examples (see Figure 1):

(a) Assume that $p_{x+2,y+1} = t_{i+x+2,j+y+1} =$ red, and $p_{x+2,y+2} = t_{i+x+1,j+y+1} =$ blue; all the other pixels in the text and pattern surrounding $t_{i+x+1,j+y+1}$, $t_{i+x+2,j+y+1}$, $p_{x+2,y+1}$, and $p_{x+2,y+2}$ are green. This example satisfies the two conditions of Claim 5.1. Draw two lines, one that concatenates $p_{x+2,y+1}$ and $p_{x+2,y+2}$; and one that concatenates $t_{i+x+1,j+y+1}$ and $t_{i+x+2,j+y+1}$. In order to satisfy Assumption A1, the intersection point of these two lines has to be colored by both blue and red.

(b) Assume that $p_{x+2,y+1} = t_{i+x+2,j+y} =$ red; $p_{x+2,y+2} = t_{i+x+2,j+y+2} =$ blue; and $t_{i+x+1,j+y+1} = t_{i+x+2,j+y+1} =$ green. Clearly, the two conditions of Claim 5.1 are satisfied. Draw the two lines again. There is no way to color The intersection point so that Assumption A1 is satisfied.   □

This NR algorithm runs in $O(n^2 m^2)$ time.

The more **intellectually challenging** problem is to design an algorithm for the SID-2-D problem that is both NR and PR. We note that traditionally the algorithms community has designated only algorithms which are both NR and PR as "algorithms for the problem". The next section presents such an algorithm.

5.2 A SID-2-D algorithm which is both NR and PR

The algorithm works on each subarray of size $(m+1)^2$ of the text, separately. Given are the pattern and a subarray of size $(m+1)^2$ of the text. The *main goal* is to check *whether there exists a continuous origin, which satisfies assumption A1 and allows the following two incarnations: (i) the pattern, and (ii) the subarray of the text.*

$p_{x,y}$		$p_{x,y+1}$		$p_{x,y+2}$	
	$t_{i+x,j+y}$		$t_{i+x,j+y+1}$		$t_{i+x,j+y+2}$
$p_{x+1,y}$		$p_{x+1,y+1}$		$p_{x+1,y+2}$	
	$t_{i+x+1,j+y}$		$t_{i+x+1,j+y+1}$		$t_{i+x+1,j+y+2}$
$p_{x+2,y}$		$p_{x+2,y+1}$		$p_{x+2,y+2}$	
	$t_{i+x+2,j+y}$		$t_{i+x+2,j+y+1}$		$t_{i+x+2,j+y+2}$
$p_{x+3,y}$		$p_{x+3,y+1}$		$p_{x+3,y+2}$	
	$t_{i+x+3,j+y}$		$t_{i+x+3,j+y+1}$		$t_{i+x+3,j+y+2}$

Fig. 1. Parts of the pattern and a subarray of the text

Figure 1 demonstrates that each pattern pixel is surrounded by four text pixels and each text pixel is surrounded by four pattern pixels. We assume that the pattern pixels lie on integers points, and each text pixel lies in the unit square that is bordered by the four pattern pixels.

The description of the algorithm starts with a reduction: The main goal (i.e., finding a continuous origin) is reduced to a problem of coloring certain $2m^2$ points subject to some rules (to be called *legal* coloring). The Reduction Lemma implies that a legal coloring of these points exists if and only if a continuous origin exists.

Next, we specify the $2m^2$ points to be colored by the algorithm. Consider a grid whose intersection points are the pattern pixels and another grid whose intersection points are the text pixels. Namely, the uppermost leftmost intersection point of the pattern grid fall inside the uppermost leftmost unit square defined by the text grid. *The $2m^2$ points are those points where the two grids intersect.* More explicitly, $p_{x+2,y+1}$ is connected to $p_{x+2,y+2}$, $p_{x+1,y+1}$, $p_{x+2,y}$, and $p_{x+3,y+1}$. For example, the line between $p_{x+2,y+1}$ and $p_{x+2,y+2}$ intersects the line between the text pixels $t_{i+x+1,j+y+1}$ and $t_{i+x+2,j+y+1}$. Define *an intermediate point* $n^p_{x+2,y+1}$ as the intersection point of these lines. Such an intermediate point will henceforth be indexed by the pixel on its left; the type of the this pixel (pattern or text) is given as postscript (see Figure 2).

Initially, all these $2m^2$ intermediate points are uncolored and the algorithm colors them.

A *legal coloring* of the these $2m^2$ intermediate points is a coloring where: *The color of each pattern pixel, each text pixel, and each intermediate point is equal to at least one of its four "corners".* For example, consider a pattern pixel $p_{x+2,y+1}$; then at least one of the following four conditions holds: (1) $p_{x+2,y+1} = t_{i+x+1,j+y+1} = n^p_{x+2,y+1} = n^t_{i+x+1,j+y}$, (2) $p_{x+2,y+1} = t_{i+x+1,j+y} = n^t_{i+x+1,j+y} = n^p_{x+2,y}$, (3) $p_{x+2,y+1} = t_{i+x+2,j+y} = n^p_{x+2,y} = n^t_{i+x+2,j+y}$, and (4) $p_{x+2,y+1} = t_{i+x+2,j+y+1} = n^t_{i+x+2,j+y} = n^p_{x+2,y+1}$.

The Reduction Lemma. Existence of a legal coloring is necessary and sufficient. That is, a legal coloring exists if and only if there is some continuous origin, which satisfies assumption A1 and whose possible incarnations include the pattern and the text.

Proof of the Reduction Lemma. Existence of a continuous origin, as required clearly implies existence of a legal coloring. Assume that a legal coloring is given. We describe a continuous origin as required. Assume that the pattern pixels lie on integer points. Let us put text pixels on half integer points (namely, points of the form x.5,y.5). Draw lines on the x.25 and x.75 columns and rows (0.25, 0.75, 1.25, 1.75 ....) these lines form a grid of squares of size 0.5×0.5 each. Each pixel (pattern and text) and intermediate point is in the middle of such a square. Color the square with the color of its pixel or intermediate point. Each pixel and each intermediate point are part of a legal coloring and there has a square of size 1×1 (unit square) of the same color. There is no point which does not lie in a unit square of the same color. □

The computation that follows is designed to satisfy two rules, R1 and R2, which are defined below. The rules are slightly weaker than legal coloring. However, the Correctness Theorem below guarantees that a legal coloring exists if and only if our algorithm succeed to satisfy the two rules.

$p_{x,y}$	$n^p_{x,y}$	$p_{x,y+1}$	$n^p_{x,y+1}$	$p_{x,y+2}$	$n^p_{x,y+2}$
$n^t_{i+x,j+y-1}$	$t_{i+x,j+y}$	$n^t_{i+x,j+y}$	$t_{i+x,j+y+1}$	$n^t_{i+x,j+y+1}$	$t_{i+x,j+y+2}$
$p_{x+1,y}$	$n^p_{x+1,y}$	$p_{x+1,y+1}$	$n^p_{x+1,y+1}$	$p_{x+1,y+2}$	$n^p_{x+1,y+2}$
$n^t_{i+x+1,j+y-1}$	$t_{i+x+1,j+y}$	$n^t_{i+x+1,j+y}$	$t_{i+x+1,j+y+1}$	$n^t_{i+x+1,j+y+1}$	$t_{i+x+1,j+y+2}$
$p_{x+2,y}$	$n^p_{x+2,y}$	$p_{x+2,y+1}$	$n^p_{x+2,y+1}$	$p_{x+2,y+2}$	$n^p_{x+2,y+2}$
$n^t_{i+x+2,j+y}$	$t_{i+x+2,j+y}$	$n^t_{i+x+2,j+y}$	$t_{i+x+2,j+y+1}$	$n^t_{i+x+2,j+y+1}$	$t_{i+x+2,j+y+2}$
$p_{x+3,y}$	$n^p_{x+3,y}$	$p_{x+3,y+1}$	$n^p_{x+3,y+1}$	$p_{x+3,y+2}$	$n^p_{x+3,y+2}$
$n^t_{i+x+3,j+y-1}$	$t_{i+x+3,j+y}$	$n^t_{i+x+3,j+y}$	$t_{i+x+3,j+y+1}$	$n^t_{i+x+3,j+y+1}$	$t_{i+x+3,j+y+2}$

Fig. 2. Parts of the pattern the text and the intermediate points

R1. *The color of each pattern pixel, and each text pixel, is equal to at least one of its four "corners".* For example, consider a pattern pixel $p_{x+2,y+1}$; then at least one of the following four conditions holds: (1) $p_{x+2,y+1} = t_{i+x+1,j+y+1} = n^p_{x+2,y+1} = n^t_{i+x+1,j+y}$, (2) $p_{x+2,y+1} = t_{i+x+1,j+y} = n^t_{i+x+1,j+y} = n^p_{x+2,y}$, (3) $p_{x+2,y+1} = t_{i+x+2,j+y} = n^p_{x+2,y} = n^t_{i+x+2,j+y}$, and (4) $p_{x+2,y+1} = t_{i+x+2,j+y+1} = n^t_{i+x+2,j+y} = n^p_{x+2,y+1}$. We sometimes refer to Rule R1 as having at least one corner satisfy its pixel, and a pixel with a satisfying corner is *satisfied*.

R2. The color of an intermediate point must be equal to at lease one pattern pixel and one text pixel among the four pixels that surround it. For example, the color of $n^p_{x+2,y+1}$ is equal to at least one of $p_{x+2,y+1}$ and $p_{x+2,y+2}$ and one of $t_{i+x+1,j+y+1}$ and $t_{i+x+2,j+y+1}$.

The algorithm works in four steps.

Step 1. For every intermediate point, check whether it can *locally* satisfy rule R2. Namely, check whether there is a color which is shared by (at least) one of the two surrounding text pixels and (at least) one of the two surrounding pattern pixels. If there is no such color, then stop the whole algorithm - conclude that no legal coloring exists. If there is one such color, assign it to the intermediate point. Nothing is done if there are two such colors.

The following two observations characterize the situation following Step 1.

Observation 1. An intermediate point that is not yet colored is surrounded by two different colors, as follows: one of the pattern pixels and one of the text pixels are equal to one color and the other two pixels are equal to the second one.

For Observation (2), and later, we classify each pattern and text pixel into four different states, based on its four surrounding points, and four surrounding pixels:
P1 a *satisfied pixel* - one that satisfies rule R1 (i.e., the pixel itself, the other pixel

and the two intermediate points in, at least, one of its four corner have all the same color).

P2 a *free pixel* - one that is not satisfied yet, but (the intermediate points in) two of its corners can still be colored to satisfy it, without changing any color that has already been assigned.

P3 a *determined pixel* - one that is not satisfied yet, but the already colored intermediate points enable to color exactly one of its corners to satisfy it.

P4 a *disappointed pixel* - one that the only way to satisfy it, is by changing colors that already have assigned to intermediate points.

Observation 2. A pattern (or text) pixel can be P2 only if exactly two of the four pixels surrounding the pixel are equal to it.

Proof of Observation 2. Consider a pattern (or text) pixel. If more than two out of its four surrounding text (pattern) pixels are equal to it, then it is satisfied (P1). If only one of the four pixels is equal then only the corner containing this pixel can satisfy it and therefore it cannot be P2. Only if exactly two of the four pixels are equal to it, the pixel can be P2. □

Examples where given a pixel, exactly two of its four surrounding ones have the same color as it. For instance, consider $p_{x+2,y+1}$; if it is equal to $t_{i+x+1,j+y}$ and $t_{i+x+1,j+y+1}$ and not equal to $t_{i+x+2,j+y}$ and $t_{i+x+2,j+y+1}$ then it can be satisfied by coloring either $n^p_{x+2,y}$ or $n^p_{x+2,y+1}$ by its color, since in Step 1, $n^t_{i+x+1,j+y}$ was already colored by the same color. (The same considerations apply if the pixel is equal to $t_{i+x+1,j+y}$ and $t_{i+x+2,j+y}$, or to $t_{i+x+2,j+y}$ and $t_{i+x+2,j+y+1}$ or to $t_{i+x+2,j+y+1}$ and $t_{i+x+1,j+y+1}$.) The other possible case is where $p_{x+2,y+1}$ is equal to $t_{i+x+1,j+y}$ and $t_{i+x+2,j+y+1}$, and not to $t_{i+x+1,j+y+1}$ and $t_{i+x+2,j+y}$; then it can be satisfied by coloring either $n^t_{i+x+1,j+y}$ and $n^p_{x+2,y}$, or $n^t_{i+x+2,j+y}$ and $n^p_{x+2,y+1}$ by its color. (The same considerations apply if it is equal only to $t_{i+x+1,j+y+1}$ and $t_{i+x+2,j+y}$.)

The remainder of the algorithm - an overview. The goal of Steps 2 and 3 is to satisfy all the text and pattern pixels, if possible. Step 2 propagates the coloring of Step 1 to other intermediate points, as long as rules R1 and R2 dictate assignment of colors in a unique way. When such propagation becomes impossible we advance to Step 3, where some P2 pixel that can be satisfied in two ways is satisfied in both. For each way separately, an iteration of Step 3 propagates the coloring to other intermediate points, in order to satisfy as many pixels as possible. This propagation is similar to Step 2. If each of the two ways leads to a P4 pixel, we conclude that there is no legal coloring exists. However, if the propagation in one of the ways does not lead to a P4 pixel, we fix the coloring of the intermediate points as in this way, and proceed to another iteration of Step 3, or to Step 4 if all pixels have become P1. *Perhaps surprisingly*, our Correctness Theorem shows that while this may lead to ignoring some legal colorings, the algorithm is still correct; namely, that if there exist several legal colorings then the algorithm will find a coloring that satisfies all rules R1 and R2. Furthermore, the theorem also guarantee that Step 4 will be able to color all remaining intermediate points so they are all satisfied, resulting in a legal coloring.

Step 2. Visit, in some order, all the pattern and text pixels. As the computation progresses some pixels will be put for revisits in a special queue Q.

Visiting a pixel:

1. If the pixel is P1 (rule R1 is satisfied) or P2: do nothing.

2. If the pixel is P3: make it P1 by coloring the required intermediate points; put in Q (for a revisit in the current Step 2) every visited pixel which is adjacent to these intermediate points (it is possible that a pixel was P2 when visited, and later became P3 or even P4).

3. If the pixel is P4: stop - conclude that no legal coloring exists.

Step 3 works in iterations till either all pixels are P1, or a conclusion that there no legal coloring exists.

Step 3. While there is a pattern (or text) pixel $p_{i,j}$, which is P2, perform the following *iteration*.

Step 3.1. Pick one of the P2 pixels. Without loss of generality, denote the picked pixel $p_{i,j}$. Being P2, the pixel $p_{i,j}$ can be satisfied in two ways. For each of the two ways separately, and in parallel (see also the implementation remark below), do the following:

Step 3.1.1. Satisfy the pixel $p_{i,j}$ by coloring the required intermediate points; enter into a queue Q every pixel which is adjacent to these intermediate points for a future visit (in the current iteration). Note that there will be two separate Q's, one for each way.

Step 3.1.2. If Q is empty proceed to Step 3.2. While the queue Q is not empty, pick one of its pixels, and apply the same routine for visiting a pixel as in Step 2 subject to some minor re-adjustments, as follows:

1. If the pixel is P1 (rule R1 is satisfied) or P2: do nothing.

2. If the pixel is P3: make it P1 by coloring the required intermediate points; put in Q (for a visit in the current iteration) every pixel which is adjacent to these intermediate points (it is possible that a pixel was P2 when visited, and later became P3 or even P4).

3. If the pixel is P4: stop - conclude that the current way in which pixel $p_{i,j}$ is satisfied cannot be extended into a legal coloring.

Step 3.2. Suppose that one of the ways reaches this step with an empty Q. Leave the colors assigned to intermediate points as set by this way. Abort Step 3.1 for the other way.

Repeat Step 3. If both ways for satisfying pixel $p_{i,j}$ lead to a P4 pixel, conclude that no legal coloring exists.

A remark on implementation. For efficiency reasons, the two ways for satisfying the pixel $p_{i,j}$ are implemented in parallel at the following pace: we alternate back and forth between the two ways performing one computation for each at a time. If one of the ways leads to an empty Q, we can charge all the operations of the current iterations, including those of the aborted way, to its pixels.

Step 4. For each intermediate point which is not yet colored, color it in a way which will satisfy it; namely, so that its color is equal to at least one of its four corners. (This is doable by Observation 3.)

Complexity: Step 1 runs in $O(m^2)$ time. Step 2 runs in $O(m^2)$ time, since when a pixel is satisfied we may change the status of at most 6 more pixels (specifically, one may check that two pixels may change to P1 and four pixels to P3 or P4). The above analysis of Step 2, together with the implementation remark of Step 3, implies that it also takes $O(m^2)$ time. Step 4 runs in $O(m^2)$ time. To sum up, the algorithm runs in time which is linear in m^2. Since we run the above algorithm n^2 time, once for each possible subarray of the text the total time solving the SID-2-D problem is $O(m^2 n^2)$. (The running time of the algorithm is independent of the size of the alphabet; that is, the number of different colors.)

Correctness Theorem. The algorithm is correct. This means that: (1) the algorithm finds a coloring which satisfies rules R1 and R2 if and only if such a coloring exists, and (2) if the algorithm finds such coloring then it extends it into a legal coloring.

Proof: We first prove item (2) of the theorem. Suppose that the algorithm provides a coloring which satisfies rules R1 and R2.

It is clear at this point that all the text and pattern pixels are satisfied (by Rule R1). The intermediate points that were colored in Steps 2 and 3 are also satisfied, since satisfying a pattern or text pixel was done by coloring the intermediate points in a way that satisfies them, as well.

It remains to check intermediate points that were colored in Steps 1 and 4.

For Step 4, we show that:

Observation 3. All intermediate points that are not colored after Step 3 can be colored, in Step 4, to obtain a legal coloring.

Proof of Observation 3. We will demonstrate the proof on an example, which is general enough. In Figure 2, consider $n^t_{i+x+1,j+y}$ and assume that $p_{x+1,y+1}$ and $t_{i+x+1,j+y+1}$ have color 1 while $t_{i+x+1,j+y}$ and $p_{x+2,y+1}$ have color 2. If $n^t_{i+x+1,j+y}$ can be satisfied by being colored with colors 1 or 2, then we are done. Therefore, we assume, in contradiction that coloring $n^t_{i+x+1,j+y}$ with either color 1 or color 2 does not satisfy it. The inability to use color 1 implies that the color of $n^p_{x+1,y+1}$ is not 1 and the inability to use color 2 implies that the color of $n^p_{x+2,y}$ is not 2. However, since $t_{i+x+1,j+y+1}$ is satisfied, the color of $n^p_{x+2,y+1}$ must be 1, and since $p_{x+2,y+1}$ is satisfied, the color of $n^p_{x+2,y+1}$ must be 2. A contradiction.   □

For Step 1, we show that:

Observation 4. All intermediate points that where colored in Step 1 are satisfied at the end of Step 3.

Proof of Observation 4. It is easy to see that any intermediate point that is surrounded by three or more pixels with the same color must be satisfied when the pixels are satisfied. The only interesting case where the intermediate point is colored in Step 1 is where it is surrounded by three colors: two pixels with color 1 one pixel with color 2 and one with color 3. The proof is the same as the proof of Observation 3.   □

To prove item (1) of the theorem, we only show that if the algorithm fails to find a coloring that satisfies rules R1 and R2, such a coloring does not exists. The proof traces the algorithm.

If in Step 1, an intermediate point cannot satisfy rule R2 then rule R2 cannot be satisfied.

The algorithm in Step 2 does not have any choice. Therefore, aborting Step 2 because a P4 pixel occurs, means that a pixel cannot be satisfied.

Next we trace Step 3. In the beginning of each iteration of Step 3 all the pixels are either P1 or P2. Consider an iteration that starts by satisfying rule R1 for pixel $p_{i,j}$ in two ways and suppose that each way led to a pixel which was changed to P4. We argue that no coloring that satisfies rules R1 and R2 exists. Consider the computation for (any) one of the two ways and denote the pixel where the P4 status occurred by x. Tracing this computation will show two directed paths leading from pixel $p_{i,j}$ to x. The first path is denoted $s_1(=p_{i,j}), s_2, ..., s_\alpha(=x)$ and the second path $t_1(=p_{i,j}), t_2, ..., t_\beta(=x)$. Each edge (s_k, s_{k+1}), $1 \le k < \alpha$ reflects the following. Prior to satisfying pixel s_k, pixel s_{k+1} was P2, and after that, pixel s_{k+1} became P3. The same applies to every edge of the form (t_k, t_{k+1}), $1 \le k < \beta - 1$. For the edge $(t_{\beta-1}, t_\beta)$, the situation is that prior to satisfying pixel $t_{\beta-1}$, pixel t_β was P3, and after that it became P4. Now, given a coloring which satisfied rules R1 and R2, we reach a contradiction. Pick one possibility in which pixel $p_{i,j}$ is satisfied (namely, included in a unit square which contains, in particular, two intermediate points). Next, follow the first path corresponding to this example, and then the second path. If all pixels are satisfied in the same way as in the two paths, Rule R1 cannot hold at the last pixel which is the same for both paths. If some pixel in any of the paths is satisfied in a different way, look at the first pixel in that path which is satisfied in a different way and get there a contradiction to Rule R1. Assume that it happened in the first path on pixel s_ℓ. Then, at least one among the pixels $s_1, s_2, ..., s_{\ell-1}$ is P4.

This completes the proof of item (1) of the Correctness Theorem. $\square$

Remark. Step 3 can be represented as a 2-SAT problem, and solved alternatively by the linear time algorithm for 2-SAT, of [EIS-76]. We sketch the basic idea for presenting Step 3 as a 2-SAT problem and leave the details to the interested reader. Each pattern pixel and text pixel which are P2 define two literals. For example, $p_{i,j}$ defines x_k and x_{k+1} - one literal for each way (i.e., corner) which can satisfy $p_{i,j}$. Namely, x_k is true if $p_{i,j}$ is satisfied in by one corner, and x_{k+1} is true if $p_{i,j}$ is satisfied by another corner. So, for $p_{i,j}$, we will have a clause $(x_k \lor x_{k+1})$. Consider now a text pixel $t_{a,b}$ which is a neighbor of $p_{i,j}$, is P2, and defines the literals x_ℓ and $x_{\ell+1}$. The definition of the 2-SAT problem may, for instance, not allow $p_{i,j}$ and $t_{a,b}$ be satisfied by their first way simultaneously. Specifically, for the pair $p_{i,j}$ and $t_{a,b}$ we will have the clause $(\neg x_k \lor \neg x_\ell)$.

6 Algorithms Using String Matching Methods

In this section, additional algorithms that solve the SID-1-D and the SID-2-D problems are given. For the results to apply we need to assume that the colors in each of these problems are taken from the set $\{c_1, c_2, ..., c_C\}$, where C is some constant..

Below, we overview only the algorithm for the SID-2-D problem, which includes all the ideas needed for the SID-1-D algorithm and more. Note that the algorithm that solves the SID-1-D problem is both NR and PR, but the one that solves the SID-2-D problem is NR only. At a high-level, the SID-1-D algorithm is similar to the algorithm of section 4 and the SID-2-D algorithm is similar to the algorithm of section 5.1. The difference, however, is in the implementation, which is more efficient.

Section 5.1 deals with the SID-2-D problem. Under the assumptions given there, we claimed that an occurrence of an incarnation of the pattern starts in location $t_{i,j}$ of the text if the following two conditions hold.

(a) For every $1 \le x, y \le m-1$, $t_{i+x,j+y}$ equals either $p_{x,y}$ or $p_{x+1,y}$ or $p_{x,y+1}$ or $p_{x+1,y+1}$.

(b) For every $1 \le x, y \le m$, $p_{x,y}$ equals to at least one of the following four characters: $t_{i+x-1,j+y-1}, t_{i+x-1,j+y}, t_{i+x,j+y-1}$, or $t_{i+x,j+y}$.

The algorithm for the SID-2-D problem has three steps. The first two are guided by the above two conditions. The third step combines the answers of steps 1 and 2.

Steps 1 and 2 are similar to one another in the following sense. In Step 1 (resp. Step 2) we check if a character of the text (resp. pattern) is equal to at least one of four characters of the pattern (resp. text). Hence, we are going to describe briefly only Step 1.

Step 1. Step 1 works in C iterations, one per color. In iteration k, we count for each $t_{i,j}$ separately, the number of characters $t_{i+x,j+y}$ that have the following two properties: (i) they are equal to c_k and (ii) they satisfy condition (a). After the last iteration C, it remains to identify all locations in the text for which the sum of counts reaches $(m-1)^2$.

For the implementation of Step 1 some folklore tricks from string matching algorithms are used. We briefly allude to some of them. In each iteration k, every character of the text, which is equal to c_k, becomes 1 and every other becomes 0; a pattern $\bar{P}$ is produced as follows: For each pair i, j $(1 \leq i, j \leq m-1)$, $\bar{p}_{i,j} = 1$ if at least one of the following four characters is equal to c_k: $p_{i,j}$, $p_{i+1,j}$, $p_{i,j+1}$, $p_{i+1,j+1}$; otherwise $\bar{p}_{i,j}$ is equal 0. Then, in order to count the number of matches, two techniques are used: (i) the convolution technique (as was suggested in [FP-74], and used in many algorithms such as [P-85], [A-87], [BP-89], [K-89], [AL-90], and [DGM-90]), combined with (ii) a "wraparound technique" given in [BP-89] and [AL-90] is used.

Step 3. Combining the answers of step 1 and 2, derive all occurrences of an incarnation of the pattern in the text.

Complexity. Each of steps 1 and 2 runs in $O(n^2 \log m)$ time. Step 3 runs in $O(n^2)$ time. Therefore the total time complexity of the algorithm is $O(n^2 \log m)$ time. (In the case of general alphabet the running time of the algorithm is $O(n^2 \log^2 m)$ time.) The new algorithm for the SID-1-D problem runs in $O(n \log m)$ time.

7 Yet Additional Discussion

While being motivated by the fact that original digitized images were derived from a continuous (i.e., Euclidean) domain, our problem definitions *do not abandon* these original digitized images in favor of continuous ones at the image formation or early image processing stages - these stages are part of what is viewed as low level image processing (see, e.g., [BB-82]). This is in contrast to the contemporary procedure for handling such a problem by computers which seems to be a roundabout way: one endeavors to transform a digitized input into a continuous domain and then discretize it again for solution by a digitized computer. It should be clear that our approach does not depart from the contemporary procedure at higher levels. The contemporary procedure places an intermediate level of representation before proceeding to "image understanding" in order to cope with issue such as: (1) Variability in the structure (geometry and radiometry) of natural objects, and the need to recognize them independent of perspective changes; this suggests the need for a symbolic representation that will be invariant to these sources of variability. (2) Noise and clutter in images also lead to variability in appearances of objects, which can often be overcome using intermediate image representations. The motivation being to extract some information in the image which is crucial to recognition (and, generally, to other vision tasks). *The contribution of our approach is at the low level.* It identifies an important gap in the fundamental theory for image formation and image processing - how to determine, even in the absence of noise, if a digitized pattern could be an image of a given scene (either specified in terms of its "continuous" properties or through other, exemplar, images)?

One can object to considering NR and PR algorithms on grounds that they do not really solve the digitized problem. However, even algorithms that solve a digitized problem are limited to the available data, and they cannot be guaranteed to solve the problem we essentially care about - the origin problem. Another justification for NR and PR algorithms is that they enable to classify heuristics, which may be relevant in practice, for a problem. Such algorithms are not that unusual: in [LLS-91], page 33, an interesting quote of Richard Lipton concerning computational biology problems which are NP-complete, advocates not only approximation algorithm and ones that work well only on the average, but also algorithms that do not work every time.

A more general discussion is given in [LV-92b], which is a comprehensive journal version of [LV-92a] and the present paper.

Acknowledgments. Helpful discussions with Neal Young are gratefully acknowledged.

References

[A-87]   K. Abrahamson, "Generalized string matching," *SIAM J. Comput.*, 17:1039-1051, 1987.

[AL-90] A. Amir and G.M. Landau, "Fast parallel and serial multi dimensional approximate array matching," *Theoretical Computer Science*, 81:97-115, 1991.

[BB-82] D.H. Ballard, and C.M. Brown, *Computer Vision*, Prentice-Hall, 1982.

[BP-89] S. Ben-Yehuda and R.Y. Pinter, "Symbolic layout improvement using string matching based local transformation," *Proc. of the Decennial Caltech Conf. on VLSI*, 227-239, 1989.

[C-71]   M.B. Clowes, "On seeing things," *Artificial Intelligence*, 2:79-116, 1971.

[DGM-90] M. Dubiner, Z. Galil, and E. Magen, "Faster tree pattern Matching," *Proc. 31th IEEE Symp. Foundations of Computer Science*, 145-150, 1990.

[EIS-76] S. Even, A Itai and A. Shamir, "On the complexity of timetables and multicommodity flow problems," *SIAM J. Computing*, 5(4):691-703, 1976.

[FP-74] M.J. Fischer and M.S. Paterson, "String Matching and Other Products," *Complexity of Computation*, R.M. Karp (editor), SIAM-AMS Proceedings, 7:113-125, 1974.

[H-71]   D.A. Huffman, "Impossible objects as nonsense sentences," *Machine Intelligence*, 6:295-323, 1971, B. Meltzer and D. Michie (Eds.), Edinburgh University Press.

[K-89]   S.R. Kosaraju, "Efficient tree pattern Matching," *Proc. 30th IEEE Symp. Foundations of Computer Science*, 178-183, 1989.

[LLS-91] E.S. Lander, R. Langridge, and D.M. Saccocio, "A report on computing in Molecular Biology: mapping and interpreting Biological information," *CACM*, 34,11: 33-39, 1991.

[LV-92a] G.M. Landau and U. Vishkin, "Pattern matching in a digitized image," *Proc. 3rd ACM-SIAM Symposium on Discrete Algorithms*, 453-462, 1992.

[LV-92b] G.M. Landau and U. Vishkin, "Pattern matching in a digitized image," *Algorithmica*, to appear. Also, Technical Report UMIACS-TR-92-132, Institute for

Advanced Computer Studies, University of Maryland, College Park, MD 20742, 1992. This paper is a journal version of [LV-92a] and the present paper.

[P-85]   R. Y. Pinter, "Efficient string matching with don't-care patterns," in A. Apostolico and Z. Galil (editors), *Combinatorial Algorithms on Words*, NATO ASI Series, Series F: Computer and System Sciences, Vol. 12, Springer-Verlag, 97-107, 1985.

[S-90]   H. Samet, *The Design and Analysis of Spatial Data Structures*, Addison-Wesley, 1990.

Analysis of a String Edit Problem in a Probabilistic Framework
(Extended Abstract)

Guy Louchard[1] and Wojciech Szpankowski[2]

[1] Lab. d'Informatique Théorique, Université Libre de Bruxelles, B-1050 Brussels, Belgium
[2] Dept. of Computer Science, Purdue University, W. Lafayette, IN 47907, USA

Abstract. We consider a string edit problem in a probabilistic framework. This problem is of considerable interest to many facets of science, most notably molecular biology and computer science. A string editing transforms one string into another by performing a series of weighted edit operations of overall maximum (minimum) cost. An edit operation can be the deletion of a symbol, the insertion of a symbol or the substitution of a symbol. We assume that these weights can be arbitrary distributed. We reduce the problem to finding an optimal path in a weighted grid graph, and provide several results regarding a typical behavior of such a path. In particular, we observe that the optimal path (i.e., edit distance) is asymptotically almost surely (a.s.) equal to αn where α is a constant and n is the sum of lengths of both strings. We also obtained some bounds on α in the so called independent model in which all weights (in the associated grid graph) are assumed to be independent. More importantly, we show that the edit distance is well concentrated around its average value. As a by-product of our results, we also present a precise estimate of the number of alignments between two strings. To prove these findings we use techniques of random walks, diffusion limiting processes, generating functions, and the method of bounded difference.

1 Introduction

We first review the string editing problem, its importance, and its relationship to the longest path problem in a special grid graph.

Let b be a string consisting of ℓ symbols on some alphabet Σ of size V. There are three operations that can be performed on a string, namely *deletion* of a symbol, *insertion* of a symbol, and *substitution* of one symbol for another symbol in Σ. With each operation is associated a *weight* function. We denote by $W_I(b_i)$, $W_D(b_i)$ and $W_Q(a_i, b_j)$ the weight of insertion and deletion of the symbol $b_i \in \Sigma$, and substitution of a_i by $b_j \in \Sigma$, respectively. An *edit script* on b is any sequence ω of edit operations, and the total weight of ω is the sum of weights of the edit operations.

The *string editing problem* deals with two strings, say b of length ℓ (for *long*) and a of length s (for *short*), and consists of finding an edit script ω_{max} (ω_{min}) of

---

* This research was partially done while the second author was visiting INRIA, Rocquencourt, France, and he wishes to thank INRIA (projects ALGO, MEVAL and REFLECS) for a generous support. Additional support was provided in part by NSF Grants CCR-9201078, NCR-9206315 and INT-8912631, in part by AFOSR Grant 90-0107, NATO Grant 0057/89, and Grant R01 LM05118 from the National Library of Medicine.

minimum (maximum) total weight that transforms **a** into **b**. The maximum (minimum) weight is called the *edit distance from* **a** *to* **b**, and its is also known as the Levenshtein distance. In molecular biology, the Levenshtein distance is used to measure similarity (homogeneity) of two molecular sequences, say DNA sequences (cf. [27]).

The string edit problem can be solved by the standard dynamic programming method (cf. [2], [24], [30]). More importantly for us, it can be modelled as a path in a special weighted grid graph. Let $C_{\max}(i, j)$ denote the maximum weight of transforming the prefix of **b** of size i into the prefix of **a** of size j. The key observation is to note that interdependency among the partial optimal weights $C_{\max}(i, j)$ induce an $\ell \times s$ grid-like directed acyclic graph, called further a *grid graph*. In such a graph vertices are points in the grid and edges go only from (i, j) point to grid points $(i, j+1)$, $(i+1, j)$ and $(i+1, j+1)$. A horizontal edge from $(i, j-1)$ to (i, j) carries the weight $W_I(b_j)$; a vertical edge from $(i, j-1)$ to (i, j) has weight $W_D(a_i)$; and finally a diagonal edge from $(i-1, j-1)$ (i, j) is weighted according to $W_Q(a_i, b_j)$.

In this paper, we analyze the string edit problem in a probabilistic framework. We adopt the Bernoulli model for a random string, that is, *all symbols of a string are generated independently with probability p_i for symbol $i \in \Sigma$.* A standard probabilistic model assumes that both strings are generated according to the Bernoulli scheme (cf. [3], [6], [7], [8], [11], [18], [29], [30]). We call it the **string model**. Such a framework, however, leads to statistical dependency of weights in the associated grid graph. To avoid this problem, most of the time we shall work within the framework of another probabilistic model which postulates that all weights in the associated grid graph are statistically independent. We call it **independent model**. This is closly related to a model in which *only* one string is random, say **b**, while the other one , say **a**, is deterministic. Indeed, in such a situation all weights in a "horizontal" strip in the associated grid graph are independent, while weights in a "vertical" strip are dependent (e.g., if **a** = 101, and **b** is random, then the "1"s in the string **a** match independently all "1"s in **b**, but clearly the first "1" and the third "1" in **a** have to match "1"s in **b** at the same places). We call such a model **semi-independent**.

Most of the results in this paper deal either with the independent model or the string model. We believe that better understanding of the independent model should be the first step to obtain valuable results for the semi-independent model. Certainly, results of the semi-independent model can be further used to deduce probabilistic behavior of the string model (cf. Corollary 2.2). In passing, we note that the semi-independent model might be useful in some applications (e.g., when comparing a given string to all strings in a data base).

Clearly, in the edit distance problem the distributions of weights $W_D(a_i)$, $W_I(b_j)$ and $W_Q(a_i, b_j)$ depend on the given string **a**. However, to avoid complicated notations we ignore this fact and consider a grid graph with weights W_I, W_D and W_Q. In other words, we concentrate on finding the longest path in a grid graph with independent weights W_I, W_D and W_Q, not necessary equally distributed. By selecting properly these distributions, we can model the string edit problem. For example, in the standard setting the deletion and insertion weights are identical, and usually constant, while the insertion weight takes two values, one (high) when matching between a letter of **a** and a letter of **b** occurs, and another value (low) in the case of a mismatch (e.g., in the *Longest Common Substring* problem, one sets $W_I = W_D = 0$, and $W_Q = 1$ when a matching occurs, and $W_Q = -\infty$ in the other case).

Our results can be summarized as follows: Applying the *Subadditive Ergodic Theorem* we note that $C_{\max} \sim \alpha n$ almost surely (a.s.), where $n = \ell + s$ (cf. Theorem 2.1 and Corollary 2.2). Our main contribution lies in establishing bounds for the constant α (cf. Theorem 2.7) for the independent model (cf. Corollary 2.2 for a possible extension to the string model). The upper bound is rather tight as verified by simulation experiments. More importantly, using the powerful and modern method of bounded differences (cf. [23]) we establish in the string model (and the other two models) a sharp concentration of $C_{\max}$ around the mean value $EC_{\max}$ under a mild condition on the tail of the weight distributions (cf. Theorem 2.3). This proves the conjecture of Chang and Lampe [10] who observed empirically such a sharp concentration of $C_{\max}$ for a version of the string edit problem, namely the approximate string matching problem.

Our probabilistic results are proved in a unified manner by applying techniques of random walks (cf. [12], [14]), generating functions (cf. [13], [20], [21]), and bounded differences (cf. [23]). In fact, these techniques allow us to establish further results of a more general interest. In particular, we present a precise asymptotic estimate for the number of paths in the grid graph (cf. Theorem 2.4), which coincides with the number of sequence alignments (cf. [15], [16]). Finally, for the independent model we establish the limiting distribution of the total weight (cf. Theorem 2.5) and the tail distribution of the total weight (cf. Theorem 2.6) of a randomly selected path (edit script) in the grid graph.

The string edit problem and its special cases (e.g., the longest-common subsequence problem and approximate pattern matching) were studied quite extensively in the past, and are subject of further vigorous research due to their vital applications in molecular biology. There are many algorithmic solutions to the problem, and we only mention here Apostolico and Guerra [1], Apostolico *et al.* [2], Chang and Lampe [10], Myeres [24], Ukkonen [28], and Waterman [30]. On the other hand, a probabilistic analysis of the problem was initiated by Chvatal and Sankoff [11] who analyzed the longest common subsequence problem. To the best of our knowledge, there is no much literature on the probabilistic analysis of the string edit problem with a notable exception of a recent marvelous paper by Arratia and Waterman [7] (cf. [29]). There is, however, a substantial literature on probabilistic analysis of pattern matching and approximate pattern matching without weights. We mention here a series of papers by Arratia and Waterman (cf. [5], [6]) and with Gordon (cf. [3], [4]), as well as papers by Karlin and his co-authors (cf. [17], [18]). Another approach for the probabilistic analysis of pattern matching with mismatches was recently reported by Atallah *et al.* in [8].

In this preliminary version of the paper, we mainly present our results delaying all necessary proofs to a journal version (cf. [22]).

2 Main Results

In this section we present our main results concerning the typical behavior of the longest (shortest) path in a grid graph, i.e., the cost of an edit scrip. We also report some findings on the probabilistic behavior of a *given* path in such a graph. We deal in this section only with the grid graph model unless otherwise stated.

To recall, we consider a grid graph of size ℓ and s ($\ell \geq s$). All of our results, however, will be expressed in terms of $n = \ell + s$ and $d = \ell - s$. We assign to every edge in such a graph a real number representing its weight. A family of such directed acyclic weighted graphs will be denoted as $\mathcal{G}(n, d)$ or shortly $\mathcal{G}(n)$. We write $\mathbf{G}(n) \in \mathcal{G}(n, d)$ for a member of such a family.

In the independent model, we assume that *weights are independent* with respect to different edges. Let $F_I(\cdot)$, $F_D(\cdot)$ and $F_Q(\cdot)$ denote distribution functions of W_I, W_D and W_Q respectively. We assume that the mean values m_I, m_D and m_Q, and the variances s_I^2, s_D^2 and s_Q^2, respectively, are finite. The distribution functions are not necessary identical.

In order to formulate our optimization problem on the grid graph, we introduce some further notation. By $\mathcal{B}(n, d)$ or shortly $\mathcal{B}(n)$ we denote the set of all directed paths from the point O of the grid graph to the end point E. The cardinality of $\mathcal{B}(n)$, that is, the total number of paths between O and E, is denoted as $L(n, d)$. Finally, a particular path from O to E is denoted as $\mathcal{P}$, i.e., $\mathcal{P} \in \mathcal{B}(n, d)$. Note that the length $|\mathcal{P}|$ of a path $\mathcal{P}$ satisfies $\ell \leq |\mathcal{P}| \leq l + r = n$.

We formulate our problem as an optimization problem on a grid graph. Let $N_I(\mathcal{P})$, $N_D(\mathcal{P})$ and $N_Q(\mathcal{P})$ denote the number of horizontal edges (say I-steps), vertical edges (say D-steps), and diagonal edges (say Q-steps) in a path $\mathcal{P}$. Then,

$$C_{\max} = \max_{\mathcal{P} \in \mathcal{B}(n)} \left\{ \sum_{i=1}^{N_I(\mathcal{P})} W_I(i) + \sum_{i=1}^{N_D(\mathcal{P})} W_D(i) + \sum_{i=1}^{N_Q(\mathcal{P})} W_Q(i) \right\}, \qquad (1)$$

$$C_{\min} = \min_{\mathcal{P} \in \mathcal{B}(n)} \left\{ \sum_{i=1}^{N_I(\mathcal{P})} W_I(i) + \sum_{i=1}^{N_D(\mathcal{P})} W_D(i) + \sum_{i=1}^{N_Q(\mathcal{P})} W_Q(i) \right\}. \qquad (2)$$

We also denote by $W_n(\mathcal{P})$, or shortly W_n, the total weight of a path $\mathcal{P}$, that is,

$$W_n(\mathcal{P}) = \sum_{i=1}^{N_I(\mathcal{P})} W_I(i) + \sum_{i=1}^{N_D(\mathcal{P})} W_D(i) + \sum_{i=1}^{N_Q(\mathcal{P})} W_Q(i). \qquad (3)$$

We write W_n to denote the total weight of a randomly selected path. More precisely, we define

$$\Pr\{W_n < x\} = \frac{1}{L(n, d)} \sum_{\mathcal{P} \in \mathcal{B}} \Pr\{W_n(\mathcal{P}) < x\}. \qquad (4)$$

Our results crucially depend on the order of magnitude of d with respect to n. We consider separately several cases, as defined below:

CASE (A): $d = O(\sqrt{n})$, and let $x = d\sqrt{\sqrt{2}/n} = \zeta d/\sqrt{n}$ where $\zeta = 2^{1/4}$.
CASE (B): $d = \Theta(n)$, and let $x = d/n$.
CASE (C): $d = n - O(n^{1-\varepsilon})$, that is, for some constant x we have $d = n(1 - x/n^{\varepsilon})$.
CASE (D): $d = O(1)$ will be reduced to Case (A).
CASE (E): $s = O(1)$ will be reduced to case (C).

Since the last two cases, as will turn out, can be reduced to the previous three ones, we shall concentrate below on the three top cases.

Now, we are in a position to present our results. To simplify further our presentation, we concentrate mainly on the longest path C_{max}. We start with a simple general result concerning the typical behavior of C_{max}. The more refined results containing a computable upper bound for EC_{max} (in the independent model) is given at the end of this section (cf. Theorem 2.7).

Theorem 2.1. *In all three probabilistic models, by the Subadditive Ergodic Theorem, the following holds*

$$\lim_{n \to \infty} \frac{C_{max}}{n} = \lim_{n \to \infty} \frac{EC_{max}}{n} = \alpha \quad (a.s.) , \tag{5}$$

provided ℓ/s has a limit as $n \to \infty$. ∎

The above result is true for all probabilistic models, however, in the string (and semi-independent) model one has to consider the fact that weights depend on the given string **a**. Let $P(\mathbf{a})$ be the probability of **a** occurrence in our standard Bernoulli model (e.g., for the binary alphabet $\Sigma = \{\alpha, \beta\}$ we have $P(\mathbf{a}) = p^{|\alpha|}(1-p)^{|\beta|}$ where p is the probability of α occurrence, and $|\alpha|$ ($|\beta|$) is the number of α's (β's) in the string **a**). Since now the weights are functions of $P(\mathbf{a})$, hence the constant α in Theorem 2.1 (cf. (5)) depends on **a**, too. We denote it by $\alpha_{\mathbf{a}}$ for the semi-independent model. The following corollary follows directly from Theorem 2.1.

Corollary 2.2. *In the string model, the edit distance C_{max} satisfies (5) with α as below*

$$\alpha = \sum_{\mathbf{a} \in \mathcal{H}} \alpha_{\mathbf{a}} P(\mathbf{a}) \tag{6}$$

*where $\mathcal{H}$ is the set of all possible strings **a** of length s over the alphabet Σ.* ∎

Finally, for all probabilistic models we can report the following finding concerning the concentration of the edit distance. It proves the conjecture of Chang and Lampe [10]. The proof of this result uses a powerful method of bounded differences (cf. [23]) (or in other words: Azuma's type inequality).

Theorem 2.3. (i) *If all weights are bounded random variables, say $\max\{W_I, W_D, W_Q\} \leq 1$, then for arbitrary $\varepsilon > 0$ and large n*

$$\Pr\{|C_{max} - EC_{max}| > \varepsilon EC_{max}\} \leq 2 \exp(-\varepsilon^2 \alpha n) . \tag{7}$$

(ii) *If the weights are unbounded but such that for large n, $W_{max} = \max\{W_I, W_D, W_Q\}$ satisfies the following*

$$n \Pr\{W_{max} \geq n^{1/2-\delta}\} \leq U(n) \tag{8}$$

for some $\delta > 0$ and a function $U(n) \to 0$ as $n \to \infty$, then

$$\Pr\{|C_{max} - EC_{max}| > \varepsilon EC_{max}\} \leq 2 \exp(-\beta n^\delta) + U(n) \tag{9}$$

for any $\varepsilon > 0$ and some $\beta > 0$.

Proof: We consider only the string model. Part (i) is a direct consequence of the following inequality of Azuma's type (cf. [23]): *Let X_i be i.i.d. random variables such that for some function $f(\cdot, \ldots, \cdot)$ the following is true*

$$|f(X_1, \ldots, X_i, \ldots, X_n) - f(X_1, \ldots, X_i', \ldots, X_n)| \le c_i , \qquad (10)$$

where $c_i < \infty$ are constants, and X_i' has the same distribution as X_i. Then,

$$\Pr\{|f(X_1, \ldots, X_i, \ldots, X_n) - Ef(X_1, \ldots, X_i, \ldots, X_n)| \ge t\} \le 2\exp(-2t^2 / \sum_{i=1}^{n} c_i) \qquad (11)$$

for some $t > 0$. The above technique is also called the method of bounded differences.

Now, for part (i) it suffices to set $X_i = b_i$ for $1 \le i \le \ell$, and $X_i = a_i$ for $\ell + 1 \le i \le n$, where a_i and b_i are the i symbols of the two strings **a** and **b**. Under our Bernoulli model, the X_i are i.i.d. and (10) holds, hence we can apply (11). Inequality (7) follows from the above and $t = \varepsilon EC_{\max} = O(n)$.

To prove part (ii), we note that for the string edit problem

$$|C_{\max}(X_1, \ldots, X_i, \ldots, X_n) - C_{\max}(X_1, \ldots, X_i', \ldots, X_n)| \le \max_{1 \le i \le n}\{X_i\} . \qquad (12)$$

Then, for some constant c

$$\begin{aligned}
\Pr\{|C_{\max} - EC_{\max}| \ge t\} &= \Pr\{|C_{\max} - EC_{\max}| \ge t , \max_{1 \le i \le n}\{X_i\} \le c\} \\
&\quad + \Pr\{|C_{\max} - EC_{\max}| \ge t , \max_{1 \le i \le n}\{X_i\} > c\} \\
&\le 2\exp(-2t^2/nc^2) + n\Pr\{X_i > c\} .
\end{aligned}$$

Set now $t = \varepsilon EC_{\max} = O(n)$ and $c = O(n^{1/2-\delta})$, then

$$\Pr\{|C_{\max} - EC_{\max}| \ge \varepsilon EC_{\max}\} \le 2\exp(-\beta n^\delta) + n\Pr\{X_i > n^{1/2-\delta}\} ,$$

for some constant $\beta > 0$, and this implies (9) provided (8) holds. ∎

Hereafter, we investigate only the independent model. For this model, we have obtained several new results regarding the probabilistic behavior of an edit script, that is , a path in a weighted grid graph. These results are of their own interests. For example, the total number of paths $L(n, d)$ in the grid graph (cf. Theorem 2.5) represents the number of ways the string **a** can be transformed into **b**, and this problem was already tackled by others (cf. [15], [16]).

Theorem 2.4. *The limiting distribution of the total weight is normal. More precisely,*

$$\frac{W_n - n\mu_W}{\sqrt{n}\sigma_W} \to \mathcal{N}(0, 1) \qquad (13)$$

where $\mathcal{N}(0, 1)$ is the standard normal distribution, and

$$\mu_W = m_I \mu_I + m_D \mu_D + m_Q \mu_Q , \qquad (14)$$
$$\sigma_W^2 = \mu_I s_I^2 + \mu_D s_D^2 + \mu_Q s_Q^2 + \tilde{\sigma}_Q^2 (m_I + m_D - m_Q)^2 \qquad (15)$$

where $\mu_I = EN_I(\mathcal{P})$, $\mu_D = EN_D(\mathcal{P})$, $\mu_Q = EN_Q(\mathcal{P})$ and $\tilde{\sigma}_Q^2 = var N_Q(\mathcal{P})$. *Explicit formulas for these quantities varies, and are given for each case (A)-(E) separately in the next section.* ∎

The next result provides information about the enumeration of paths in a grid graph $\mathbf{G} \in \mathcal{G}$. In fact, we count the number of ways a string **a** can be transformed into string **b** (i.e., number of alignments between **a** and **b**). Its formulation depends on a parameter u that takes different values for every case (A)-(E), however, (D) and (E) can be reduced to the other three cases. Let us define $u = d/n$, and then x becomes

CASE (A): Set $d = x\sqrt{n/\sqrt{2}}$. Then, $u = x/\sqrt{\sqrt{2}n} = x/(\zeta\sqrt{n})$.

CASE (B): Set $u = x$.

CASE (C): Set $d = n(1 - x/n^\varepsilon)$. Then, $u = 1 - x/n^\varepsilon$.

Theorem 2.5. *Let $L(u) = L(n,d)$ be the number of paths in a grid graph $\mathbf{G} \in \mathcal{G}(n)$, where u is defined for each cases (A)-(E) as above. Then,*

$$L(u) = \frac{C\psi_2(\beta_2(u))^n}{\beta_2(u)^{n(1+u)/2}\sqrt{2\pi n V(u)}}(1 + O(1/n)) \tag{16}$$

where

$$\beta_2(u) = \frac{1 + 3u^2 + u\sqrt{8(u^2+1)}}{1 - u^2}, \tag{17}$$

$$\psi_2(u) = \psi_2[\beta_2(u)] = \frac{2u\beta_2(u)}{\beta_2(u) - 1 - u(1 + \beta_2(u))}, \tag{18}$$

and C can be computed explicitly. In the above, $V(u)$ is the variance obtained from the generating function $h(z)$ defined as $h(z) = \psi_2(z\beta_2(u))/\psi_2(\beta_2(u))$, that is, $V(u) = h''(1) - 0.25(1 - u^2)$ where $h''(z)$ is the second derivative of $h(z)$. ∎

For most of our computations, we only need the asymptotics of $L(u)$ in the following, less precise, form

$$\log L(u) = n\rho(u) - 0.5\log n + O(1), \tag{19}$$

where $\rho(u)$ differs for every case (A)-(E). In particular, for cases (A), (B) and (C) we have respectively

$$\rho(u) = -\log(\sqrt{2} - 1), \tag{20}$$

$$\rho(u) = \log\psi_2(\beta_2(u)) - \frac{1+u}{2}\log\beta_2(u), \tag{21}$$

$$\rho(x) = \frac{x\log n}{4\sqrt{n}} + \frac{x(1 + \log 4)}{2\sqrt{n}} - \frac{x\log x}{2\sqrt{n}} + O(\frac{1}{n}). \tag{22}$$

and x is defined above for the case (C).

Finally, in order to obtain an upper bound for the cost $C_{\max}$, we need an estimate on the tail distribution of the total weight W_n along a random path (in the independent model). Formula (3) suggests to apply Cramer's large deviation result (cf. Feller [12]) with some modifications (due to the fact that the total weight W_n as in (3) is a sum of *random* number of weights). To avoid unnecessary complications, we consider in details only two cases, namely:

(a) all weights are *identically* distributed with mean $m = m_I = m_D = m_Q$ and the *cumulant function* $\Psi(s) = \log E e^{s(W-m)}$ for the common weight $W - m$;

(b) insertion weight and deletion weight are constant, say all equal to -1 (e.g., $W_I = W_D = -1$), and the substitution weight $W_Q - m_Q$ has the cumulant function $\Psi_Q(s) = \log E e^{s(W_Q - m_Q)}$. Such an assignment of weights is often encountered in the string edit problem.

We prove the following result.

Theorem 2.6. (i) *In the case (a) of all identical weights, define s^* as the solution of*

$$a = \Psi'(s^*) , \qquad (23)$$

for a given $a > 0$, and let

$$Z_0(a) = s^* \Psi'(s^*) - \Psi(s^*) , \qquad (24)$$

$$E_1(a) = -(s^* m + \Psi(s^*)) , \qquad (25)$$

$$E_2^2(a) = \frac{\tilde{\sigma}_Q^2 m^2 + 2\tilde{\sigma}_Q^2 ma + \tilde{\sigma}_Q^2 (\Psi'(s^*))^2 + (1 - \mu_Q)\Psi''(s^*)}{2(1 - \mu_Q)\tilde{\sigma}_Q^2 \Psi''(s^*)} , \qquad (26)$$

where $\mu_Q = EN_Q$ and $\tilde{\sigma}_Q^2 = var\, N_Q$. Then,

$$\Pr\{W_n > (1 - \mu_Q)(a + m)n\} \sim \qquad (27)$$

$$\frac{1}{2s^* E_2(a)\tilde{\sigma}_Q \sqrt{\pi(1 - \mu_Q)n\Psi''(s^*)}} \exp\left(-n(1 - \mu_Q)Z_0(a) + n\frac{E_1^2(a)}{4E_2^2(a)}\right) .$$

(ii) *In case (b) of constant I-weights and D-weights, we define s^* as a solution of*

$$a = \Psi_Q'(s^*) , \qquad (28)$$

and let

$$Z_0(a) = s^* \Psi_Q'(s^*) - \Psi_Q(s^*) ,$$

$$E_1(a) = s^*(m_Q + 2) + 2s^* a^* - \Psi(s^*) ,$$

$$E_2^2(a) = \frac{\tilde{\sigma}_Q^2 (m_Q + 2)(m_Q + 2 + 2a^* + 4s^* \Psi_Q''(s^*)) + \tilde{\sigma}_Q a^*(a^* + 4s^* \Psi_Q''(s^*)) + \mu_Q \Psi_Q''(s^*)}{2\mu_Q \tilde{\sigma}_Q^2 \Psi_Q''(s^*)}$$

Then,

$$\Pr\{W_n > \mu_Q(a + \beta/\mu_Q)n\} \sim$$

$$\frac{1}{2s^* E_2(a)\tilde{\sigma}_Q \sqrt{\pi\mu_Q n\Psi_Q''(s^*)}} \exp\left(-n\mu_Q Z_0(a) + n\frac{E_1^2(a)}{4E_2^2(a)}\right) . \qquad (29)$$

where $\beta = 2\mu_Q + m\mu_Q - 1$. ∎

Having the above estimates on the tail of the total cost of a path in the grid graph $\mathbf{G} \in \mathcal{G}(n)$, we can provide a more precise information about the constant α in our Theorem 2.1, that is, we compute an upper bound $\overline{\alpha}$ and a lower bound $\underline{\alpha}$ of

α for the independent model. We prove below the following result, which is one of our main finding.

Theorem 2.7 *Assume the independent model.*
(i) *Consider first the identical weights case (cf. case (a) above). Let a^* be a solution of the following equation*

$$(1 - \mu_Q)Z_0(a^*) = \rho + \frac{E_1^2(a^*)}{4E_2^2(a^*)} \;, \tag{30}$$

where ρ is defined in (20)-(22), and Z_0, E_1 and E_2^2 are defined in (24)-(26). Then, the upper bound $\overline{\alpha}$ of α becomes

$$\overline{\alpha} = (1 - \mu_Q)(a^* + m) + O(\log n/n) \;. \tag{31}$$

In the case of constant I and D weights, let a^* be a solution of the equation

$$\mu_Q Z_0(a^*) = \rho + \frac{E_1^2(a^*)}{4E_2^2(a^*)} \;, \tag{32}$$

where Z_0, E_1 and E_2^2 are as before. Then,

$$\overline{\alpha} = \mu_Q(a^* + \beta/\mu_Q) + O(\log n/n) \;, \tag{33}$$

where β is defined in Theorem 2.6(ii).

(ii) *The lower bound $\underline{\alpha}$ of α can be obtained from a particular solution to our optimization problem (1). In particular, we have*

$$\underline{\alpha} = \max\{\mu_W, \ell m_D + s m_I, \alpha_{gr}\} \;, \tag{34}$$

where α_{gr} is constructed from a greedy solution of the problem, that is,

$$n\alpha_{gr} = (\ell + sp)m_{max} \tag{35}$$

where $p = \Pr\{W_Q > W_I \text{ and } W_Q > W_D\}$, and $m_{max} = E\max\{W_I, W_D, W_Q\}$.

Proof. We first prove part (i) provided Theorem 2.6 is *granted*. Observe that by Boole's inequality we have for any real x

$$\Pr\{C_{max} > x\} \leq \sum_{\mathcal{P} \in \mathcal{B}} \Pr\{W_n(\mathcal{P}) > x\} = L(u)\Pr\{W_n > x\}$$

where the last equality follows from (4). Let now a_n be such that

$$L(u)\Pr\{W_n > a_n\} = 1 \;. \tag{36}$$

Then, in view of Theorem 2.6 one immediately proves that for $a_n \sim (1 + \varepsilon)n\overline{\alpha}$ for any $\varepsilon > 0$, with $\overline{\alpha}$ as in (31) and (33), we have $\Pr\{C_{max} > a_n\} = o(1)$, as needed for (31).

The lower bound can be established either by considering some particular paths $\mathcal{P}$ or applying a simple algorithm like a greedy one. The greedy algorithm selects in every step the most expensive edge, that is, the average cost per step is $m_{max} =$

$E \max\{W_D, W_I, W_Q\}$. Let $p = \Pr\{W_Q > W_I, W_Q > W_D\}$. Then, assuming we have k D-steps, the number of Q-steps is binomially distributed with parameters p and $s - k$. Ignoring boundary conditions, we conclude that the average total cost for the greedy algorithm is

$$n\alpha_{gr} = m_{max} \sum_{k=1}^{s} (\ell + k)\binom{s}{k} p^{s-k}(1-p)^k = \ell + sp .$$

This formula should be modified accordingly if some boundary conditions must be taken into account. This can be accomplished in the same manner as discussed in Section 3. ∎

We compared our bounds for $C_{\max}$ with some simulation experiments. In the simulation we restricted our analysis to uniformly and exponentially distributed weights, and here we only report the latter results.

Table 1. Simulation results for exponentially distributed weights with means $m_I = m_D = m_Q = 1$ for case (B) with $d = 0.6n$.

ℓ	s	α	α_{sim}	$\overline{\alpha}$
200	50	1.588	1.909	2.45
400	100	1.588	1.808	2.45
600	150	1.5888	1.899	2.45
800	200	1.588	1.926	2.45
1000	250	1.588	1.922	2.45

From Table 1 one concludes that the upper bound is quite tight. It is plausible that the normalized limiting distribution for $C_{\max}$ is double exponential (i.e., $e^{-e^{-x}}$), however, the normalizing constants are quit hard to find.

The edit problem can be generalized, as it was recently done by Pevzner and Waterman [26] for the longest common subsequence problem. In terms of the grid graph, their generalization boils down to adding new edges in the grid graph that connect *no-neighboring* vertices. In such a situation our Theorem 2.1 may not hold, as easy to see. In fact, based on recent results of Newman [25] concerning the longest (unweighted) path in a general acyclic graph, we predict that a phase transition can occur, and $C_{\max}$ may switch from $\Theta(n)$ to $\Theta(\log n)$. This was already observed by Arratia and Waterman [7] for another string problem, namely, for the score in matching.

Acknowledgement

We would like to thank Professors Luc Devroye, McGill University, and Michel Talagrand, Ohio State University and Paris VI, for discussions that led to our Theorem 2.3.

162

References

1. A. Apostolico and C. Guerra, The Longest Common Subsequence Problem Revisited, *Algorithmica*, 2, 315-336, 1987.
2. A. Apostolico, M. Atallah, L. Larmore, and S. McFaddin, Efficient Parallel Algorithms for String Editing and Related Problems, *SIAM J. Comput.*, 19, 968-988, 1990.
3. Arratia, R., Gordon, L., and Waterman, M., An Extreme Value Theory for Sequence Matching, *Annals of Statistics*, 14, 971-993, 1986.
4. Arratia, R., Gordon, L., and Waterman, M., The Erdös-Rényi Law in Distribution, for Coin Tossing and Sequence Matching, *Annals of Statistics*, 18, 539-570, 1990.
5. R. Arratia and M. Waterman, Critical Phenomena in Sequence Matching, *Annals of Probability*, 13, 1236-1249, 1985.
6. Arratia, R., and Waterman, M., The Erdös-Rényi Strong Law for Pattern Matching with a Given Proportion of Mismatches, *Annals of Probability*, 17, 1152-1169, 1989.
7. R. Arratia and M. Waterman, A Phase Transition for the Score in Matching Random Sequences Allowing Deletions, *Annals of Applied Probability*, to appear.
8. M. Atallah, P. Jacquet and W. Szpankowski, A Probabilistic Approach to Pattern Matching With Mismatches, *Random Structures & Algorithms*, 4, 1993.
9. Z. Galil and K. Park, An Improved Algorithm for Approximate String Matching, *SIAM J. Computing*, 19, 989-999, 1990.
10. W. Chang and J. Lampe, Theoretical and Empirical Comparisons of Approximate String Matching Algorithms, *Proc. Combinatorial Pattern Matching*, 172-181, Tuscon 1992.
11. V. Chvatal and D. Sankoff, Longest Common Subsequence of Two Random Sequences, *J. Appl. Prob.*, 12, 306-315, 1975.
12. W. Feller *An Introduction to Probability Theory and its Applications*, Vol.II, John Wiley & Sons, 1971
13. D.H. Greene and D.E. Knuth, *Mathematics for the Analysis of Algorithms*, Birkhauser, 1981
14. D.L. Iglehart, Weak Convergence in Applied Probability, *Stoch. Proc. Appl.* 2, 211-241, 1974.
15. J. Griggs, P. Halton, and M. Waterman, Sequence Alignments with Matched Sections, *SIAM J. Alg. Disc. Meth.*, 7, 604-608, 1986.
16. J. Griggs, P. Halton, A. Odlyzko and M. Waterman, On the Number of Alignments of *k* Sequences, *Graphs and Combinatorics*, 6, 133-146, 1990.
17. S. Karlin and A. Dembo, Limit Distributions of Maximal Segmental Score Among Markov-Dependent Partial Sums, *Adv. Appl. Probab.*, 24, 113-140, 1992.
18. S. Karlin and F. Ost, Counts of Long Aligned Word Matches Among Random Letter Sequences, *Adv. Appl. Prob.*, 19, 293-351 (1987).
19. J.F.C. Kingman, *Subadditive Processes*, in Ecole d'Eté de Probabilités de Saint-Flour V-1975, Lecture Notes in Mathematics, 539, Springer-Verlag, Berlin (1976).
20. G. Louchard, Random Walks, Gaussian Processes and List Structures, *Theor. Comp. Sci.*, 53, 99-124, 1987.
21. G. Louchard, R. Schott and B. Randrianarimanna, Dynamic Algorithms in D.E. Knuth's Model : A Probabilistic Analysis, *Theor. Comp. Sci.*, 93, 201-225, 1992.
22. G. Louchard and W. Szpankowski, A Probabilistic Analysis of a String Edit Problem, INRIA Rapports de Recherche, No. 1814, 1992.
23. C. McDiarmid, On the Method of Bounded Differences, in *Surveys in Combinatorics*, J. Siemons (Ed.), vol 141, pp. 148-188, London Mathematical Society Lecture Notes Series, Cambridge University Press, 1989.

24. E. Myeres, An $O(ND)$ Difference Algorithm and Its Variations, *Algorithmica*, 1, 251-266, 1986.
25. C. Newman, Chain Lengths in Certain Random Directed Graphs, *Random Structures & Algorithms*, 3, 243-254, 1992.
26. P. Pevzner and M. Waterman, Matrix Longest Common Subsequence Problem, Duality and Hilbert Bases, *Proc. Combinatorial Pattern Matching*, 77-87, Tuscon 1992.
27. D. Sankoff and J. Kruskal (Eds.), *Time Warps, String Edits, and Macromolecules: The Theory and Practice of Sequence Comparison*, Addison-Wesley, Reading, Mass., 1983.
28. E. Ukkonen, Finding Approximate Patterns in Strings, *J. Algorithms*, 1, 359-373, 1980.
29. M. Waterman, L. Gordon and R. Arratia, Phase Transitions in sequence matches and nucleic acid structures, *Proc. Natl. Acad. Sci. USA*, 84, 1239-1242, 1987.
30. M. Waterman, (Ed.) *Mathematical Methods for DNA Sequences*, CRC Press Inc., Boca Raton, (1991).

Detecting False Matches
in String Matching Algorithms *

S. Muthukrishnan [†]

Courant Institute, New York University
251 Mercer Street, New York, NY 10012.
Internet: muthu@cs.nyu.edu

Abstract

Consider a text string of length n, a pattern string of length m and a match vector of length n which declares each location in the text to be either a mismatch (the pattern does not occur beginning at that location in the text) or a potential match (the pattern may occur beginning at that location in the text). Some of the potential matches could be *false*, i.e., the pattern may not occur beginning at some location in the text declared to be a potential match. We investigate the complexity of two problems in this context, namely, *checking* if there is any false match, and *identifying* all the false matches in the match vector.

We present an algorithm on the CRCW PRAM that checks if there exists any false match in $O(1)$ time using $O(n)$ processors. Since string matching takes $\Omega(\log \log m)$ time on the CRCW PRAM, checking for false matches is *provably* simpler than string matching. As an important application, we use this simple algorithm to convert the Karp-Rabin Monte Carlo type string matching algorithm into a Las Vegas type algorithm without asymptotic loss in complexity. We also present an efficient algorithm for identifying all the false matches and as a consequence, show that string matching algorithms take $\Omega(\log \log m)$ time even given the flexibility to output a few false matches.

In addition, we give a sequential algorithm for checking using three heads on a 2-way deterministic finite state automaton (DFA) in linear time and another on a 1-way DFA with a fixed number of heads.

1 Introduction

Given a pattern string of length m and a text string of length n, the problem of *string matching* is to output a match vector which indicates, for every location

---

*This research was supported in part by NSF/DARPA under grant number CCR-89-06949 and by NSF under grant number CCR-91-03953.

†The author sincerely thanks Ravi Boppana, Richard Cole, Babu Narayanan and Krishna Palem for very helpful discussions.

in the text, whether or not the pattern occurs beginning at that location in the text. Several efficient deterministic sequential and parallel algorithms (See [Ah89] for a survey, [KMP77,BM77,GS83,We73,CP91] for sequential algorithms and [CG+93,Ga92,BG90,Ga85a,Vi85,Vi90] for parallel algorithms) are known for string matching. These algorithms commit no errors in outputting all the locations where the pattern matches the text. Karp and Rabin [KR87] presented a probabilistic string matching algorithm which is simple and efficient; however it commits errors. In the match vector output by their algorithm, the matches indicated might be *false*, i.e., the pattern might not occur beginning at a location in the text indicated as a match. However, it is guaranteed that the pattern does not occur beginning at those locations in the text which are indicated as mismatches. In this paper, we investigate the complexity of detecting the false matches, if any, in the given match vector for a given text and pattern.

The motivation for detecting false matches arises partly from the task of converting the Monte Carlo type string matching algorithm of Karp and Rabin [KR87] into a Las Vegas type algorithm. Given a Monte Carlo type algorithm for a problem $\mathcal{P}$ (string matching, in our case), an algorithm that detects errors (false matches, in our case) in the output of the Monte Carlo type algorithm and that which is provably simpler than any algorithm for $\mathcal{P}$, can be utilized to derive a Las Vegas type algorithm for $\mathcal{P}$. Also, checking for errors in string matching algorithms has inherent interest in view of the recently formalized notion of program checking [BK89]. In addition, as we shall show, it helps address a notion of approximate string matching. In the standard notion of approximate string matching, all those positions in the text are sought, where the pattern mismatches the text by at most k character-by-character mismatches, for a parameter k. An alternate notion of approximate string matching is to allow a few false matches in the output. This leads to a question partly addressed in this paper – can string matching be performed more efficiently, if a few false matches are tolerated?

First, we consider the problem of *checking* if there exists a false match in the match vector. Checking can be done in linear work by simply running any optimal parallel string matching algorithm [CG+93,BG90] on the given text and pattern and comparing the output with the given match vector. However, in the spirit of converting a Monte Carlo algorithm to a Las Vegas one, we would like a procedure for checking which is provably simpler than string matching itself. Utilizing the crucial observation that a series of "structured" matches can be checked quickly without explicitly considering each of them, we present a parallel algorithm which runs in $O(1)$ time on the CRCW PRAM using $O(n)$ processors. This algorithm is simple and it does not involve preprocessing the pattern. Note that string matching takes at least $\Omega(\log \log m)$ time on the CRCW PRAM with $O(n)$ processors [BG91]. Hence, checking for the occurrence of a false match in string matching algorithms is *provably* simpler than string matching itself.

Using this algorithm that checks for false matches, we convert the Monte Carlo type string matching algorithm of Karp and Rabin [KR87] to a Las Vegas type algorithm. Karp and Rabin gave a parallel algorithm for string matching that performs linear work, denoted $\mathcal{A}$, which may output false matches with a small probability. To remove the occurrences of the false matches in the output, they suggested naively verifying every match indicated by their algorithm. Thus the resultant algorithm $\mathcal{B}$ that does not commit errors in the form of false matches, performs $O(n + m + (\#M + \#F)m)$ work (informally, for sequential algorithms, the total time taken is the work performed) where $\#M$ denotes the number of the matches of the pattern in the text and $\#F$ denotes the number of the false matches in the output of the algorithm $\mathcal{A}$. In the worst case, this is no better than the naive string matching algorithm. By combining our algorithm that checks for an occurrence of a false match with their linear work algorithm $\mathcal{A}$ which might output false matches, we derive a string matching algorithm that, on any input, performs $O(n + m)$ work with a high probability, never outputs false matches and detects all the occurrences of the pattern in the text.

Next, we consider the problem of identifying all the occurrences of the false matches in the match vector. We present a parallel algorithm for this problem which works in $O(\#F_m)$ time using $O(n)$ processors on the CRCW PRAM where $\#F_m$ denotes the maximum number of consecutive false matches in any substring of length $m/2$. We use this as a reduction to show that any string matching algorithm which outputs $o(\log \log m)$ consecutive false matches using $O(n)$ processors while making no errors in finding the locations where the pattern does not match, has to take $\Omega(\log \log m)$ time on the CRCW PRAM.

It has been of considerable theoretical interest to determine the inherent complexity of string matching. In particular, there have been several investigations on the complexity of string matching on simple devices like finite state automaton (See [Ga85b,GS83]). We investigate the relative complexity of checking for false matches and string matching using deterministic finite state automaton (DFA). Such an investigation gives intuition into the complexity of string matching on DFAs. We show that checking for any false matches can be accomplished using three heads on a 2-way DFA in linear time. Note that string matching is not known to be performable with fewer than six heads on a 2-way DFA in linear time [GS83]. We also demonstrate that checking can be performed by a 1-way DFA with a fixed number of heads when the text size and the pattern size are comparable and the heads can not sense if they coincide. It was a longstanding conjecture that a 1-way DFA can not perform string matching with a fixed number of heads [Ga85b]. Recently, Jiang and Li affirmed this conjecture when the text size n is significantly larger than the pattern size m, say $n = \Omega(m^3)$ and the heads can not sense if they coincide. Hence, the following version of the original conjecture remains open and is of considerable interest:

Can string matching be performed on a 1-way DFA when $n \leq 2m$ and the heads can not sense if they coincide?

If the heads can sense when some of them coincide, then the original conjecture of Galil that string matching can not be performed by a 1-way DFA with fixed number of heads, remains open. We show that 1-way DFA with fixed number of heads can perform checking for arbitrary n and m if the heads can sense when they coincide.

The rest of the paper is organized as follows. In Section 2, the preliminaries and two main observations are presented. In Section 3, a simple algorithm for checking is outlined which is used in Section 4 to provide Las Vegas type string matching. In Section 5, the complexity of identifying all false matches is considered. Finally, algorithms for checking on finite state machines are presented in Section 6.

2 Problem Definition And Preliminaries

Let t be the text string of length n, p the pattern string of length m and M a binary vector of length n called the *match vector*. $M(i) = 0$ if p does not match t beginning at the location i. If $M(i) = 1$, i is a *potential match* location. A potential match i at which p does not match t, is called a *false match*. Note that $M(i) = 0$ for $n - m + 2 \leq i \leq n$. We consider two problems:

Checking: Given t, p and M, determine if there exists a false match.
Identifying: Given t, p and M, output $N[1 \ldots n - m + 1]$ such that $N[i] = 1$ if i is a false match and 0 otherwise.

Consider the auxiliary problem of **verifying**: Given a location i in the text, determine if the pattern matches the text beginning at i. Our algorithms use algorithms for verifying as "black boxes".

The combinatorial structure of the strings is utilized. A period length of a string s is k if $s[i + k] = s[i]$ for $i = 1, \ldots, m - k$. The minimal such k that is less than $m/2$ is henceforth called the *period length* [Ga85a,Vi85]. The *period* of the string is $s[1] \ldots s[k - 1]$. The period length and the period are also defined alternately (and equivalently) as follows [Ga85a,Vi85]: u is the period of s if $s = u^k \overline{u}$, $k > 1$, $\overline{u}$ is a prefix of u and $|u|$, the period length, is minimal. The following basic lemmas are proved elsewhere [Ga85a,Vi85].

Lemma 1 (GCD-lemma) *[LS62] If $l1$ and $l2$ are two period lengths of a string s such that $|s| \geq l1 + l2$, then $\gcd(l1, l2)$ must be a period length of s.*

Lemma 2 *If p occurs in t beginning at i and at $i+d$, $1 \leq d < m/2$, and nowhere in between, then d is the period length of p.*

Lemma 3 *If p, with a period length d, occurs in t beginning at i and $i + j$ such that $j < m/2$, then j is a multiple of d.*

We divide the text into disjoint segments of length $m/2$ and each segment is called a *block*. Assume that $n = k(3m/2)$, $k \geq 1$. Unless otherwise stated, our algorithms can be modified to handle the other values of n. The following observation would be often used.

Lemma 4 *[FRW88] Given a boolean vector of length l, the first, second and the last 1 in this vector can be found in $O(1)$ time using $O(l)$ processors on a CRCW PRAM.*

Now we make two observations that are useful in checking and identifying false matches quickly. Consider two potential matches in the text at i and $i + kd$ such that d is the period length of the pattern and $kd < m/2$.

Lemma 5 (Quick-Match Lemma) *If the pattern matches at i and $i + kd$, it matches at $i + ld$ for $1 \leq l \leq k - 1$.*

Proof: Let the pattern be $u^j \overline{u}$ as in the definition. Since it matches at i and $i + kd$, the text string at i contains $u^k u^j \overline{u}$ which is $u^l u^j u^{k-l} \overline{u}$ for any l, $1 \leq l \leq k - 1$. Thus the pattern matches at $i + ld$. $\square$

Let the pattern occur beginning at i in the text and let d be the period length of p. Also, let $i + kd + f$ be the smallest position in which the pattern placed on the text beginning at $i + kd$, $kd < m/2$, does not match the text.

Lemma 6 (Quick-Identify Lemma) *The pattern matches the text beginning at a position $i+ld$ in the text for $1 \leq l \leq k$, if and only if $i+ld+m-1 < i+kd+f$.*

Proof: From the definition of the period, it follows that when copies of the pattern are placed on the text beginning at $i + ld$, $1 \leq l \leq k$, all the positions which fall on each other contain the same symbol. Hence, none of the copies which fall on t_{i+kd+f} match the text. Any copy on $i + ld < i + kd + f - (m - 1)$ does not fall on t_{i+kd+f} and every other copy falls on t_{i+kd+f} (See Figure 1). $\square$

Given a sequence of "structured" potential matches, the Quick-Match Lemma implies that it is sufficient to verify the occurrence of the pattern at the last location in the sequence to ascertain if the entire sequence contains a false match. The Quick-Identify Lemma implies something stronger: by looking at the first position, where the pattern placed beginning at the last location in the sequence does not match the text, we can *precisely* determine all the false matches in the sequence.

3 Parallel Checking

Assume the common CRCW PRAM model of computation, i.e., all the processors which attempt to write into a memory location, write the same value [Ja91]. Note that the problem of verifying the occurrence of the pattern at a location in the text can be performed using m processors in $O(1)$ time on a common CRCW PRAM. Naively, the problem of checking can be solved by verifying each potential match in parallel and thereby performing $O(nm)$ work in the worst case. We perform the checking with linear work by avoiding verifying every potential

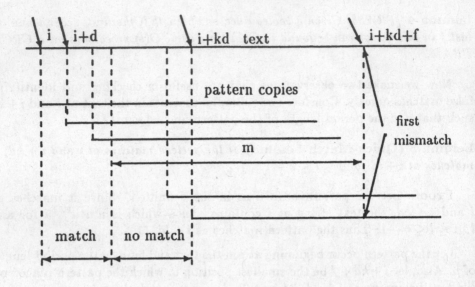

Figure 1: Quick-Identify

match in the match vector. Henceforth assume that each block under consideration has at least two potential matches. Otherwise, we can verify each potential match in the block naively. Recall that a block is a substring of length $m/2$.

Algorithm Par-Check /* Output is set to 1 if a false match is found and 0 otherwise */

Set Output to 0. Divide the text into $2n/m$ disjoint blocks. For each block do in parallel:

1. Let $M[i]$ and $M[j]$ be the first and the second 1 in the block respectively. If the pattern does not match the text at either i or j, set Output to 1 and exit.

2. Let $d = j - i$. If the distance of every potential match from i is not a multiple of d, set Output to 1 and exit.

3. Let $M[k]$ be the last 1 in the block. If the pattern does not match the text beginning at k, set Output to 1.

Lemma 7 *Algorithm Par-Check correctly detects the occurrence of any false match.*

Proof: Within each block, if the first two potential matches i and j are indeed matches, then the period length is $d = j - i$ (Lemma 2). The match positions are multiples of the period length away from i (Lemma 3 and Quick-Match Lemma). Consider the series of potential matches satisfying this property in which the first occurrence i is indeed a match. By the Quick-Match Lemma, if the last occurrence is a match, every potential match in the series is a match. Note that if there is a failure in any of the steps, a false match is found. □

Theorem 1 *There exists an algorithm for checking that takes $O(1)$ time using $O(n)$ processors on the Common CRCW PRAM.*

Proof: Consider Algorithm Par-Check. Each of the three verifications per block takes $O(1)$ time using $O(m)$ processors. Other steps take $O(1)$ time using m processors (Lemma 4). Hence each block can be checked using $O(m)$ processors in $O(1)$ time. There are $O(n/m)$ such blocks in all. □

Note that, to perform checking in constant time optimally, we have only used the information about the period length of the pattern gleaned from matching the pattern at various positions in the text. In particular, our algorithm does not involve preprocessing the pattern. Let $n = 3m/2$. Breslauer and Galil [BG91] prove that any string matching algorithm takes $\Omega(\log \log m)$ time with $O(n)$ processors on the CRCW PRAM. That Algorithm Par-Check takes a constant time with optimal work implies that checking is *provably* simpler than string matching on the CRCW PRAM.

Note that the bound of Breslauer and Galil [BG91] includes the cost of preprocessing the pattern. Following preprocessing which takes $O(\log \log m)$ time, text processing can be done in $O(1)$ time optimally [CG+93]. However, even with this preprocessing, deterministic optimal parallel algorithms in $O(\log m)$ time are not known for text processing on the Exclusive-Read-Exclusive-Write (EREW) PRAM [Ja91].

Theorem 2 *There exists an algorithm that checks for false matches in $O(\log m)$ time and $O(n)$ work on the EREW PRAM.*

Proof: In Algorithm Par-Check, all the tasks that take $O(1)$ time on the CRCW PRAM can be replaced by $O(\log m)$ time algorithms on the EREW PRAM with optimal work by using the standard binary tree method [Ja91]. □

4 Las Vegas String Matching

Karp and Rabin [KR87] present a Monte Carlo algorithm (referred to as Algorithm MC henceforth) for string matching.

Lemma 8 *[KR87] By randomly choosing a prime $\leq mn^k$, Algorithm MC finds all the matches in $O(\log m)$ time and $O(n)$ work on the EREW PRAM with the probability that a false match occurs on any input instance being at most c/n^{k-1} for a small constant c.*

Our Las Vegas string matching algorithm (referred to as Algorithm LV henceforth) works as follows: it repeatedly runs Algorithm MC checking the output using Algorithm Par-Check till the output contains no false matches.

Theorem 3 *Algorithm LV correctly finds all the matches with high probability in $O(\log m)$ time and $O(n)$ work on the EREW PRAM.*

Proof: Let $k = 2$ in Lemma 8. The probability of occurrence of a false match is at most c/n. The probability that Algorithm MC is run exactly $r + 1$ times is at most c/n^r. Thus the probability that MC is run 3 or more times is at most $\sum_{r \geq 2} c/n^r$. This probability is at most $c/n(n-1)$ which tends to 0 as n tends to infinity. Hence with a high probability, Algorithm MC is run $O(1)$ times. The theorem follows from Lemma 8. □

A sequential simulation of Algorithm LV on a RAM gives a string matching algorithm which works in linear time with a high probability and detects all the occurrences of the pattern in the text without outputting any false matches. Note that Karp and Rabin [KR87] had suggested naively verifying all the potential matches so as to avoid outputting false matches. Let $\#M$ be the number of occurrences of the pattern in the text and $\#F$ be the number of false matches in the output of Algorithm MC. Their algorithm takes $O(n + m + (\#M + \#F)m)$ time to find all the occurrences of the pattern in the text. In the worst case, $\#M = O(n)$ and their algorithm is no better than the naive algorithm for string matching even though the probability of Algorithm MC outputting false matches is provably low. The sequential simulation of our Algorithm LV works in linear time with a high probability irrespective of $\#M$. Furthermore, it retains the appealing simplicity of Algorithm MC.

5   Identifying in Parallel

In this section, we consider the problem of identifying all the false matches. The Quick-Identify Lemma is used crucially.

Algorithm Par-Identify: For each block in parallel do:

1. Verify each of the potential match locations in succession till the first two matches i and j are found.

2. Let $d = j - i$. As in Algorithm Par-Check determine that every potential match is a multiple of d away from i. Every other potential match is false.

3. Let k be the last potential match in the $m/2$ block. Find the smallest position $k + f$ where the pattern placed at k does not match the text. Every potential match l such that $l + m - 1 \geq k + f$ is false.

Theorem 4 *Let $\#F_m$ be the maximum number of consecutive false matches in any block. There exists an algorithm that detects all the false matches in $O(\#F_m)$ time using $O(n)$ processors on the common CRCW PRAM.*

Proof: At most $2(\#F_m)$ potential matches are verified in Algorithm Par-Identify before the first two matches in a block are found. Rest of the algorithm, including finding the first mismatch in Step 3, takes $O(1)$ time using $O(n)$ processors (Lemma 4). Correctness follows from the Quick-Identify Lemma. □

Theorem 5 *Let* $n = 3m/2$. *Any string matching algorithm* $\mathcal{A}$ *which outputs at most* $\#F_m$ *consecutive false matches where* $\#F_m = o(\log \log m)$, *takes* $\Omega(\log \log m)$ *time using* $O(n)$ *processors on the CRCW PRAM.*

Proof: Let algorithm $\mathcal{A}$ take $o(\log \log m)$ time using $O(m)$ processors on the CRCW PRAM. Given any input text and pattern, run algorithm $\mathcal{A}$ first. Subsequently run Algorithm Par-Identify on the text, the pattern and the output and remove all the false matches thereby achieving a string matching algorithm in time $o(\log \log m)$ using $O(m)$ processors on the CRCW PRAM. But since it is known that string matching takes $\Omega(\log \log m)$ time using $O(m)$ processors on the CRCW PRAM [BG91], we derive a contradiction. □

Thus the claim that the false matches in the output of a string matching algorithm can be detected and removed efficiently implies string matching takes $\Omega(\log \log m)$ time even when given the flexibility to output a few false matches.

6 Sequential Checking

The complexity of string matching on finite state machines has been studied extensively [Ga85b] in an attempt to determine the "simplest" machine that can perform string matching efficiently. In this section, we study the complexity of checking for false matches using deterministic finite state automaton (DFA). Depending on whether the heads in the DFAs can sense if they coincide or not, we have, respectively, *sensing* and *nonsensing* DFAs. Note that sensing DFAs are at least as powerful as nonsensing DFAs.

6.1 Checking on a 2-Way DFA

Assume that input is given as follows: $p_1 p_2 \ldots p_m @M(1)t_1 M(2)t_2 \ldots M(n)t_n \&$ where @ and & do not appear in the text or the pattern. We simulate the sequential version of the Algorithm Par-Check using 2-way DFA. Verifying the occurrence of p at i in t is done as follows: let h_1 and h_2 be at t_i and p_1 initially. Move the heads h_1 and h_2 on text and pattern respectively and perform character by character comparisons. When the occurrence is verified, both the heads are dragged back to their initial positions. Thus, verifying a location takes $O(m)$ time. At various places in our algorithm, we would be required to move some head say h_1 a distance at most $m/2$ to the right. This is ensured by using a "count" head say h_2 which scans two cells on the pattern for every move to the right made by h_1.

The 2-way DFA which checks the given input has three heads denoted by $\mathcal{H} = \{h_a, h_b, h_p\}$. The two heads h_a and h_b moving on the text are referred to as text heads. The head h_p moving on the pattern is referred to as the pattern head. The pattern head is used in two ways: first, to scan the pattern symbols when text locations are verified, and second, to count the displacement of the text heads. The text head h_a moves on the text and the head h_b usually trails behind. The movements of these heads are now described. To begin with, h_p scans p_1 and, h_a and h_b scan $M(1)$.

1. Move h_a and h_b to the first potential match f in the text. Verify f using h_a and h_p.

2. Move h_a to second potential match s at most $m/2$ to the right of f. Use h_p to count the displacement of h_a. If no potential matches are found within a displacement of $m/2$, h_p is brought back to p_1 and go to Step 1 to continue searching other blocks. If s is found, move h_p back to p_1. Verify s using h_a and h_p.

3. The heads h_a and h_b scan s and f respectively. The head h_p scans p_1. Move h_a and h_b across the text simultaneously such that h_a does not move more than $m/2$ locations to the right of f. This is ensured using h_p.

   (a) Before $m/2$ text symbols are scanned by h_a: If h_a and h_b scan a potential match simultaneously, continue. If h_a scans a potential match while h_b does not, signal a false match. If h_b scans a potential match while h_a does not, move h_a past the end of the $m/2$th text symbol using h_p and insure that no potential matches are encountered enroute. Goto Step 4.

   (b) Assume h_a has passed scanning the text symbol $m/2$ locations to the right of f. Move h_b right till a potential match is encountered. It is easily seen that such a potential match will be encountered before h_b hits h_a, i.e., within $m/2$ locations to the right of f.

4. The head h_b is left scanning the last potential match l in the block of size $m/2$ from f. Move h_p back to p_1 and verify l using h_b and h_p.

5. If no false matches have been detected so far, the head h_b scans l, h_p scans p_1 and the head h_a scans the location $m/2+1$ to the right of f. Move h_b to coincide with h_a as follows: first, move h_a left to l while simultaneously moving h_p forward. When h_a scans l, h_p is to the right of p_1 by an amount which represents the distance of h_b and h_a from the location $m/2+1$ to the right of f. Move h_a and h_b right while simultaneously moving h_p left till h_p hits p_1. Now, h_a and h_b scan $m/2+1$th symbol to the right of f. Go to Step 1 and continue checking the remaining text locations.

Theorem 6 *Checking can be performed by a 2-way DFA using three heads in linear time.*

Proof: It is easily seen that within a block of length $m/2$ from the first potential match location, the first, the second and the last potential matches are verified. Also, h_a and h_b move simultaneously to check that successive potential match locations are separated by the period length. Hence, the block beginning at the first potential match in the text is checked correctly. Now, the argument is applied to subsequent blocks. In each block, the work done is $O(m)$ and there are $O(n/m)$ blocks in all. □

Note that linear time string matching is not known to be performable with fewer than six heads on a 2-way DFA [GS83]. The above theorem states that checking can be done on a 2-way DFA with substantially fewer number of heads. However, nothing can be said about the relative complexity of checking and string matching on a 2-way DFA without stronger bounds for string matching and checking on 2-way DFA.

It follows from Theorem 6 that three extra storage locations are sufficient to perform checking on a RAM in linear time: each head is simulated at a separate storage location. Again by analogy to string matching on 2-way DFA, string matching is not known to be performable on a RAM in linear time with fewer than six extra memory locations.

Remarks:

1. We have assumed that the match vector and the text are interleaved in the input. Now assume otherwise: let the input be presented as $p@t@M$. In this case, four heads are sufficient to perform the checking. Essentially use three heads as described earlier except that the heads moving on the text, now move on the match vector. Use an additional head moving on the text to verify locations.

2. Note that our algorithm works on a *nonsensing* 2-way DFA.

6.2   Checking on a 1-Way DFA

Assume $n = 3m/2$. We can simulate the 2-way DFA described in the previous section which performs the checking, using several additional heads on a 1-way DFA. Recall verifying a location i using a 2-way DFA. Verifying the location i by a 1-way DFA is performed similarly except that the heads are not moved back to their initial position:

1. Move h_1 to the first potential match location i while counting its movements using h_2. Verify i using h_1 and h_3. Use three more heads to do the same for the second potential match.

2. Move h_f and h_b to the first and the second potential match location respectively. Subsequently move h_f and h_b simultaneously as with 2-way DFA. Use h_c to count the movement of h_f.

3. When h_b is scanning the last potential match in the first block, use h_b and an extra head h_e to verify that location.

Theorem 7 *Checking can be performed by a 1-way DFA using ten heads.*

Note that the various heads in our algorithm are *nonsensing*. Since the heads moving on the pattern can not be reused for subsequent blocks by moving them back to the first pattern symbol, this theorem is true only when $n \leq 3m/2$. It is not clear if checking can be performed by a 1-way DFA when n is significantly larger than m and the heads are nonsensing.

We remark that by carefully simulating the algorithm in Section 6.1 on a 1-way DFA, the number of heads in the above theorem can be decreased substantially. However, it is not relevant to our goal of comparing the complexity of string matching and that of checking given our present knowledge. It was a longstanding conjecture that string matching can not be performed by a 1-way DFA using a fixed number of heads [Ga85b]. Jiang and Li [JL93] have recently proved that if the text length n is significantly larger than m (say, $n = \Omega(m^3)$) and the heads are nonsensing, then in fact, this conjecture is true. The following two questions remain open in this context:

Q1. Can string matching be performed on a 1-way DFA when $n \leq 3m/2$ and the heads are nonsensing?

Q2. Can checking be performed on a 1-way DFA when $n >> 3m/2$ and the heads are nonsensing?

We believe that the answer to both these questions is "no" which would imply that checking is provably simpler than string matching on 1-way DFAs. If the heads are sensing, then the original conjecture of Galil and in particular, the analog of $Q1$ remains open. However, the analog of $Q2$ when the heads are sensing is easily settled. We claim that an algorithm on a 1-way DFA for checking when $n \leq 3m/2$ can be immediately converted to one for any n provided the heads are sensing. The intuition behind this claim is that if the heads are sensing, then copies of the pattern in a portion of the text can be used to match subsequent portions of the text. We formalize this claim below.

Consider an algorithm $\mathcal{A}$ that takes a portion of a string of length m denoted by s_1 and another portion s_2 of length at most $3m/2$ and finds all the occurrences of s_1 in s_2 using $\mathcal{H}(m)$ heads on a 1-way DFA. At the completion of this algorithm, all the heads on s_1 scan the symbol past the last symbol in s_1 and all the heads on s_2 scan the symbol past the last symbol in s_2. Then,

Theorem 8 *There exists an algorithm that performs checking on a text of length n, a pattern of length m and a match vector of length n using $O(\mathcal{H}(m))$ heads for any n.*

Proof: The algorithm $\mathcal{A}$ is used repeatedly. Consider three disjoint sets of heads of size $\mathcal{H}(m)$ denoted S^1, S^2 and S^3. The heads in S^1, S^2 and S^3 check the potential matches in the following blocks respectively,

1. $1 \cdots m/2, \quad 3m/2 \cdots 3m/2 + m/2, \quad \cdots \quad , \quad 3mk'/2 \cdots 3mk'/2 + m/2$

2. $m/2 \cdots m, 2m \cdots m/2+2m, \cdots, m/2+3mk''/2 \cdots m/2+3mk''/2+m/2$

3. $m \cdots 3m/2, m+3m/2 \cdots m+2m, \cdots, m+3mk'''/2 \cdots m+3mk'''/2+m/2$

where k', k'' and k''' are integers such that the blocks are within the text. We now describe how checking is performed using the set S^1; checking with S^2 and S^3 are similar.

We loosely maintain the following invariant at each stage of the algorithm: two heads h_1 and h_2 are maintained at the beginning and at the end of a substring of the input such that the string between the two heads is a copy of the pattern. We also maintain three heads h_3, h_4, h_5 on a substring of length $3m/2$ such that h_3 and h_4 are at the beginning and h_5 is at the end. These three heads mark off portions of the text in which the occurrences of the pattern marked off by heads h_1 and h_2 will be found. Initially, h_1 and h_2 scan p_1 and p_m respectively. The heads h_3, h_4, h_5 scan $M(1), M(1)$ and $M(3m/2)$ respectively.

Let the pattern heads in S^1 be S_p^1 and the text heads be S_t^1. Position S_p^1 at h_1 and S_t^1 at h_3. Check if there exist any potential matches within first $m/2$ locations of h_3. If there exist none, move heads in S_t^1 to h_5 and move h_3, h_4 and h_5 such that the heads h_3 and h_4 scan $M(3m/2)$ and the head h_5 scans $M(3m/2 + 3m/2)$. This represents the next block to the checked by S^1. Thus, repeat the steps from the beginning on this block.

If there exists a potential match within $m/2$ locations to the right of h_3, the following series of steps are performed. Using S^1 as in Algorithm $\mathcal{A}$, find occurrences of p in the portion of t contained between h_3 and h_5. Any false matches are signaled appropriately. If there exist no false matches, at the end of this step, the heads S_p^1 and h_2 coincide and the heads S_t^1 and h_5 coincide. Note that initially h_2 is to the left of h_3. This invariant will be maintained at the end of following two steps. Move h_1 and h_2 such that h_1 scans the first potential match within $m/2$ locations to the right of h_3 and h_2 is m symbols to the right of h_1. Since no false matches were detected when Algorithm $\mathcal{A}$ was used, the string between h_1 and h_2 represents a copy of p. Now, move h_3 and h_4 to the location scanned by h_5 and move h_5 to a location $3m/2$ to the right. The three heads h_3, h_4 and h_5 represent the next block to be checked using S^1. Move the heads S_p^1 to h_1. At the end of invocation of Algorithm $\mathcal{A}$, the heads S_t^1 are left scanning the current position of h_3. Now, the entire procedure is repeated with the new substring of text marked off by the heads h_3, h_4 and h_5.

For each set of $\mathcal{H}(m)$ heads, a constant number of extra heads is used. There are three such sets in all. In total, $O(\mathcal{H}(m))$ heads are used. $\square$

It follows from Theorem 7 and 8 that checking a text of size n, a pattern of size m and a match vector of length n can be done using a fixed number of heads on a 1-way DFA provided the heads are sensing. As mentioned earlier, the analogous question for string matching remains open.

7 Discussion

We have shown that in the CRCW PRAM, checking for false matches in the output of string matching algorithms is *provably* simpler than string matching itself. We have also presented efficient algorithms on 2-way DFA and 1-way DFA that check for false matches. However, in the absence of appropriate lower bounds on the number of heads required on these models for string matching, we can not claim that checking for false matches is simpler than string matching on these models.

We have converted the Karp-Rabin Monte Carlo type string matching algorithm to a Las Vegas string matching algorithm. The fingerprinting technique of Karp and Rabin has several other applications [KR87] yielding Monte Carlo algorithms. It would be of interest to convert these into Las Vegas type algorithms.

References

[Ah89]   A. Aho. Algorithms for finding patterns in strings. *Handbook of theoretical computer science*, Vol 1, Van Leeuwen Ed., 1989.

[BG90]   D. Breslauer and Z. Galil. An optimal $O(\log \log n)$ time parallel string matching algorithm. *SIAM J. Computing*, 19:6, 1051-1058.

[BG91]   D. Breslauer and Z. Galil. A lower bound for parallel string matching. *Proc 23rd Annual ACM Symposium on Theory of Computation*, 1991, 439-443.

[BK89]   M. Blum and S. Kannan. Designing Programs That Check Their Work. *Proc 21st Annual Annual ACM Symposium on Theory of Computation*, 1989, 86-97.

[BM77]   R. Boyer and S. Moore. A fast string matching algorithm. *Communications of ACM*, 20(1977), 762-772.

[CG+93]   M. Crochemore, Z. Galil, L. Gasieniec, S. Muthukrishnan, K. Park, H. Ramesh and W. Rytter. Fast two dimensional/string pattern matching. *Manuscript*, 1993.

[CP91]   M. Crochemore and D. Perrin. Two-way pattern matching. *Journal of ACM*, Vol 38, 1991, 651-675.

[FRW88]   F. E. Fich, R. L. Ragde, and A. Wigderson. Relations between concurrent-write models of parallel computation. *SIAM J. Computing*, 17:1988, 606-627.

[Ga85a]   Z. Galil. Optimal Parallel Algorithms for String Matching. *Information and Control*, Vol. 67, 1985, 144-157.

[Ga85b]   Z. Galil. Open Problems in Stringology. *Combinatorial Algorithms on Words*, A. Apostolico and Z. Galil Eds, NATO ASI Series, Springer Verlag, 1985, 1-8.

[Ga92]   Z. Galil. Hunting lions in the desert optimally or a constant time optimal parallel string matching algorithm. *Proc. 24th Annual ACM Symposium on Theory of Computation*, 1992.

[GS83]   Z. Galil and J. Seiferas. Time space optimal string matching. *Journal Comput. Syst. Sci.* 26(1983), 280-294.

[HU79]   J. Hopcroft and J. Ullman. *Introduction to automata theory, languages and computation*. Addison-Wesley, 1979.

[Ja91]   J. JaJa. *Introduction to Parallel Algorithms*. Addison-Wesley, 1991.

[JL93]   T. Jiang and M. Li. *k* one way heads can not perform string matching. To appear in *Proc ACM Symposium on Theory of Computation*, 1993.

[KMP77] D.E. Knuth, J. Morris, V. Pratt. Fast pattern matching in strings. *SIAM J. Computing*, 6:1973, 323-350.

[KR87]   R. Karp and M.O. Rabin. Efficient randomized pattern matching algorithms. *IBM Journal of Research and Development*, 31(2), 1987, 249-260.

[LS62]   R. Lyndon and M. Schutzenberger. The equation $a^M = b^N c^P$ in a free group. *Michigan Math. J.* 9, 1962, 289-298.

[Vi85]   U. Vishkin. Optimal pattern matching in strings. *Information and Control*, Vol. 67, 1985, 91-113.

[Vi90]   U. Vishkin. Deterministic sampling – A new technique for fast pattern matching. *Proc 22nd Annual ACM Symposium on Theory of Computation*, 1990, 170-180.

[We73]   P. Weiner. Linear Pattern Matching Algorithms. *Proc. 14th IEEE Ann. Symp. on Switching and Automata Theory*, 1973. 1-11.

On Suboptimal Alignments of Biological Sequences

Dalit Naor * and Douglas Brutlag

Department of Biochemistry
Stanford University Medical Center
Stanford, California 94305-5307

Abstract. It is widely accepted that the optimal alignment between a pair of proteins or nucleic acid sequences that *minimizes the edit distance* may not necessarily reflect the correct *biological* alignment. Alignments of proteins based on their structures or of DNA sequences based on evolutionary changes are often different from alignments that minimize edit distance. However, in many cases (e.g. when the sequences are close), the edit distance alignment is a good approximation to the biological one. Since, for most sequences, the true alignment is unknown, a method that either assesses the significance of the optimal alignment, or that provides few "close" alternatives to the optimal one, is of great importance.

A suboptimal alignment is an alignment whose score lies within the *neighborhood* of the optimal score. Enumeration of suboptimal alignments [Wa83, WaBy] is not very practical since there are many such alignments. Other approaches [Zuk, Vi, ViAr] that use only partial information about suboptimal alignments are more successful in practice.

We present a method for representing all alignments whose score is within any given delta from the optimal score. It represents a large number of alignments by a compact graph which makes it easy to impose additional biological constraints and select one desirable alignment from this large set. We study the combinatorial nature of suboptimal alignments. We define a set of "canonical" suboptimal alignments, and argue that these are the essential ones since any other suboptimal alignment is a combination of few canonical ones. We then show how to efficiently enumerate suboptimal alignments in order of their score, and count their numbers. Examples are presented to motivate the problem.

Since alignments are essentially (s, t)-paths in a directed acyclic graph with (possibly negative) weights on its edges, our solution gives an extremely simple method to enumerate all K shortest (or longest) paths from s to t in such graphs in increasing order, as well as all (s, t) paths that are within δ of the optimum, for any δ. We compare this solution with known algorithms that find the K-best shortest paths in a graph.

* Supported by a Postdoctoral Fellowship from the Program in Mathematics and Molecular Biology of the University of California at Berkeley, under National Science Foundation Grant DMS-9720208

1 Introduction

1.1 Motivation

Protein and Nucleic Acid sequences are regularly compared agains one another, since it is assumed that sequences with similar biological function have similar physical characteristics, and therefore are similar, or homologous, at the amino or nucleic acid sequence level [D1, D2]. Since a good similarity measure between sequences that predicts the structural or evolutionary similarity is not yet known, biologists resort to the classical string comparisons methods [Sel, SK]. The most commonly used measure of similarity between strings is the well known edit distance measure, which is the minimum-weight set of edit operations (substitutions or insertions/deletions of gaps) that are needed to transform one sequence to another. The minimum edit distance alignment can also be expressed as the best scored alignment; throughout the paper we use the maximization terminology. Biological constraints are imposed via the weights that are assigned to the different edit operations (i.e., the PAM matrix for proteins). It is still controversial how to correctly derive these weights. With any set of weights, the edit distance measure optimizes a single simple objective function, and there is no evidence to believe that this function is being optimized by nature.

Despite these limitations, this technique has become the method of choice in molecular biology for aligning sets of sequences. The main reason is that structural or evolutionary alignments are very hard to find; they involve the determination of the three-dimensional structures of molecules (via crystallography, NMR techniques or phylogenies), which are extremely laborious tasks. String comparison techniques, on the other hand, are much faster and in many cases have proven to be good predictors of the correct alignment. However, lacking experimental data, it is difficult to assess the significance of an alignment that was obtained purely by a computational method.

In summary, the edit distance measure is a well-defined, rigorous combinatorial measure that can be efficiently optimized, and which can be justified biologically (by a set of parameters, whose values are determined empirically). However, the edit distance does not always yield the biological alignment when optimized. It is therefore useful to intelligently explore a larger set of solutions. One approach is the *Parametric* approach, which looks for different optimal solutions obtained for different sets of parameters [FS, GBN, WEL, Wa92]. The second is to consider a larger, but still a manageable, set of alignments that are in the vicinity of the optimum. These are "Suboptimal Alignments" [Zuk, ViAr, WaBy, Wa83, WaEg].

1.2 Summary of Results

In this paper we study the combinatorial nature of suboptimal alignments under a simple scoring function. For every Δ, we define the *minimum* set of edges E_Δ in the edit distance graph that includes *every* suboptimal alignment in the Δ neighborhood, and then show that this set of edges is *exactly* the union of certain types of alignments which we call "canonical" alignments. We argue that these alignments can be viewed as a *canonical* set for all suboptimal alignments, since any suboptimal alignment is a "combination" of few canonical ones. We denote by $\Delta_0 < \Delta_1 < \Delta_2 < \ldots$ the maximal sequence of values such that

$$E_{\Delta_0} \subset E_{\Delta_1} \subset E_{\Delta_2} \subset \ldots$$

and show that the information revealed by this set of graphs is much richer than merely a single optimal alignment, or a list of few suboptimal ones. It allows a *graphical* view of:

- suboptimal alignments with almost optimal score that are *substantially different* than the optimal alignment.
- the regions that are *common to many* suboptimal alignments, which may indicate that these regions are significant [Zuk, Vi, ViAr].
- the best alignment which satisfies additional biological constraints.

We developed a program, called SUBOPT, that computes and displays these graphs, and counts the number of δ-suboptimal alignments for each δ.

We also define a transformation on the weights of the edit distance graph with which suboptimal alignments can be output efficiently (ordered or unordered). Let n and m be the lengths of the sequences, $m \geq n$. Given the transformed weights, the next best alignment (canonical or non-canonical) can be output in $O(m)$ time. If alignments are to be enumerated in increasing order, then the space requirement is $O(Km+mn)$; otherwise, $O(mn)$ space suffices. We also show how to efficiently *count* the number of suboptimal alignments, rather than enumerate them. This method is extremely simple. It requires only one additional edit distance computation between the reversed strings. Computing the edit distance between the sequences, and between the reversed sequences, was used in the algorithms of [KIM, Vi, ViAr, Wa83, WaBy, Zuk]. This information is sufficient to represent canonical as well as non-canonical suboptimal alignments and can be obtained in $O(nm)$ time and space.

Examples To motivate this problem, we show two examples. The first demonstrates that the biologically correct alignment is not necessarily the optimal one. The second example shows how a compact representation of a *set* of alignments between a *single* pair of sequences may reveal a wealth of information that is traditionally obtained from a multiple alignment of a large set of sequences. In both examples an alignment is represented by a path in a grid graph that starts at the upper-left corner and ends at the bottom-right corner. A diagonal edge corresponds to a pair of aligned characters, and a horizontal or a vertical line corresponds to a gap in one of the two sequences.

To demonstrate how misleading the optimality criteria can be, consider the example in Figure 2. Here, two 25-long amino acid sequences are compared against each other. The two are prealigned substrings of longer "Leucine Zippers" sequences, in which amino acid Leucine (L) appears every 7 positions. In this example, L appears at the 4th, 11th, 18th and 25th positions of the two sequences, so we expect a good alignment to pick up this periodicity. Hence, we define a biologically "correct" alignment as one that coincides with the diagonal at the 4th, 11th, 18th and 25th positions.

When the two sequences were compared (with a PAM80, gap score -1) the best score was 43. All alignments with this optimal score are represented as paths in the top part of Figure 2. Note that no optimal alignment is the biologically "correct" one under our definition. The bottom part of Figure 2 shows that when the score drops by 2, another large set of paths which are closer to the diagonal are introduced, among them a biologically "correct" one.

In the second example, the variable regions of the heavy and the light chain of a Human Immunoglobulin F_{ab} (the first 120 amino acids of PDB sequence 2fb4_H and 2fb4_L) have been aligned (Figure 3). The graph that represents optimal alignments reveals that all optimal alignments agree on three well-aligned regions ("diagonals"). This phenomena is reinforced by the consideration of suboptimal alignments (for $\delta = 0, 1, 2$), as they all share the same well-aligned regions, and introduce more alternatives at the remaining parts of the alignment. The fact that these two sequences are conserved in three region and are variable between them is the well known fact of "hypervariable regions" in Immunoglobulins. Hence, the hypervariable regions which were discovered in the '70s by aligning *multiple* immunoglobulin sequences can be obtained from this *single* pair of sequences and their suboptimal alignments. Since the Human F_{ab} has been structured, we located the turns that correspond to the hypervariable regions, and correlated them with those observed in the suboptimal graphs. In the heavy chain, the turns occur at positions 27-32, 52-55, 72-75 and 99-111, and the corresponding regions revealed by the suboptimal graphs are 24-32, 47-60, 77-79 and 97-111.

1.3 Problem Definition

Let A and B be two sequences (over a finite alphabet) of lengths m and n respectively, $m \geq n$. Let A_i (B_i) denote the first i characters in A (B). An alignment between A and B introduces spaces into the sequences such that the lengths of the two resulting sequences are identical, and places these spaced sequences one upon the other so that no space is ever above another space. The length of an alignment is between m and $n + m$. A column that contains two identical characters is called a *match* and is assigned a high positive weight *mat*. A column that contains two different characters a and b is called a *mismatch*, and is assigned a weight w_{ab} ($w_{aa} = mat$). A column that contains a space is called a *space*, and is assigned a small weight of *sp*. We define the score of an alignment as the sum over all weights assigned to its columns. (There are many possible definitions for the alignment score – in this paper we only consider this simple definition [2]). The edit distance problem is to find the alignment that maximizes its score.

---

[2] our results hold for any scoring function with the property that $D(m,n) - D(i,j)$ is the score of optimal alignment between the last $m - i$ characters from A and the last $n - j$ characters from B

If $D(i,j)$ is the score of the best alignment between A_i and B_j, and if the ith and j^{th} characters of A and B are a and b respectively, then the following dynamic programming formulation will correctly compute $D(i,j)$ [SK]:

$$D(i,j) = \max\{D(i-1,j) + sp, D(i-1,j-1) + w_{ab}, D(i,j-1) + sp\}$$

All values $D(i,j)$ can be computed in $O(nm)$ time, as well as $D(m,n)$, which is the desired score of the optimal alignment. Also, **all** optimal alignments can be found by backtracing through the matrix of the $D(i,j)$ values.

This problem can be viewed as a longest path problem on a directed acyclic graph G with nm nodes and $3nm$ edges. A node in G corresponds to some cell (i,j) in an $(m \times n)$ table, and each node (i,j) has three edges coming into it from adjacent cells $(i-1,j), (i-1,j-1), (i,j-1)$. The weight of a horizontal or a vertical edge is sp, whereas the weight of a diagonal edge is w_{ab}, a and b are the ith and jth characters of A and B respectively. This graph is called "the edit distance graph of A and B". If node s corresponds to the cell $(0,0)$, and node t corresponds to the cell (m,n), then there is a 1:1 correspondence between alignments and $s - t$ paths in G. Hence, suboptimal alignments are essentially suboptimal paths in this simple graph.

By the graph analogue of the problem, it is well known that all optimal alignments between A and B (that is, optimal paths from s to t) have a compact representation: it is the backtrace graph, which can be constructed in $O(nm)$ time. (This property does not directly carry over for other definitions of scores, e.g. affine gap weights [Go, AlEr]). However, no such elegant representation exists for suboptimal paths, and therefore alignments. Throughout the paper we use the graph notation, where a node (i,j) is simply denoted by u, s is the node $(0,0)$ and t is (m,n). An edge, directed form u to v, is denoted by $\langle u, v \rangle$.

1.4 Previous Work

The problem of enumerating all K-best shortest paths (or longest paths in a DAG) in increasing order was addressed very early on (in the 60's and 70's) in the context of general graphs, and it is a well studied problem in combinatorial optimization. In general, the problem may or may not allow paths with loops. We do not attempt to give a complete overview of this problem, but rather to summarize some relevant ideas that have been suggested. Three main approaches have been suggested. The first is an iterative procedure, that generalizes the Bellman-Ford algorithm for optimal paths, suggested by [BK, Dr] and improved by [Fox, Law76]. The second employs a general scheme for enumerating the K best objects [Yen, Law72, Law76, Pol]. The third exploits the structural relations between suboptimal paths. It is based on the observation of [HP] that the kth-best path P is a "deviation" of some j^{th}-best path Q, $j < k$, i.e. there is an edge $e = \langle u, v \rangle$ on P such that P's segment up to u is the same as Q, $e \notin Q$, and the portion of P from v to t is an optimal path from v to t. The algorithms of [CKR, HP, KIM, Per, Yen] all show how to store the $k - 1$ best paths efficiently so that the kth best, which must be a deviation from one of them, can be found easily. All of these approaches do not require the knowledge of K in advance; the fourth approach of [Sh] relies on a known K.

The edit distance graph is a very regular graph. It is a directed (acyclic) grid of degree 3, and its edge-weights may be negative. For this type of graph, the algorithm of [BK, Dr] as implemented by [Fox, Law76] enumerates the K-best suboptimal paths in $O(mn)$ time per path, where K need not be specified in advance. In fact, it finds the K-best suboptimal paths from any node in the graph to t. The space required is $O(Kmn)$. The algorithm of [Yen,Law72] generates the next best path in $O(Knm)$ time, and $O(Km + nm)$ space.

A more relaxed goal is suggested in [Wa83, WaBy]: enumerate all shortest paths (not necessarily in order) that are within δ of the optimum, where δ is known in advance. For this problem, each path can be enumerated in $O(m)$ time, but only $O(nm)$ space is needed. The requirement to know δ in advance is a limitation, since it is not possible to know a priori which level of significance is the interesting one. Another variant, considered in [WaEg] and used in [ScWa], is to list all K-best non-overlapping local alignments. These enumeration methods (of K-best or δ-suboptimal) turn out not to be practical in the biological application, since the number of suboptimal paths grows very fast, and explicit listing of them provides too much information.

In realizing that explicit enumeration is not the desired representation, both [Zuk] and [Vi, ViAr] suggested building $(m \times n)$ 0-1 matrices S, T that store some *partial* information about all δ-suboptimal alignments. S and T are easily computable in $O(nm)$ time, but in both methods δ needs to be specified in advance. $T(i, j) = 1$ iff there is a δ-suboptimal alignment in which the ith character of A and the jth character of B are aligned, and $S(i, j) = 1$ iff in *every* δ'-suboptimal alignment, $\delta' < \delta$, the ith character of A and the jth character of B are aligned. These matrices are then used to assess significance of optimal alignments [ViAr, Zuk], to detect alternatives to the optimal one and to construct multiple alignments [Vi]. This representation, although powerful, looses "connectivity information" since, clearly, not every legal path through the 1-entries of the matrices is a suboptimal alignment.

It is therefore clear that it is not the actual enumeration of suboptimal alignments, but rather a representation of their common or uncommon features, that is needed. This is the motivation of the representation suggested in this paper, which contains more information than the partial 0-1 matrices of [Zuk, Vi, ViAr], and yet is compact. We believe that this additional information, which is manageable, can be valuable in many cases. Also, the transformation of the edge weights suggested in the paper provides an alternative intuitive way to explore the search space of suboptimal alignments.

2 The Combinatorial Structure of Suboptimal Paths

Notation - Let G be the edit distance graph between two sequences A and B, and let E be its set of edges. The weight of an edge $e = \langle u, v \rangle$ is denoted by $w(e)$. Let $d(x, y)$ be the length of the optimal (maximal) path from node x to node y in G. If P is a path that goes through nodes x and y, then $d_P(x, y)$ is the length of the portion of P from x to y. For a path P from s to t (an (s, t)-path), define

$\delta(P) = d(s,t) - d_P(s,t)$. If $\delta(P) = \delta$ ($\delta \geq 0$) then P is called a δ-path (or a δ-suboptimal path). Throughout the paper only simple paths are considered (since G is a DAG, it contains only simple paths).

Given a pair of sequences, two edit distance computations (between the sequences and between their reverse) produce all values $d(s,u)$ and $d(u,t)$ for every node u, as well as the optimal path from s to u and from u to t for every node u.

Definition – For an edge $e \in E$, define $\delta(e) = d(s,t) - (d(s,u) + w(u,v) + d(v,t))$. $\delta(e)$ is therefore the difference between the length of the best path from s to t that uses e and the optimal (s,t) path.

Definition – Define E_Δ, a subset of the edge set E in G, as:

$$E_\Delta = \{e \mid \delta(e) \leq \Delta\}$$

Claim 1. E_Δ is the smallest set of edges such that every δ' suboptimal path for some $\delta' \leq \Delta$ is a path in E_Δ.

Proof. Note first that every δ'-suboptimal path, $\delta' \leq \Delta$, is a path in E_Δ. Let P be a δ'-suboptimal path for some $\delta' \leq \Delta$. Assume that P is not a path in E_Δ, so there must be some edge $e = \langle u,v \rangle \notin E_\Delta$ on P. But since P uses e, $\delta(e) = \delta'' \leq \delta' \leq \Delta$. Hence, $e \in E_\Delta$, a contradiction. E_Δ is the smallest such set since every $e \in E_\Delta$ is on some δ' suboptimal path, $\delta' \leq \Delta$. $\qquad\square$

Definition – A δ-path P from s to t is called *canonical* (for e) if there exists an edge $e = \langle u,v \rangle \in P$ such that $\delta(P) = \delta(e)$. That is, a canonical path consists of the best path from s to u, followed by e and further followed by the best path from v to t. Note - there is a canonical path for every edge e; however, a path P can be canonical for more than one edge.

We next argue that the set of canonical paths are the essential ones among all suboptimal paths since all non-canonical suboptimal paths can be derived from them.

Lemma 2. Let $P = e_1, e_2 \ldots, e_k$ be a canonical Δ-suboptimal path. Then
(1) $\forall e \in P$, $\delta(e) \leq \Delta$,
(2) there is a contiguous segment $e_l, \ldots e_r$ ($l \leq r$) along P such that $\delta(e_i) = \Delta$ for $l \leq i \leq r$ and $\delta(e_j) < \Delta$ for $j < l$ or $j > r$. Hence, P is a canonical path for $e_l, \ldots e_r$.
(3) $\delta(e_{i-1}) \leq \delta(e_i)$ for all $i \leq l$, and $\delta(e_i) \geq \delta(e_{i+1})$ for all $i \geq r$.

Proof. For any $e \in P$, $\delta(e) = d(s,t) - d_{P'}(s,t)$, where P' is the best (s,t)-path that uses e. Since P uses e, $\Delta = \delta(P) = d(s,t) - d_P(s,t) \geq d(s,t) - d_{P'}(s,t) = \delta(e)$, so $\delta(e) \leq \Delta$ and (1) is proven.

To show (2), let e_l be the first edge on P for which $\delta(e_l) = \Delta$ (since P is a canonical path, there must be such an edge), and let e_r be the first edge on P such that $\delta(e_r) = \Delta$ but $\delta(e_{r+1}) < \Delta$, or $e_r = e_k$ if no such r exists. By definition, $\delta(e_i) < \Delta$ for $i < l$. If $r = k$, then we are done. Otherwise, it is left to be shown that $\delta(e_i) < \Delta$ for $i > r$. Suppose $e_l = \langle u, v \rangle$. Since P is canonical for e_l, $d_P(v, t) = d(v, t)$, and this holds for all nodes that occur after v on P. Let $e_{r+1} = \langle x, y \rangle$. P is not canonical for e_{r+1}, but since y occurs after v on P we know that $d_P(y, t) = d(y, t)$, hence $d_P(s, x) < d(s, x)$, so there is a better path to reach x from s than via P. Take now any edge $e_i = \langle x', y' \rangle$, $i > r$. P is not canonical for e_i since there is always a better way to reach x' from s than along P: reach x optimally, then follow the portion from x to x' on P. Hence $d_P(s, x') < d(s, x')$, so $\delta(e_i) < \delta(P) = \Delta$ for all $i > r$.

Recall that P is optimal for $e_l = \langle u, v \rangle$. Consider some $i \leq l$, and let $e_{i-1} = \langle p, q \rangle$ and $e_i = \langle q, r \rangle$. e_{i-1} and e_i are on the best path from s to u, and therefore e_{i-1} is on the best path from s to q. Hence, e_{i-1} is on the canonical path for e_i and by (2) $\delta(e_{i-1}) \leq \delta(e_i)$. A similar argument shows that for every $i \geq r$, e_{i+1} is on the canonical path for e_i, hence $\delta(e_i) \geq \delta(e_{i+1})$. (3) is therefore proven. $\square$

A canonical path is depicted in Figure 1.

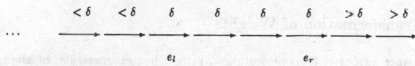

$$< \delta \qquad < \delta \qquad \delta \qquad \delta \qquad \delta \qquad \delta \qquad > \delta \qquad > \delta$$

$$e_l \qquad\qquad\qquad\qquad e_r$$

Fig. 1. A Canonical Path

Lemma 3. *If $P = e_1, e_2 \ldots, e_k$ is a non-canonical Δ-suboptimal path, then $\delta(e) < \Delta$ $\forall e \in P$.*

Proof. For any $e \in P$, the length of the best path that uses e is $d(s, t) - \delta(e)$. Since P uses e but is not canonical (therefore not the best) for e, $d_P(s, t) = d(s, t) - \Delta < d(s, t) - \delta(e)$, so $\delta(e) < \Delta$. $\square$

Denote by $\Delta_0 < \Delta_1 < \Delta_2 < \ldots$ the maximal sequence of values such that

$$E_{\Delta_0} \subset E_{\Delta_1} \subset E_{\Delta_2} \subset \ldots$$

Lemma 4. *E_{Δ_i}, which is the smallest set of edges that contains every δ' suboptimal path ($\delta' \leq \Delta_i$), is also the union of all Δ_j canonical paths for $j \leq i$.*

Proof. Recall that a canonical path must be Δ_j-suboptimal for some i. If P is a canonical Δ_j-suboptimal path, then it will first appear as a path in E_{Δ_j} since by Lemma 2 $\delta(e) \leq \Delta_j$ for all $e \in P$, and $\delta(e') = \Delta_j$ for some $e' \in P$. Also, every edge in E_{Δ_i} belongs to some canonical Δ_j-suboptimal path, where $\Delta_j \leq \Delta_i$ ($j \leq i$). $\square$

Lemma 4 implies that the union of all **canonical suboptimal** paths within Δ of the optimum contain **all** suboptimal paths in this neighborhood. Hence, any non-canonical δ-suboptimal path is a combination of few segments from canonical paths that are δ'-suboptimal, $\delta' < \delta$. In general, let P_1 and P_2 be δ_1 and δ_2 suboptimal paths, respectively, which intersect at some node u. Then, any path P_3 obtained by concatenating the (s, u) segment from one path and the (u, t) from the other is δ_3 suboptimal, $\delta_3 \le \delta_1 + \delta_2$. In the special case where P_1 and P_2 are both optimal, P_3 is also optimal.

In terms of alignments, in order to cope with this combinatorial explosion, we suggest the subset of alignments that correspond to the canonical paths as the representatives for the entire set of suboptimal alignments, which can be very large. This set fairly represents the entire set of alignments since any non-canonical alignments can be obtained by recombining few canonical ones.

Remark - Define $G_\Delta \equiv (V, E_\Delta)$. The above discussion also implies that any task that involves alignments that are at most Δ suboptimal (such as counting and enumeration) can be done on G_Δ instead of G, taking advantage of its sparsity. No theoretical bounds are known on the ratio of $|E_\Delta|/3nm$. However, it is typically the case that the interesting Δ is the one for which $|E_\Delta| = O(n + m) = O(|E|^{0.5})$, to avoid combinatorial explosion.

3   Transformation of Weights

This paper advocates for representation, rather than enumeration, of alignments. However, in some cases it may be desirable to enumerate alignments. Also, recall that E_Δ may also contain non-canonical paths that are worse than Δ. In this section we show that a simple transformation on the weights of E provides a powerful tool to manipulate suboptimal paths and to accomplish the tasks mentioned above. This transformation turns out to be related to the method of Edmonds and Karp [EK72] (see also [Law76] that transforms general weights of a graph to all non-negative weights such that the order of paths are preserved. Our transformation specializes the Edmonds-Karp transformation to (s, t)-paths; it produces a set of non-negative weights (recall that the original set of weights may be negative) which preserves the order of the (s, t) paths, and makes the enumeration extremely easy.

Definition: For every $e = \langle u, v \rangle \in E$ define

$$\epsilon(e) = \delta(e) - \min_{e' = \langle x, u \rangle} \{\delta(e')\}$$

We now prove (Lemma 5[3]) that $\epsilon(e)$ can be interpreted as the "additional penalty for using e on the path from u to t rather than following the optimal path from u to t directly". Theorem 6 shows how these transformed weights can be used.

Lemma 5. *For any $e = \langle u, v \rangle$*
(1) $\epsilon(e) = \delta(e) - \delta(e')$, where e' is the edge preceding e on the canonical path for e.
(2) $\epsilon(e) \ge 0$
(3) $w(e) + d(v, t) = d(u, t) - \epsilon(e)$.

Proof. Let $e_i = \langle x_i, u \rangle$ be the set of edges entering u and assume without loss of generality that $\delta(e_1) \leq \delta(e_2) \leq \ldots$ Note that $\delta(e_i) = d(s,t) - (d(s,x_i) + w(e_i) + d(u,t))$; hence $d(s,x_1) + w(e_1)$ is the optimal way to reach u from s since $d(s,x_1) + w(e_1) \geq d(s,x_2) + w(e_2) \geq \ldots$ This implies that the canonical path for e enters u via edge $e_1 = \langle x_1, u \rangle$, so $\epsilon(e) = \delta(e) - \delta(e_1)$ and (1) follows.

From Lemma 2 we know that if P is a Δ-suboptimal path that is canonical for e, then $\delta(e) = \Delta$ and $\delta(e') \leq \Delta$ for any edge e' preceding e on P. Hence (2) follows.

Let $e_1 = (x_1, u)$ from above. We know
(i) $\delta(e_1) = d(s,t) - (d(s,x_1) + w(e_1) + d(u,t))$
(ii) $\delta(e) = d(s,t) - (d(s,x_1) + w(e_1) + w(e) + d(v,t))$
hence
$$\epsilon(e) = \delta(e) - \delta(e_1) = d(u,t) - w(e) - d(v,t)$$
or $w(e) + d(v,t) = d(u,t) - \epsilon(e)$. □

Theorem 6. *For any path P from s to t, $\delta(P) = \sum_{e \in P} \epsilon(e)$.*

Proof. Let $P = e_1, e_2 \ldots, e_k$, where $e_i = \langle u_i, v_i \rangle$ ($u_1 = s, v_k = t$ and $v_i = u_{i+1}$). Define the path P_i as the path obtained by concatenating the edges $e_1, \ldots, e_i$ with the optimal path from v_i to t. We claim, by induction on i, $i = 1, \ldots, k$, that $\delta(P_i) = \sum_{j \leq i} \epsilon(e_j)$. Since $P_k = P$, the theorem follows.

Note that P_1 is canonical for e_1; therefore $\delta(P_1) = \delta(e_1)$. Furthermore, $\epsilon(e_1) = \delta(e_1)$ as there are no edges into s. Hence, for $i = 1$, $\delta(P_1) = \epsilon(e_1)$ as claimed.

Assume correctness for $j \leq i$. P_i and P_{i+1} share the first i edges. Since $d_{P_i}(s,t) = d_{P_i}(s,v_i) + d(v_i,t)$ we have

$$d_{P_{i+1}}(s,t) = d_{P_i}(s,v_i) + w(e_{i+1}) + d(v_{i+1},t) = \tag{1}$$
$$= d_{P_i}(s,t) - d(v_i,t) + w(e_{i+1}) + d(v_{i+1},t) \tag{2}$$

Recall that $v_i = u_{i+1}$. From Lemma 5 we have

$$w(e_{i+1}) + d(v_{i+1},t) = d(u_{i+1},t) - \epsilon(e_{i+1}) = d(v_i,t) - \epsilon(e_{i+1}) \tag{3}$$

Substituting (3) in (2) we get
$$d_{P_{i+1}}(s,t) = d_{P_i}(s,t) - \epsilon(e_{i+1})$$
By induction $d_{P_i}(s,t) = d(s,t) - \sum_{j \leq i} \epsilon(e_j)$, hence
$$d_{P_{i+1}}(s,t) = d(s,t) - \sum_{j \leq i+1} \epsilon(e_j)$$
so $\delta(P_{i+1}) = \sum_{j=1}^{i+1} \epsilon(e_j)$. □

The proof of Theorem 6 implies that after $\epsilon(e)$ has been computed for every edge e, the "goodness" of a path can be computed "on the fly" as follows. Take a path from s to u for some node u: if δ is the sum of the ϵ's along that path, then there is always a δ-suboptimal path from s to t that begins with this segment from s to u; namely, the one that proceeds with the optimal path from u to t. If an edge e is followed after u, then *any* path to t that uses the segment up to u, followed by e, will be at least $(\delta + \epsilon(e))$-suboptimal.

3.1 Enumerating Suboptimal Paths

A natural enumeration procedure for all suboptimal paths is now readily available. First, compute $\delta(e)$ and $\epsilon(e)$ for every edge in G. This preprocessing takes $O(nm)$ time and space. Then, using the new set of weights $\epsilon(e)$, enumerate the (s,t) paths in order. This can be done as follows: Build a "search tree" that explores all the paths in this graph, starting at s. An internal node in the tree represents a partial path that starts at s and ends at some node u; the leaves represent complete (s,t)-paths. Each internal node is expanded via all edges eminenting from the last node u on the partial path.

With each internal node in the tree we associate a cost δ, which is to the sum of $\epsilon(e)$ of all edges e on the path that is represented by this node. This path can always be extended to an (s,t) path P with $\delta(P) = \delta$. When an internal node is expanded via an edge e, the new node receives the cost of its parent $+ \epsilon(e)$.

It is now straightforward to observe (from Theorem 6) that if the tree is expanded in a "best first search" manner (i.e. the next node to expand is the internal node with the minimum δ), then paths are enumerated in increasing order. Hence, to enumerate the K-best paths, simply stop after the Kth leaf in the search tree has been reached. A priority queue that maintains the best K costs of the nodes in the search tree needs to be maintained. The size of the tree is at most $O(Km)$ nodes, as every leaf is preceded by at most $m + n$ nodes. After the $O(nm)$ time preprocessing stage, a new path P is enumerated in this manner in $O(|P|) = O(n + m)$ time. The priority queue maintenance takes $O(\log K)$. Since the entire search tree needs to be kept throughout, the space requirement is $O(K(n + m))$ (plus an additional $O(nm)$ space to store the $\epsilon(e)$'s).

The same method can be employed to list all δ'-suboptimal paths for $\delta' \leq \Delta$, where Δ is a specified threshold. First build E_Δ; then enumerate paths in E_Δ by building the search tree, while pruning any extension that leads to a node whose cost exceeds Δ. This is basically the method of [WaBy]. Since any extended node eventually leads to a legal leaf (whose cost is within the threshold), and since at any node there is a constant number of extensions to check, the search tree can be explored in any order. Only the current path needs to be maintained, so the space required is $O(m + n)$, and the time is $O(|P|)$ per path.

A More Efficient Enumeration Algorithm Using the idea of "deviations", together with the transformed weights, another algorithm that outputs suboptimal paths in increasing order can be suggested. It requires only $O(K)$ space and $O(|P|)$ time per path. For simplicity, assume that there are no ties between path lengths (e.g. impose lexicographic order in addition to the length).

The algorithm builds a search tree (which is different from the search tree described above). A node in the tree represents a solution (a path), and its cost is the path length. A node b is a child of a node a in the tree if the path represented by b is a deviation of the path represented by a. Specifically, let a be a node in the tree that represents a path $P = x_1, \ldots x_k$ which deviates from its parent by the edge

$\langle x_i, x_{i+1} \rangle$. Let the cost of a (i.e. $\delta(P)$) be δ. Note that $\epsilon(\langle x_j, x_{j+1} \rangle) = 0$ for all $j > i$. The children of a are found as follows: for every vertex x_j on P, $j > i$, let e_1, e_2 be the two eges that are incident at x_j but **not** on P (i.e. $\langle x_j, x_{j+1} \rangle \neq e_1, e_2$). Create two new nodes, b_1 and b_2, with costs $\delta + \epsilon(e_1)$ and $\delta + \epsilon(e_2)$ respectively, and make them the children of a. b_1 and b_2 share the prefix $x_1, \ldots, x_j$ with a, and deviate from it on the edges e_1 and e_2 respectively.

The algorithm, which outputs the first, second, third ... paths in order, proceeds at follows: Initially, the tree consist of a root, where the root of the tree is the optimal path from s to t, and its cost is 0. To find the next best path, find the leaf a in the tree of minimal cost, and ouput the path represented by a. Expand a by attaching all deviations of it as its children.

Since expanding a node requires $O(n+m)$ time (need to check at most two edges at each vertex of the path), the time to output the next path is $O(n+m)$. The size of the search tree is $O(Km)$ since every new solution creats at most $2(n+m)$ new nodes. However, this space requirement can be improved by using the technique of [Law72]: when a leaf a is first expanded, only its best child b^* needs to be explicitly stored in the tree. When b^* is eventually used at some iteration (i.e it is the next best path), b^* is expanded as before, but also the **next** best child of a is found again. The is repeated for the third, fourth .. child of a. This assures that after K solutions have been output, there are only $2K$ nodes in the tree. Moreover, at each iteration, two nodes (instead of one node) are expanded, hence the time requitement is only doubled whereas the space requirement is reduced by $O(m)$.

4 Counting Suboptimal Alignments

Simulations show that the number of suboptimal alignments grows rapidly as the threshold increases. This behavior depends on the scoring system, i.e. on the weights of the edges in the graph, as well as on the actual sequences. Hence, the significance of an alignment can be assessed not only by the distance between its score and the optimal score, but also by its **rank**, i.e. the *number* of alignments with better scores than its own score. Computing the rank of a given alignment is closely related to the problem of *counting* the number of suboptimal alignments. The number of suboptimal alignments can clearly be computed by enumerating them using the methods of Section 3.1. In this section we are interested in counting methods that are more efficient than the corresponding enumeration solution. We show that by using the transformed set of weights $\epsilon(e)$ instead of the original weights, better bounds can be achieved. As before, we use the graph notation, where alignments correspond to paths from s to t in the edit distance graph.

The number of *optimal* paths from s to t is found in time and space $O(|E_0|)$ as follows. Let $\mathcal{N}(v)$ be the number of optimal paths from s to some node v. Then,

$$\mathcal{N}(v) = \sum_{u | (u,v) \in E_0} \mathcal{N}(u)$$

$\mathcal{N}(t)$ is the desired count. A natural generalization of this method to count all suboptimal paths within Δ of the optimum is to maintain, at each node v, a list of all possible lengths of paths from s to v, as well as their count. A node creates its own list by inheriting and merging the lists and the counts of its predecessors [Wa92]. This evaluation requires $O(C|E_\Delta|) = O(Cnm)$ time and space, where C is the maximum list size (C can be very large for arbitrary weights).

Let Δ be a specified threshold. If the number of paths which are δ suboptimal for $\delta \leq \Delta$ is sought, then the $O(Cnm)$ bound can be dramatically reduced by using the transformed set of weights. Recall that if the weights on the edges of the graph are transformed as suggested in Section 3, then the paths which are δ suboptimal, $\delta \leq \Delta$, are exactly the paths whose transformed length does not exceed Δ. Hence, if we compute $\epsilon(e)$ for each edge $e \in E_\Delta$, then we can count the number of these paths in the transformed graph as follows when the weights are all integers: Let $C(v, k)$, $k = 0, 1, \ldots, \Delta$, be the number of paths of length k from s to v in the transformed graph. Then,

$$\mathcal{N}(v, k) = \sum_{u | \langle u, v \rangle \in E_\Delta} \mathcal{N}(u, k - \epsilon(\langle u, v \rangle))$$

This recursive formula can be evaluated in $O(\Delta |E_\Delta|) = O(\Delta nm)$ time and space if we use the transformed weights $\epsilon(e)$. A similar solution on the original set of weights requires $O((d(s, t) + \Delta) nm)$ time and space. Similarly, for non-integral weights, the space requirement will be reduced.

Counting Canonical Suboptimal Path

The set of *canonical* suboptimal paths is a smaller, restricted and more structured set of suboptimal paths. To count the number of canonical δ suboptimal paths we need the following lemma:

Lemma 7. $P = e_1, e_2, \ldots, e_k$ *is a canonical path iff there is some l, $1 \leq l \leq k$, such that*
(1) for any $i \leq l$, $\sum_{j \leq i} \epsilon(e_j) = \delta(e_i)$, and
(2) $\epsilon(e_i) = 0$ for all $i > l$.

Proof. Suppose first that (1) and (2) hold. Then, $\sum_{j \leq k} \epsilon(e_j) = \delta(e_l)$, and from Theorem 6 $\delta(P) = \delta(e_l)$, hence P is canonical for e_l.

We prove the only if part by induction on Δ. Assume that for any canonical δ-suboptimal path, $\delta < \Delta$, (1) and (2) hold. Let P be a canonical Δ-suboptimal path, and let $e_l = \langle v, w \rangle$ be the first edge on P such that $\delta(P) = \delta(e_l) = \Delta$. We will show that (1) and (2) hold for this l. Suppose that $e_{l-1} = \langle u, v \rangle$ and let P' be a canonical path for e_{l-1}. Consider the segment on P from s to v; it is the best path from s to v and it also goes through u, hence its portion up to u is the best path from s to u. Therefore, P' coincides with P on the segment up to u. Lemma 2 implies that $\delta(P') = \delta(e_{l-1}) < \delta(e_l) = \Delta$. Hence, by induction, for any $i \leq l - 1$, $\sum_{j \leq i} \epsilon(e_j) = \delta(e_i)$. e_{l-1} precedes e_l on the canonical path for e_l, so from Lemma 5 $\epsilon(e_l) = \delta(e_l) - \delta(e_{l-1})$. Hence, (1) holds since

$$\sum_{j \leq l} \epsilon(e_j) = \delta(e_{l-1}) + \epsilon(e_l) = \delta(e_{l-1}) + \delta(e_l) - \delta(e_{l-1}) = \delta(e_l)$$

Now (2) follows directly, since $\delta(P) = \delta(e_l)$ by definition, and $\delta(P) = \sum_{j \leq k} \epsilon(e_j) = \delta(e_l) + \sum_{l < j \leq k} \epsilon(e_j)$ which implies $\epsilon(e_i) = 0$ for all $i > l$. $\qquad\square$

The above lemma implies that every canonical path $P = e_1, e_2, \ldots, e_k$ can be decomposed into two parts, $P_1 = e_1, e_2, \ldots, e_l$ and $P_2 = e_{l+1}, \ldots, e_k$, such that in $P1$ the property $\sum_{j \leq i} \epsilon(e_j) = \delta(e_i)$ holds for any i, and in P_2 $\epsilon(e_i) = 0$ for any i. If we choose P_1 as the maximal segment with this property, then this decomposition is unique. We therefore associate with each canonical path P a pair of edges (e, f), where e is the last edge on the P_1 (maximal) segment, and f is the first edge on the P_2 segment. Note that $\delta(P) = \delta(e)$, and that from the maximality of P_1, $\delta(e) + \epsilon(f) > \delta(f)$.

Many canonical paths may be associated with the same pair of edges (e, f). Denote by $\mathcal{N}(e, f)$ the number of canonical ($\delta(e)$-suboptimal) paths that are associated with the pair (e, f). Since each canonical path is associated with a unique pair, all we need is to count how many canonical paths are associated with each pair of adjacent edges (e, f).

Define $\mathcal{N}_{can}(e)$ as the number of paths that start at s and end with the edge e, which have the property that $\sum_{j \leq i} \epsilon(e_j) = \delta(e_i)$ for any edge e_i along the path. Also, define $\mathcal{N}_{opt}(f)$ as the number of paths that start with the edge f and end at t such that $\epsilon(e) = 0$ for any edge e along the path. Then,

$$\mathcal{N}(e, f) = \begin{cases} \mathcal{N}_{can}(e) + \mathcal{N}_{opt}(f) & \text{if } \delta(e) + \epsilon(f) > \delta(f) \\ 0 & \text{otherwise} \end{cases}$$

$\mathcal{N}_{can}(e)$ and $\mathcal{N}_{opt}(f)$ are computed as follows. Let e_1, e_2, e_3 be the three edges that are immediate predecessors of e, and f_1, f_2, f_3 be the three edges that are immediate successors of f. Set $I_j = 1$, $j = 1, 2, 3$, if $\delta(e_j) + \epsilon(e) = \delta(e)$, otherwise $I_j = 0$. Then, initially, $\mathcal{N}(e) = 1$ for $e = \langle s, u \rangle$, and

$$\mathcal{N}_{can}(e) = \sum_{j=1,2,3} I_j \mathcal{N}_{can}(e_j)$$

To compute $\mathcal{N}_{opt}(f)$, initially set $\mathcal{N}_{opt}(f) = 1$ if $f = \langle u, t \rangle$ and $\epsilon(f) = 0$, and

$$\mathcal{N}_{opt}(f) = \begin{cases} \sum_{j=1,2,3} \mathcal{N}_{opt}(f_j) & \text{if } \epsilon(f) = 0 \\ 0 & \text{otherwise} \end{cases}$$

There are $3nm$ pairs of edges to consider. The recursive evaluation of $\mathcal{N}_{can}(e)$ and of $\mathcal{N}_{opt}(f)$ requires $O(nm)$ time and space. Hence the total time and space is $O(nm)$. This can be reduced to $O(|E_\Delta|)$ if we are only concerned with canonical paths that are δ-suboptimal for some $\delta \leq \Delta$.

5   Acknowledgements

We would like to thank Dan Gusfield, Gene Lawler, Frank Olken and Martin Vingron for valuable discussions, and Tod Klingler for many helps, and in particular in interpreting the graphs of Figure 3.

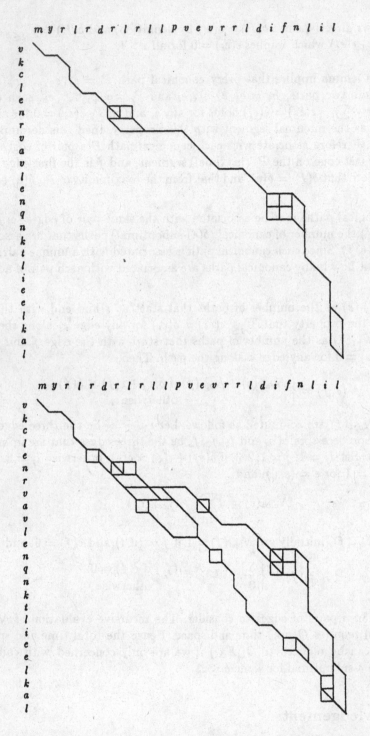

Fig. 2. Aligning two Leucine Zippers: the top figure depicts E_0, and the bottom figure depicts E_2. Note that the biologically "correct" alignment, the one that coincides with the diagonal at the 4th, 11th, 18th and 25th positions, is in E_2.

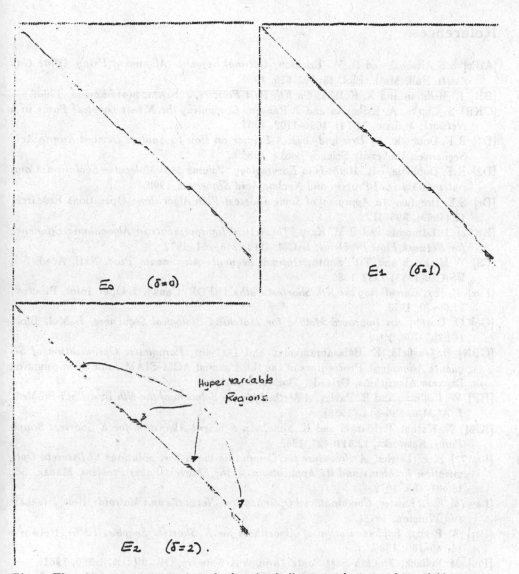

Fig. 3. The compact representation of suboptimal alignments between the variable regions of a heavy and a light chain of a Human Immunoglobulin reveals their hypervariable regions.

References

[AlEr] S.F Altschul and B. W. Erickson, *Optimal Sequence Alignment Using Affine Gap Costs*, Bull. Math. Biol. 48: 603–616, 1986.

[BK] R. Bellman and R. Kalaba, *On Kth Best Policies*, J. SIAM, 8(4):582–588, 1960.

[CKR] S. Clarke, A. Krikorian and J. Rausen, *Computing the N best Loopless Paths in a Network*, J. Siam, 11 (4): 1096–1102, 1963.

[D1] R.F. Doolittle, *Of Urfs and Orfs: A Primer on How to Analyze Derived Amino Acid Sequences*, University Science Books, 1986.

[D2] R.F. Doolittle, edt. *Methods in Enzymology, Volume 183: Molecular Evolution: Computer Analysis of Protein and Nucleic Acid Sequences*, 1990.

[Dr] S.E. Dreyfus, *An Appraisal of Some Shortest-Path Algorithms*, Operations Research, 17 (1969), 395–412.

[EK72] J. Edmonds and R.M. Karp, *Theoretical Improvements in Algorithmic Efficiency for Network Flow Problems*, JACM, 19(2):248–264, 1972.

[FS] W.M. Fitch and T.F. Smith, *Optimal Sequence Alignments*, Proc. Natl. Acad. Sci. USA 80(1983) 1382–1386.

[Fox] B. Fox, *Calculating the Kth Shortest Paths*, INFOR, Canad. J. Oper. Infor. Process., 11:66–70, 1973.

[Go] O. Gotoh, *An Improved Method for Matching Biological Sequences*, J. Mol. Biol., 162:705–708, 1982.

[GBN] D. Gusfield, K. Balasubramanian and D. Naor, *Parametric Optimization of Sequence Alignment*, Proceedings of the third annual ACM–SIAM Joint Symposium on Discrete Algorithms, Orlando, Florida, Jan. 1992.

[HP] W. Hoffman and R. Pavley, *A Method for the Solution of the Nth Best Path Problem*, J. ACM 6, 506–514 (1959).

[KIM] N. Katoh, T. Ibaraki and H. Mine, *An Efficient Algorithm for K Shortest Simple Paths*, Networks, 12:411–427, 1982.

[Law72] E. L. Lawler, *A Procedure for Computing the K-best Solutions to Discrete Optimization Problems and its Applications to the Shortest Paths Problem*, Manag. Sci. 18:401–405, 1972.

[Law76] E. L. Lawler, *Combinatorial Optimization, Networks and Matroids*, Holt, Rinehart and Winston, 1976.

[Per] A. Perko, *Implementation of Algorithms for K Shortest Loopless Paths*, Networks 16:149–160, 1986.

[Pol] M. Pollack, *The kth Best Route Through A Network*, OR, 9 (4):578–580, 1961.

[SK] D. Sankoff and J. Kruskal, Editors, *Time Warps, String Edits, and Macromolecules: The Theory and Practice of Sequence Comparison*, Addison-Wesley, 1983.

[ScWa] M. Schoniger and M.S. Waterman, *A Local Algorithm for DNA Sequence Alignment with Inversions*, Bull. Math. Bio. 54(4):521–536, 1992.

[Sel] P. H. Sellers, *An Algorithm for the Distance Between Two Finite Sequences*, J. Comb. Theory. (A), 16:253–258, 1974.

[Sh] D.R. Shier, *On Algorithms for Finding the K Shortest Paths in a Network*, Networks, 9:195–214, 1979.

[Vi] M. Vingron, *Multiple Sequence Alignment and Applications in Molecular Biology*, Preprint 91–12, Universitat Heidelberg, 1991.

[ViAr] M. Vingron and P. Argos, *Determination of Reliable Regions in Protein Sequence Alignments*, Protein Engin., 7: 565–569, 1990.

[Wa83] M.S. Waterman, *Sequence Alignments in the Neighborhood of the Optimum with General Application to Dynamic Programming*, Proc. Nat, Acad, Sci. USA. 80:3123–3124, 1983.

[Wa92] M.S. Waterman, *Parametric and Ensemble Sequence Alignment Algorithms,* manuscript.

[WaBy] M.S. Waterman and T.H. Byers, *A Dynamic Programming Algorithm to Find All Solutions in a Neighborhood of the Optimum,* Math, Biosci. 77:179–188, 1985.

[WaEg] M.S. Waterman and M. Eggert, *A New Algorithm for Best Subsequence Alignments With Applications to tRNA-rRNA Comparisons,* J. Mol. Biol. 197:723–728, 1987.

[WEL] M.S. Waterman, M. Eggert and E. Lander, *Parametric Sequence Comparisons,* PNAS 89:6090–6093, July 1992.

[Yen] J.Y. Yen, *Finding the K Shortest Loopless Paths in a Network,* Management Science, 17 (11):712–716, 1971.

[Zuk] M. Zuker, *Suboptimal Sequence Alignment in Molecular Biology, Alignment with Error Analysis,* J. Mol. Biol., 221:403–420, 1991.

A Fast Filtration Algorithm for the Substring Matching Problem *

Pavel A. Pevzner[1] and Michael S. Waterman[2]

[1] Computer Science Department
The Pennsylvania State University
University Park, PA 16802

[2] Departments of Mathematics and of Molecular Biology
University of Southern California
Los Angeles, California 90089-1113

Abstract. Given a text of length n and a query of length q we present an algorithm for finding all locations of m-tuples in the text and in the query that differ by at most k mismatches. This problem is motivated by the dot-matrix constructions for sequence comparison and optimal oligonucleotide probe selection routinely used in molecular biology. In the case $q = m$ the problem coincides with the classical *approximate string matching with k mismatches* problem. We present a new approach to this problem based on multiple filtration which may have advantages over some sophisticated and theoretically efficient methods that have been proposed. This paper describes a two-stage process. The first stage (multiple filtration) uses a new technique to preselect roughly similar m-tuples. The second stage compares these m-tuples using an accurate method. We demonstrate the advantages of multiple filtration in comparison with other techniques for approximate pattern matching.

1 Introduction

Suppose we are given a string of length n, $T[1 \ldots n]$, called the *text*, a shorter string of length q, $Q[1 \ldots q]$, called the *query*, and integers k and m. The *substring matching problem with k-mismatches* ([CL90]) is to find all *"starting"* locations $1 \leq i \leq q - m + 1$ in the query and $1 \leq j \leq n - m + 1$ in the text, such that the substring of the query $Q[i, i+1, \ldots, i+m-1]$ matches the substring of the text $T[j, j+1, \ldots, j+m-1]$ with at most k mismatches. In the case $q = m$ the substring matching problem yields the *approximate string matching problem with k-mismatches*.

The approximate string matching problem with k-mismatches has been intensively studied in computer science. The naive brute-force algorithm for approximate string matching runs in $O(nm)$ time. Landau and Vishkin ([LV86])

---

* The research was supported in part by the National Science Foundation (DMS 90-05833) and the National Institute of Health (GM-36230). This paper was written when P.A.P. was at the Department of Mathematics, University of Southern California.

gave a $O(kn + km\log m)$ approximate pattern matching algorithm. Galil and Giancarlo ([GG86]) improved the Landau-Vishkin algorithm, achieving a time performance $O(kn + m\log m)$. All these algorithms and their improved versions (see [LV89], [TU90]) are based on the preprocessing of the pattern/text

Recently several approaches emphasizing expected running time have appeared in contrast to earlier results ([BG89], [CL90], [GL90], [TU90], [HS91], [WM92a], [WM92b], [BP92]). In particular, Grossi and Luccio ([GL90]) demonstrated that although earlier algorithms yield the best performance in the worst cases, they are far from being the best in practice. In particular, a simple *filtration* algorithm from [GL90] runs approximately 10 times faster than the algorithm from [GG86] for a wide range of k and m.

The idea of filtration algorithms for approximate matching involves a two-stage process. The first stage preselects a set of positions in the text that are *potentially* similar to the pattern. The second stage verifies each potential position using an accurate method rejecting potential matches with more than k mismatches. Denote p the number of potential matches found at the first stage of the algorithm. Preselection is usually done in $O(n + p)$ time where the coefficient of n is much smaller than for the algorithms based on the preprocessing of the pattern/text. If the number of potential matches is small and the accurate method for potential match verification is not too slow, this idea brings a significant speed up in comparison to the algorithms based on the preprocessing of the pattern/text.

The idea of filtration for information retrieval/pattern matching goes back to early 70's [H71]. The idea of filtration for string matching problems first was described by Karp and Rabin [KR87] for the case $k = 0$. Notice that the idea of filtration in computational molecular biology for related alignment problems was stated even earlier (see [DN82], [WL83], [LP85] for *l-tuple filtration*, [B86], for *filtration by composition*).

For $k > 0$ Owolabi and McGregor [OM88] used an idea of *l-tuple filtration* based on a simple observation that if a pattern approximately matches a substring of the text then they share at least one l-tuple for sufficiently large l. Finding all l-tuples shared by pattern and text can be easily done by hashing. If the number of shared l-tuples is relatively small they can be verified and all *real* matches with k mismatches can be rapidly located. The theoretical analysis of the expected running time of this approach has been recently done by Kim and Shawe-Taylor [KS92]. The idea of l-tuple filtration has been significantly developed by Baeza-Yates and Perleberg [BP92] and by Wu and Manber ([WM92a], [WM92b]).

Grossi and Luccio ([GL90]) observed that if a pattern approximately matches a substring of the text then they have similar letter compositions. This observation leads to a simple algorithm running in $O(n\log|A| + pm)$ time, where A is the alphabet of pattern P and $p < n$ is the number of m-substrings of text T with letter composition having at most k differences with the letter composition of the pattern. Computational experiments with such *filtration by composition* show that $pm < nk$ for a wide range of parameters thus making the Grossi-

Luccio algorithm important in practice. Recently Ukkonen ([U92]) generalized the Grossi-Luccio algorithm taking an advantage of l-tuple composition (*filtration by l-tuple composition*) instead of letter (1-tuple) composition.

The complexity of filtration methods depends critically on the ratio $\frac{r}{p}$ (*filtration efficiency*) between r, the number of *real* matches with k mismatches and p, the number of *potential* matches found on the first stage of the algorithm. The larger this ratio the smaller the running time of the second stage of filtration algorithm. In the case $\frac{r}{p} = 1$ we would have an *ideal filtration* but none of the mentioned algorithms provides an ideal filtration or even lower bounds for filtration efficiency. Moreover the filtration algorithms described above do not provide a method for increasing filtration efficiency even at the expense of spending more time on the first (filtration) stage of the algorithm. Also filtration by composition does not allow an efficient implementation for the substring matching problem. We give an algorithm that allows exponential reduction of the number of potential matches at the expense of a linear increase of the filtration time. Therefore we drastically reduce the time of the 2nd stage of the algorithm (potential match verification) for the cost of linearly increased time of the first stage (filtration). Taking into account that the 2nd stage is frequently more time-consuming than the first one, the technique provides a trade-off for an optimal choice of filtration parameters.

Methods described in this paper can be applied to optimal oligonucleotide probe selection ([DMDC87]) and efficient algorithms for dot-matrices ([ML81]) in molecular biology applications. (See Landau et al. [LVN88] for a dynamic programming algorithm for substring matching problem and dot-matrix applications). Some of the described techniques have been implemented in the *OligoProbeDesignStation* software package. (Mitsuhashi M., Cooper A., Waterman M., Pevzner P. *OligoProbeDesignStation: a computerized method for designing optimal DNA probes.* Pending application for United States Letters Patent (1992).)

2 Filtration methods for approximate pattern matching

The following simple observation (compare with Theorem 5.1 from [U92]) provides a basis for *l-tuple filtration* and *filtration by l-tuple composition*.

Lemma 1. *A boolean word $v[1, \ldots, m]$ with at most k zeros contains at least $m - (k + 1)l + 1$ l-runs of ones.*

Substituting $l = \lfloor \frac{m}{k+1} \rfloor$ in Lemma 1 we derive

Lemma 2. *A boolean word $v[1, \ldots, m]$ with at most k zeros contains at least one l-run of ones with $l = \lfloor \frac{m}{k+1} \rfloor$.*

Notice that every match with at most k mismatches between strings $P[1, \ldots, m]$ and $S[1, \ldots, m]$ corresponds to a boolean word $v[1, \ldots, m]$ by the rule

$$v[i] = \begin{cases} 1, & \text{if } P[i] = S[i] \\ 0, & \text{otherwise} \end{cases}$$

This remark and lemma 2 imply the following observation of Baeza-Yates and Perleberg [BP92] and Wu and Manber [WM92b]

Lemma 3. *Let the strings* $P[1, \ldots, m]$ *and* $S[1, \ldots, m]$ *match with at most* k *mismatches and* $l = \lfloor \frac{m}{k+1} \rfloor$. *Then the strings* P *and* S *share a* l-*tuple, i.e.* $\exists i :$ $P[i, i+1, \ldots, i+l-1] = S[i, i+1, \ldots, i+l-1]$.

Lemma 3 motivates a simple two-stage l-*tuple filtration* algorithm for approximate substring matching with k mismatches between a query $Q[1, \ldots, q]$ and a text $T[1, \ldots, n]$:

Algorithm 1. Detection of all m-matches between Q and T with up to k mismatches.

- *Potential match detection.* Find all occurrences of l-tuples in both the pattern and the text.
- *Potential match verification.* Verify each potential match by extending it to the left and to the right until either the first $k+1$ mismatches are found, or the beginning/end of Q or T is found   □

Lemma 3 guarantees that Algorithm 1 finds *all* matches of length m with k or fewer mismatches between Q and T if $l \leq \lfloor \frac{m}{k+1} \rfloor$. Stage 1 (potential match detection) of the Algorithm 1 can be implemented by hashing or by building the trie ([K73]). The running time of Algorithm 1 is $O(n + p_1 m)$ where p_1 is the number of potential matches detected at the first stage of the algorithm (see [BP92] and [WM92b] for details of the implementation). For a Bernoulli text with A equiprobable letters the expected number of potential matches equals

$$E(p_1) = \frac{(n-l+1)(q-l+1)}{A^l}$$

yielding a fast algorithm for large A and l.

For Bernoulli texts with equiprobable A letters define the *filtration efficiency* e of a filtration algorithm to be the ratio of the expected number of matches with k mismatches $E(r)$ to the expected number of potential matches $E(p)$. For example for $k = 1$ the efficiency of the l-tuple filtration, (see Wu and Manber, [WM92b]) $e \approx \frac{A-1}{A^{\lfloor \frac{m}{2} \rfloor}}$ is rapidly decreasing with m and A increasing. This observation raises the question of devising a filtration method with a larger filtration efficiency.

3   The idea of multiple filtration

A set of positions $i, i+t, i+2t \ldots, i+jt, \ldots, i+(l-1)t$ is called a *gapped* l-*tuple* with *gapsize* t and *size* $1+t(l-1)$ (Fig.1). Continuous l-tuples are simply gapped l-tuples with gapsize 1 and size l.
Similarly to lemmas 1 and 2 we derive

Lemma 4. *A boolean word $v[1, \ldots, m]$ with at most k zeros contains at least $m - s + 1 - kl$ gapped l-tuples with gapsize t of size s containing only ones.*

Lemma 5. *A boolean word $v[1, \ldots, m]$ with at most k zeros contains at least one gapped $\lfloor \frac{m}{k+1} \rfloor$-tuple with gapsize $k + 1$ containing only ones.*

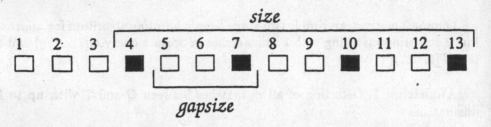

Fig. 1. Gapped 4-tuple with gapsize 3 and size 10 starting at position 4.

If an l-tuple shared by the pattern and the text starts at position i of the pattern and at position j of the query, we call (i, j) the *coordinate* of l-tuple. Define the *distance* $d(v_1, v_2)$ between l-tuples with coordinates (i_1, j_1) and (i_2, j_2) as

$$d(v_1, v_2) = \begin{cases} i_1 - i_2, & \text{if } i_1 - i_2 = j_1 - j_2 \\ \infty, & \text{otherwise} \end{cases}$$

Combining lemma 2 and lemma 5 we derive

Lemma 6. *Let the strings $P[1, \ldots, m]$ and $S[1, \ldots, m]$ match with at most k mismatches and $l = \lfloor \frac{m}{k+1} \rfloor$. Then the strings P and S share both a continuous l-tuple and a gapped l-tuple with gapsize $k + 1$ with distance d between them satisfying the condition*

$$-k \leq d \leq m - l$$

Lemma 6 is the basis of a two-stage *double filtration* algorithm for approximate string matching with k mismatches between a query $Q[1, \ldots, q]$ and a text $T[1, \ldots, n]$:

Algorithm 2. Detection of all m-matches between Q and T with up to k mismatches.

- *Potential match detection.* Find all such occurrences of continuous l-tuples from the pattern in the text where there exists a gapped l-tuple with gapsize $k + 1$ of the distance $-k \leq d \leq m - l$ from the continuous l-tuple.
- *Potential match verification.* Verify each potential match by extending it to the left and to the right until either the first $k + 1$ mismatches are found or the beginning/end of Q or T is found. □

Lemma 6 guarantees that Algorithm 2 finds *all* matches between P and T with k mismatches. Stage 1 (potential match detection) of Algorithm 2 can be implemented by hashing. The running time of algorithm 2 is $O(n + p_2 m)$ where

p_2 is the number of potential matches detected at the first stage of the algorithm (the details of an implementation are given in section 5). Define $\delta = \lceil \frac{l}{k+1} \rceil$. For a Bernoulli text with equiprobable A letters the expected number of potential matches can be roughly estimated by

$$E(p_2) \leq \frac{(n-l+1)(m-l+1)}{A^l} \cdot \frac{m}{A^{l-\delta}},$$

thus yielding better filtration than l-tuple filtration when $m < A^{l-\delta}$. The efficiency of double filtration is at least $\frac{A^{l-\delta}}{m}$ better than the efficiency of l-tuple filtration. For typical parameters of oligonucleotide probe selection ($A = 4, m = 25, k = 2$) double filtration is at least 40 times more efficient than l-tuple filtration.

In the next section we estimate the efficiency of double filtration.

4 Efficiency of double filtration

According to lemma 6 every match with k mismatches corresponds to both a continuous l-tuple and a gapped l-tuple located close to each other that contain only ones. In this section we estimate the expected number of such occurrences in a random Bernoulli boolean word.

Fix m and k and let $l = \lfloor \frac{m}{k+1} \rfloor$. We say that position j is in the *vicinity* of position i if $-k \leq i - j \leq m - l$ (see lemma 6).

A position i in a boolean word $v[1, \ldots, n]$ is a *potential match* if

(i) $v[i, \ldots, i + l - 1]$ is a run of ones,
  and
(ii) there exists a gapped l-tuple with gapsize $k + 1$ starting in the vicinity of i that contains only ones.

We denote a continuous l-tuple starting at position i as $c(i)$, and a gapped l-tuple of gapsize $k + 1$ starting at position j as $g(j)$. Notice that a continuous l-tuple $c(i)$ and a gapped l-tuple $g(j)$ of gapsize $k + 1$ can share at most $\delta = \lceil \frac{l}{k+1} \rceil$ positions. If $g(j)$ contains a position $i + s$ ($0 \leq s \leq k$), then $c(i)$ and $g(j)$ can share at most $\delta(s) = \lceil \frac{l-s}{k+1} \rceil$ positions (Fig.2).

Lemma 7. *Let $v[1, \ldots]$ be a Bernoulli boolean word with the probabilities of letters $p(1) = p$ and $p(0) = 1 - p = q$. Then the probability of a potential match at position $i > m - l$ equals*

$$p^l \cdot \{(1 - \prod_{s=0}^{s=k}(1 - (l - \delta(s))p^{l-\delta(s)}q - p^{l-\delta(s)}))\}$$

Proof. For a l-tuple $c(i)$ starting at i define $G_s(i) = \{g(t)\}$ to be the set of gapped l-tuples with gapsize $k + 1$ starting in vicinity of $c(i)$ and fulfilling the condition: $i - t = s \mod k + 1$ (Fig.2). Let $P_s(i)$ be the probability that at least one l-tuple in $G_s(i)$ contains only ones given that $c(i)$ contains only ones. Let

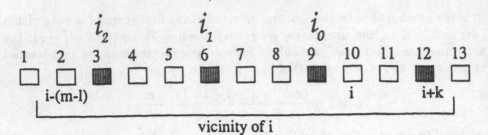

vicinity of i

Fig. 2. Vicinity of the position $i = 10$ ($m = 12, k = 2, l = \lfloor \frac{m}{k+1} \rfloor = 4, gapsize = k + 1 = 3$. Solid boxes indicate the starting positions of gapped l-tuples from $G_2(i) = \{g(3), g(6), g(9), g(12)\}$. A gapped 4-tuple and a continuous 4-tuple can share at most $\delta = \lceil \frac{l}{k+1} \rceil = \lceil \frac{4}{2+1} \rceil = 2$ positions. Gapped 4-tuples from $G_2(i)$ and continuous 4-tuple $c(i)$ can share at most $\delta(s) = \lceil \frac{l-s}{k+1} \rceil = \lceil \frac{4-2}{2+1} \rceil = 1$ positions ($l' = l - \delta(s) = 3$).

$P(i)$ be the probability to have a potential match at position i given that $c(i)$ contains only ones. As the sets $G_s(i)$ are non-overlapping for $0 \le s \le k$,

$1 - P(i) = P\{\text{there is no gapped } l\text{-tuple } g(t) \text{ in the vicinity of } i \text{ containing only ones}\} =$

$$\prod_{s=0}^{s=k} P\{\text{there is no gapped } l\text{-tuple } g(t) \epsilon G_s(i) \text{ in the vicinity of } i \text{ containing only ones}\} =$$

$$\prod_{s=0}^{s=k} (1 - P_s(i)).$$

Each l-tuple from $G_s(i)$ shares at most $\delta(s)$ positions with $c(i)$. Denote $l' = l - \delta(s)$. Fix i and consider the following positions of v to the left of i (Fig.2):

$$i_0 = i + s - (k+1), \quad i_1 = i + s - 2(k+1), \dots, \quad i_{l'-1} = i + s - l'(k+1)$$

and let $left$ be the minimum index such that $v[i_{left}] = 0$ (we assume $left = l'$ if $v[i_0] = v[i_1] = \dots = v[i_{l'-1}] = 1$). Similarly consider the positions of v to the right of $i + l - 1$

$$j_0 = i+s+(\delta(s))(k+1), j_1 = i+s+(\delta(s)+1)(k+1), \dots, j_{l'-1} = i+s+(\delta(s)+l'-1)(k+1)$$

and let $right$ be the minimum index such that $v[j_{right}] = 0$ (we assume $right = l'$ if $v[j_0] = v[j_1] = \dots = v[j_{l'-1}] = 1$).

The positions $i_{l'-1}, \dots, i_0, j_0, \dots j_{l'-1}$ represent possible positions of gapped l-tuples from $G_s(i)$. We denote $P^*(i_l, j_r)$ the probability that $left = i_l$ and $right = j_r$ in a random word v. Obviously $G_s(i)$ contains a l-tuple with only ones if and only if $left + \delta(s) + right \ge l$). Therefore

$$1 - P_s(i) = \sum_{0 \le i_l \le l'-1, 0 \le j_r \le l'-1} P^*(t_l, t_r)$$

where the product is taken over all values i_l and j_r fulfilling the conditions $i_l + j_r < l'$. As $P(left = t) = P(right = t) = qp^t$, the probabilities $P\{left + right = t\}$ constitute the *negative binomial distribution* ([F70]) and

$$1 - P_s(i) = \sum_{i_l + j_r < l'} qp^{i_l} \cdot qp^{j_r} = q^2 \sum_{t=0}^{l'-1} \binom{t+1}{t} p^t = q^2 \cdot (\sum_{t=0}^{l'-1} p^{t+1})' = q^2 \cdot (\frac{p - p^{l'+1}}{1 - p})' =$$

$$1 - l'p^{l'}q - p^{l'}$$

□

Denote $\delta_{min} = \min_s \delta(s) = \lceil \frac{l-k}{k+1} \rceil$. Lemma 7 implies

Lemma 8. *Let $Q[1,\ldots]$ and $T[1,\ldots]$ be random query and text and let p be the probability that arbitrary letters from the query and from the text are equal. Then the probability of potential match between the query and the text at position (i,j) is less than or equal $p^{2l-\delta_{min}}((m-l+k)(1-p)+k+1)$*

Proof. Let $P(i,j)$ be the probability of a potential match at (i,j) given the condition that the continuous l-tuples of Q and T starting at positions i and j are equal. Without loss of generality assume that $i - j = \Delta > 0$ and consider a boolean word $v[1,\ldots]$ corresponding to a diagonal Δ:

$$v[t] = \begin{cases} 1, & \text{if } Q[t + \Delta] = T[t] \\ 0, & \text{otherwise} \end{cases}$$

Applying lemma 7 to a word v with $p(1) = p$ and taking into account that $\delta_{min} \le \delta(s) \le \delta$ we derive

$$1 - P(i,j) = \prod_{s=0}^{k} (1 - (l - \delta(s))p^{l-\delta(s)}q - p^{l-\delta(s)})$$

$$\ge \prod_{s=0}^{k} (1 - (l - \delta_{min})p^{l-\delta}q - p^{l-\delta}) \ge 1 - (k+1)(l-\delta_{min})p^{l-\delta}q - (k+1)p^{l-\delta}$$

Therefore

$$P(i,j) \le ((k+1)l - (k+1)\delta_{min})p^{l-\delta}q + (k+1)p^{l-\delta} \le$$

$$((k+1)\lfloor \frac{m}{k+1} \rfloor - (k+1)\lceil \frac{l-k}{k+1} \rceil)p^{l-\delta}q + (k+1)p^{l-\delta} \le p^{l-\delta}((m-l+k)(1-p)+k+1).$$

□

Lemma 8 demonstrates that the efficiency of double filtration is approximately $\frac{A^{l-\delta}}{m(1-\frac{1}{A})}$ times larger than the efficiency of l-tuple filtration for a wide range of parameters m and k. Fig.3 presents the results of comparison of the efficiency of double filtration with the efficiency of l-tuple filtration for a 4-letter DNA alphabet.

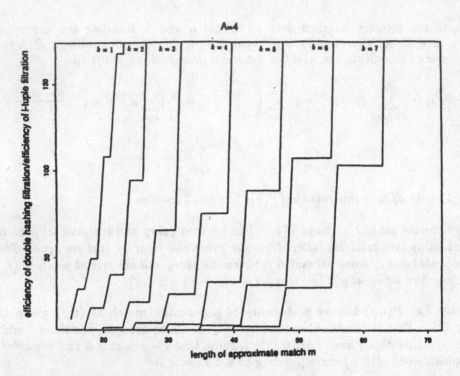

Fig. 3. The comparison of the efficiency of double filtration and l-tuple filtration. The plot shows the ratio of the efficiency of double filtration and l-tuple filtration in 4-letter alphabet for different parameters m and k.

Comment: The definition of filtration efficiency when applied to comparison of l-tuple and double filtration should be taken with caution. The definition does not take into account the number of potential matches relative to the size of the text. When l-tuple filtration is already very efficient there is no reason to apply further filtration. In other words, if the total number of expected potential matches is, say, 1.3 for the whole text, vs. 0.013 for true matches, the ratio is large but is meaningless in practice.

5 Double filtration for approximate substring matching

In this section we present a sketch of the implementation of double filtration for approximate substring matching problem. For simplicity we concentrate on double filtration described by Algorithm 2 and consider the alphabet $\mathcal{A} = \{0, \ldots, A-1\}$.

Let p be the number of potential matches between the query and the text found at the filtration stage of the Algorithm 2, and p_c (p_g) be the number of continuous (gapped) l-tuples shared by the query and the text. It is not difficult to see that the filtration stage of Algorithm 2 can be implemented in $O(q + n + p_c + p_g)$ time by hashing (compare with [U92]).

Query hashing

We need an encoding of every l-tuple v as an integer. A natural encoding is to interpret each l-tuple as an A-ary integer. For a l-tuple $v[1, \ldots, l]$ a *hash value* of v is

$$\hat{v} = v[1]A^{l-1} + v[2]A^{l-2} + \ldots v[l]A^0 \tag{1}$$

For query $Q[1, \ldots, q]$ define $v_i = Q[i, \ldots, i+l-1]$, $1 \leq i \leq m-l+1$, be the l-tuple of Q starting at position i. Obviously

$$\hat{v}_{i+1} = (\hat{v}_i - Q[i] \cdot A^{l-1}) \cdot A + Q[i+l] \tag{2}$$

By setting v_1 and then applying (2) for $1 \leq i \leq m-l$, we get the hash values for all l-tuples of Q. Assuming that each application of (2) takes constant time (we consider relatively small A and l) we can build hash table H_1 for continuous l-tuples in $O(q)$ time. Continuous l-tuples from the query with the same hash value h are put in a linked list pointed by $H_1[h]([K73])$.

Similarly we can build a hash table H_2 for gapped l-tuples with gapsize gap in $O(q)$ time. Denote $w_i = Q[i, i+gap, i+2\cdot gap, \ldots, i+j\cdot gap, \ldots, i+(l-1)\cdot gap]$. Using the same hash function (1) for w_i we get

$$\hat{w}_{i+gap} = (\hat{w}_i - Q[i] \cdot A^{l-1}) \cdot A + Q[i+l\cdot g] \tag{3}$$

By setting $\hat{w}_1, \ldots, \hat{w}_{gap}$ and then applying 3 we get hash values for all gapped l-tuples with gapsize gap. Gapped l-tuples from the query with the same hash value h are put in a linked list pointed by $H_2[h]$. Note that with such an implementation memory requirements of double filtration are doubled in comparison with l-tuple filtration.

Text scanning with double filtration

Figure 4 presents a sketch of the filtration stage for approximate substring matching with k mismatches by double filtration. We assume $l = \lfloor \frac{m}{k+1} \rfloor$ and $size = (k+1)(l-1)+1$. Given a $q \times n$ matrix we number its $q+n-1$ diagonals assigning number $j-i+q$ to a diagonal containing position (i,j). To implement double filtration we have to test efficiently if there exists a gapped l-tuple in the vicinity of a continuous l-tuple. To provide this test we use an array $diag[1, \ldots, n+q]$ and assign

$$diag[j-i+q] = j$$

every time we find a gapped l-tuple starting at (i,j). Therefore for each of $n+q$ diagonals of the $q \times n$ matrix representing all possible coordinates, $diag[t]$ equals the starting position in T of the *last* gapped l-tuple found at this diagonal. On the preprocessing stage of the algorithm we put a dummy value $diag[t] = -1$ for $1 \leq t \leq n+q$. Although memory requirements of substring matching problem are not crucial in many applications notice that to reduce memory requirements $diag$ can be actually implemented as an array of size q that is scanned in a circular manner (not shown at Fig.4).

Algorithm Text scanning with double filtration

```
for (j = 1; j < n − l + 1; j + +) /*n is the length of the text*/
{
    if (j < n − size + 1 − k) /*size is the size of gapped l-tuple with gapsize k + 1*/
    {
        compute the hash value ŵ_{j+k} of the gapped l-tuple starting at j + k
        if (linked list H_2[ŵ_{j+k}] is not empty)
        {
            for all gapped l-tuples from H_2[ŵ_{j+k}] find their starting positions
                i[1], . . . , i[t1] in the query
            for (t = 1; t <= t1; t + +)
                diag[(j + k) − i[t] + q] = j + k;
        }
    }
    compute the hash value v̂_j of the continuous l-tuple starting at j
    if (linked list H_1[v̂_j] is not empty)
    {
        for all continuous l-tuples from H_1[v̂_j] find their starting positions
            i[1], . . . , i[t2] in the query
        for (t = 1; t <= t2; t + +)
            if(j − diag[j − i[t] + q] <= m − l) /*see lemma 8*/
                report potential match (i[t], j)
    }
}
```

Fig. 4. Sketch of the filtration stage of the double filtration approximate substring matching algorithm.

Potential match verification

Brute-force implementation of the verification stage of the algorithm adds $O(m)$ time for verifing of each potential match. For the approximate pattern matching problem it leads to an algorithm with linear expected running time for $k = O(\frac{m}{\log m})$ (see [BP92] for details).

6 Overlapping potential matches and fast dot-matrix drawing

Several optimizations are included that deviate from the simple description of the algorithm given in Fig.4. We observe that m-matches with k-mismatches frequently overlap. To exclude redundant output we use an *extended potential match* data structure.

Let (i, j) be a potential match on the diagonal $j − i + q$ where i (j) is the starting position of the l-tuple representing potential match in the query (text).

Consider the positions on the diagonal $j - i + q$ *behind* (i, j) and define an array $b[1, \ldots]$:

$$b[t] = \begin{cases} 1, & \text{if } Q[i - t] = T[j - t] \\ 0, & \text{otherwise} \end{cases}$$

Similarly, consider the positions on the diagonal $j - i + q$ *ahead* (i, j) and define an array $a[1, \ldots]$:

$$a[t] = \begin{cases} 1, & \text{if } Q[i + l - 1 + t] = T[j + l - 1 + t] \\ 0, & \text{otherwise} \end{cases}$$

(for the sake of simplicity we neglect border effects when, for example $i - t < 0$). Let $behind[0, \ldots, k]$ be an array with the positions of the first $(k + 1)$ zeros in b. Similarly let $ahead[0, \ldots, k]$ be an array with the positions of the first $(k + 1)$ zeros in a. We call the structure

$$(i, j), \quad behind[0, \ldots, k] \quad ahead[0, \ldots, k]$$

an *extended potential match* starting at (i, j).

Let (i, j) be a potential match and $Q[i+l] = T[j+l]$ (it means that $(i+1, j+1)$ is a potential match also). Notice that in this case an extended potential match $(i+1, j+1)$ does not provide any additional information in comparison with (i, j) and we can exclude such overlapping extended potential matches from further consideration.

Arrays $behind[0, \ldots, k]$ and $ahead[0, \ldots, k]$ can be easily derived by simply scanning diagonal $j - i + q$ behind (i, j) and ahead $(i + l - 1, j + l - 1)$ or by faster methods (see, for example Wu and Manber [WM92b]). We say that an approximate match with k-mismatches starting at (i', j') is *generated* by a potential match (i, j) if it belongs to the same diagonal $j' - i' = j - i$ and

$$i - behind[k] < i' \text{ and } i + l - 1 + ahead[k] > i' + m - 1$$

Lemma 6 guarantees that each approximate match is generated by at least one potential match. On the other hand a potential match (i, j) generates an approximate match with k mismatches if and only if there exists $0 \leq t \leq k$: $ahead[t] + behind[k - t] + l \geq m$. This condition gives an efficient algorithm for potential match verification. Notice that for biological applications extended potential match data structure provides a useful tool for dot-matrices drawing without looking at all approximate matches generated by a given potential match.

7 Computational experiments

We have implemented the double filtration (Algorithm 3) and compared its performance with l-tuple filtration (Algorithm 1). Recent studies ([WM92b]) demonstrate that l-tuple filtration runs much faster than other approximate pattern matching algorithms. Our study indicates that double filtration outperforms l-tuple filtration for approximate substring matching in a wide range of parameters. We presented the results of the computational experiments with the parameters $l = \lfloor \frac{m}{k+1} \rfloor$ and k as they are more convenient for comparison of running times than usual parameters m and k.

Algorithm 1 (l-tuple filtration) and Algorithm 2 (double filtration) were implemented in 'C' and all tests have been run on a SUN SparcStation 2 running UNIX. Stage 2 (potential match verification) was implemented in the same straightforward way in both Algorithm 1 and Algorithm 3. Our primary interest was to reveal the advantages and disadvantages of the filtration stage; that's why we ignored fast implementations of the verification stage. The numbers given in Figure 5 should be taken with caution. They depend on our program implementation, the architecture, the operating system, and the compilers used. However we tried to avoid optimizations and fancy programming implementations which might give an advantage to the double filtration over l-tuple filtration. The only difference between two programs was the implementation of the filtration stage.

Let $t_{fil}(t_{ver})$ be a running time of the filtration (verification) stage of the l-tuple filtration algorithm. Denote $e = \frac{E(p_1)}{E(p_3)}$ the ratio the filtration efficiency of double filtration to the filtration efficiency of l-tuple filtration. Roughly speaking a running time of double filtration algorithm will be $2 \cdot t_{fil} + \frac{t_{ver}}{e}$. In the case

$$t_{fil} + t_{ver} > 2 \cdot t_{fil} + \frac{t_{ver}}{e}$$

double filtration is faster than l-tuple filtration. It means that in the case $e > \frac{t_{ver}}{t_{ver} - t_{fil}}$ double filtration might be better than l-tuple filtration. Figure 3 indicates that this is the case for various m and k as e is very large for a wide range of parameters. Figure 5 presents the results of comparisons for $q = 10000$ and $n = 100000$ indicating that double filtration might be better for a range of parameters frequently used for dot-matrices constructions and optimal oligonucleotide probes selection ($m = 14, \ldots, 30$, $k = 1, \ldots, 5$). Note that the ratio of the running time of the l-tuple filtration algorithm to the running time of the double filtration algorithm depends on $\frac{n}{q}$ (data are shown only for $\frac{n}{q} = 10$)

8 Other filtration techniques

The basic idea of all l-tuple filtration algorithms suggested to date is to reduce a (m, k) approximate pattern matching problem to a $(m', 0)$ exact pattern problem and to use a fast exact pattern matching algorithm on the filtration stage. The drawback of such approachs is relatively low filtration efficiencies. In this section we suggest reducing (m, k) approximate pattern matching to (m', k') approximate matching with $m' < m$ and $0 < k' < k$ and application of the fast approximate pattern matching technique with small k' on the filtration stage. We demonstrate that this allows an increase of filtration efficiency without significant slowing down the filtration stage. For the sake of simplicity we illustrate this idea on a simple example reducing a (m, k) problem to a $(m', 1)$ problem.

Knuth [K73] has suggested a method for approximate pattern matching with 1 mismatch based on the observation that strings differing by a single error must

k\l	2	3	4	5	6	7	8	9	10
1		0.98	1.32	1.25	1.09	0.75	0.62	0.62	0.72
		89.6	*17.0*	*5.1*	*2.1*	*1.6*	*1.6*	*1.9*	*2.5*
2	0.87	1.27	1.55	1.58	1.23	0.75	0.62	0.62	0.72
	493.3	*84.8*	*17.6*	*4.8*	*2.1*	*1.6*	*1.6*	*1.9*	*2.5*
3	0.89	1.30	1.90	1.87	1.38	0.81	0.62	0.62	0.69
	574.5	*100.1*	*17.1*	*4.7*	*2.1*	*1.6*	*1.6*	*1.9*	*2.6*
4	0.90	1.31	1.92	2.27	1.52	0.87	0.62	0.62	0.69
	658.4	*114.5*	*19.4*	*4.4*	*2.1*	*1.6*	*1.6*	*1.9*	*2.6*
5	0.90	1.31	2.04	2.47	1.66	0.93	0.68	0.62	0.69
	746.0	*128.6*	*20.8*	*4.6*	*2.1*	*1.6*	*1.6*	*1.9*	*2.6*
6	0.91	1.31	2.14	2.62	1.85	1.00	0.68	0.62	0.65
	834.6	*144.2*	*22.2*	*4.8*	*2.1*	*1.6*	*1.6*	*1.9*	*2.7*
7	0.91	1.31	2.12	2.78	2.00	1.06	0.68	0.62	0.65
	921.4	*159.8*	*24.8*	*5.0*	*2.1*	*1.6*	*1.6*	*1.9*	*2.7*
8	0.91	1.31	2.15	2.82	2.19	1.12	0.75	0.62	0.70
	1011.1	*175.3*	*26.8*	*5.4*	*2.1*	*1.6*	*1.6*	*1.9*	*2.0*
9	0.91	1.29	2.25	3.13	2.28	1.12	0.75	0.62	0.70
	1098.9	*193.2*	*28.3*	*5.3*	*2.1*	*1.6*	*1.6*	*1.9*	*2.8*
10	0.91	1.29	2.20	3.12	2.47	1.18	0.75	0.62	0.70
	1189.5	*210.7*	*30.9*	*5.7*	*2.1*	*1.6*	*1.6*	*1.9*	*2.8*

Fig. 5. The comparison of the running time of the double filtration (Algorithm 3) and l-tuple filtration (Algorithm 1) for random Bernoulli words in 4-letter alphabet with $q = 10000$ and $n = 100000$ for different parameters k (number of mismatches) and $l = \lfloor \frac{m}{k+1} \rfloor$ (size of l-tuple). Lower cell on the intersection of k-th row and l-column represents the running time of the double filtration algorithm (in seconds). Upper cell on the intersection of k-th row and l-column represents the ratio of the running time of the l-tuple filtration algorithm to the running time of the double filtration algorithm. The area shown by a solid line represent the set of parameter (k, l) for which double filtration outperforms l-tuple filtration.

match exactly in either the first or the second half. For example, $(9,1)$ approximate pattern matching problem can be reduced to $(4,0)$ exact pattern matching problem. This provides an opportunity for 4-tuple filtration algorithm. In this section we demonstrate how to reduce $(9,1)$ approximate pattern matching to a 6-tuple filtration algorithm thus increasing the filtration efficiency by a factor of $\frac{A^2}{2}$.

Let $(l_1, g_1, l_2, g_2, \ldots, l_t, g_t, l_{t+1})$-tuple be an tuple having l_1 positions followed by a gap of length $g_1 + 1$, then l_2 positions followed by a gap of length $g_2 + 1$,

Fig. 6. Example of $(4,1,2,3,3)$-tuple.

..., then l_t positions, followed by a gap of length $g_t + 1$ and finally l_{t+1} positions (Fig.6). Fig.7 demonstrates that every boolean word of length 9 with at most 1 zero contains either a continuous 6-tuple or a $(3,3,3)$-tuple containing only ones. Two 6-tuples and one $(3,3,3)$-tuple shown in Fig.7 are *packed* into 9-letter word v so that every position in v belongs to exactly two of these tuples. Therefore the only zero in v belongs to two of these tuples leaving the third. In Fig.7 the $(3.3,3)$-tuple contains only ones.

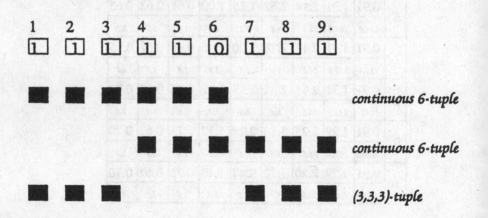

continuous 6-tuple

continuous 6-tuple

(3,3,3)-tuple

Fig. 7. A boolean word of length 9 with only zero contains either continuous 6-tuple or $(3,3,3)$-tuple containing only ones.

The following lemma generalizes the example above and allows one to perform $\lfloor \frac{2}{3}m \rfloor$-tuple filtration instead of $\lfloor \frac{1}{2}m \rfloor$-tuple filtration in Algorithm 1, thus increasing filtration efficiency approximately $4\frac{\frac{\pi}{6}}{2}$ times.

Lemma 9. *A boolean word $v[1, \ldots, m]$ with at most 1 zero contains either a continuous $\lfloor \frac{2}{3}m \rfloor$-tuple or a $(\lfloor \frac{1}{3}m \rfloor, \lfloor \frac{1}{3}m \rfloor, \lfloor \frac{1}{3}m \rfloor)$-tuple containing only ones.*

The following lemma generalizes lemma 2 and reduces the (m, k) approximate pattern matching problem to the (m', k') problem with $m' < m, k' < k$.

Lemma 10. *A boolean word $v[1, \ldots, m]$ with at most k zeros contains a subword of length m' with at most $k' < k$ zeros for $m' = \lfloor \frac{(k'+1) \cdot m + k'}{k + k' + 1} \rfloor$.*

Proof. Fix $0 < t < m$ and consider all $m - t + 1$ subwords of v of length t. Every position in v belongs to at most t of these t-words. Therefore the total number of zeros in these t-words is $z \leq k \cdot t$.

If all t-subwords of v contain at least $k' + 1$ zeros then the total number of zeros in these t-words is $z \geq (k' + 1) \cdot (m - t + 1)$ and therefore

$$k \cdot t \geq z \geq (k' + 1) \cdot (m - t + 1)$$

If this inequality fails then there exists a t-subword of v containing less than $k' + 1$ zeros. Therefore the maximum t fulfilling the inequality

$$k \cdot t < (k' + 1) \cdot (m - t + 1)$$

provides the upper bound for the length of subword containing at most k' zeros.

$$t < \frac{(k' + 1) \cdot (m + 1)}{k + k' + 1}.$$

□

Substituting $k' = 1$ in the last lemma provides a reduction of (m, k) approximate pattern matching problem to $(\lfloor \frac{2 \cdot m + 1}{k + 2} \rfloor, 1)$ approximate pattern matching problem. Lemma 9 allows further implementation of filtration with $\lfloor (\frac{2}{3} \lfloor \frac{2 \cdot m + 1}{k + 2} \rfloor) \rfloor$-tuples. For large m lemmas 9 and 10 allows one to implement l-tuple filtration with $l \approx \frac{4}{3} \frac{m}{k}$ which improves the filtration of Algorithms 1 and 3 with $l \approx \frac{m}{k}$.

Finally, there is no approximate pattern/substring matching algorithm that is the best for all possible cases. It is an open problem to find the optimal filtration techniques depending on the parameters and applications. Note that the proposed methods does not support insertions and deletions. This motivates the problem of finding an efficient filtration technique for approximate pattern matching with k differences. To solve this problem Myers, 1990 ([M90]) proposed a related method based on a reduction of the $(m, \epsilon m)$ approximate pattern matching problems with a database of length n to the $(\log n, \epsilon \log n)$ pattern matching problems. The method requires a prebuilt inverted index and so is an off-line algorithm while all the others mentioned here are on-line. This technique provides approximate pattern matching with k differences in sublinear time and gives 50- to 500-fold improvement over dynamic programming algorithms for approximate pattern matching ([U85], [MM86]).

9   Acknowledgements

We are grateful to William Chang, Udi Manber and Gene Myers for useful suggestions.

References

[BG89] Baeza-Yates R.A., Gonnet G.H. A new approach to text searching. in *Proc. of the 12th Annual ACM-SIGIR conference on Information Retrieval*, Cambridge, MA, (1989), 168-175

[BP92] Baeza-Yates R.A., Perleberg C.H. Fast and practical approximate string matching. In A.Apostolico, M.Crochermore, Z.Galil, U.Manber (eds.) Combinatorial Pattern Matching 92, Tucson, Arizona, *Lecture Notes in Computer Science*, 644, Springer-Verlag, (1992) , 185-192

[B86] Blaisdell B.E. A measure of the similarity of sets of sequences not requiring sequence alignment. *Proc. Nat. Acad. Sci. U.S.A.*, 83, (1986), 5155-5159.

[CL90] Chang W.I., Lawler E.L. Approximate string matching in sublinear expected time. *Proceedings of 31st IEEE FOCS*, (1990), 116-124

[DMDC87] Danckaert A., Mugnier C., Dessen P., and Cohen-Solal M. A computer program for the design of optimal synthetic oligonucleotides probes for protein coding genes. *CABIOS*, 3, (1987) 303-307.

[DN82] Dumas, J.P., Ninio, J. Efficient algorithms for folding and comparing nucleic acid sequences. *Nucl. Acids Res.*, 10, (1982), 197-206.

[F70] Feller W. *An introduction to probability theory and its applications*. John Wiley & Sons, New York, (1970)

[GG86] Galil, Z. and Giancarlo, R. Improved string matching with k mismatches. *SIGACT News*, April, (1986), 52-54.

[GL90] Grossi R., Luccio F. Simple and efficient string matching with k mismatches. *Information Processing Letters*, 33, (1990), 113-120

[H71] Harrison M.C. Implementation of the substring test by hashing. *C.ACM*, 14, (1971), 777-779

[HS91] Hume A., Sunday D. Fast string searching. *Software - Practice and Experience*, 21, (1991), 1221-1248

[KR87] Karp R.M., Rabin M.O. Efficient randomized pattern-matching algorithms. *IBM J. Res. Develop.*, 31, (1987), 249-260

[KS92] Kim J.Y. Shawe-Taylor J. An approximate string matching algorithm. *Theoretical Computer Science*, 92, (1992), 107-117

[K73] Knuth D.E. *The art of computer programming, vol.III: sorting and searching*. Addison-Wesley, Reading, Mass., (1973)

[LV86] Landau G.M., Vishkin U. Efficient string matching with k mismatches, *Theoret. Computer Sci.*, 43, (1986), 239-249

[LV89] Landau G.M., Vishkin U. Fast parallel and serial approximate string matching. *J. of Algorithms*, 10, (1989), 157-169

[LVN88] Landau, G.M., Vishkin, U., and Nussinov, R. Locating alignments with k differences for nucleotide and amino acid sequences. *CABIOS*, 4, (1988), 19-24.

[LP85] Lipman, D.J., Pearson, W.R. Rapid and sensitive protein similarity searches. *Science*, 227, (1985), 1435-1441.

[ML81] Maizel, J. V., Jr. and Lenk, R.P. Enhanced graphic matrix analysis of nucleic acid and protein sequences. *Proc. Nat. Acad. Sci. USA*, 78, (1981), 7665-7669.

[MM86] Myers E.W., Mount D. (1986) Computer program for the IBM personal computer that searches for approximate matches of short oligonucleotide sequences in long target DNA sequences. *Nucleic Acids Research*, 14, 501-508

[M90] Myers E.W. (1990) A sublinear algorithm for approximate keyword searching. Technical Report TR-90-25, Department of Computer Science, Thwe University of Arizona, Tucson, Arizona. (to appear in *Algorithmica*)

[OM88] Owolabi O., McGregor D.R. Fast approximate string matching. *Software-Practice and Experience*, 18, (1988), 387-393

[TU90] Tarhio J., Ukkonen E. Boyer-Moore approach to approximate string matching *Lecture Notes in Computer Science*, *447*, Springer, Berlin, (1990), 348-359

[U85] Ukkonen U. Finding approximate patterns in strings. *Journal of Algorithms*, 6, (1985), 132-137

[U92] Ukkonen U. Approximate string-matching with q-grams and maximal matches. *Theoretical Computer Science*, 92, (1992), 191-211

[WL83] Wilbur W. J., Lipman D.J.. Rapid similarity searches of nucleic acid and protein data banks. *Proc. Nat. Acad. Sci. USA* , 80, (1983), 726-730.

[WM92a] Wu S., Manber U. Agrep - A Fast Approximate Pattern-Matching Tool. *Usenix Winter 1992 Technical Conference*, San Francisco (January 1992), (1992), 153-162.

[WM92b] Wu S., Manber U. Fast Text Searching Allowing Errors. *Comm. of the ACM*, 35, No.10 (1992),83-90

A Unifying Look at d-dimensional Periodicities and Space Coverings

Mireille Régnier[1] and Ladan Rostami[2]

[1] INRIA, 78153 Le Chesnay, France,
regnier@margaux.inria.fr
[2] LITP, 2 place Jussieu,
75 251 Paris Cedex 05, France,
rostami@litp.ibp.fr

Abstract. We propose a formal characterization of d-dimensional periodicities. We show first that any periodic pattern has a canonical decomposition and a minimal generator, generalizing the 1D property. This allows to classify the d-dimensional patterns in $2^{d-1} + 1$ classes, according to their periodicities, each class having subclasses. A full classification of the coverings of a 2-dimensional space by a pattern follows. These results have important algorithmic issues in pattern matching. First, the covering classification allows an efficient use of the now classical "duel" paradigm. Second, d-dimensional pattern matching complexity is intrinsically different for each class.

1 Introduction and State of the Art

A lot of attention has been given in the last decade to string searching in a text. String searching can be generalized to "multidimensional search" or "multidimensional pattern matching". A multidimensional pattern, p, most often an array and usually connex and convex, is searched in a multidimensional array, the text, t. It remains a widely open research area, although a strong interest appeared recently [ZT89, BYR90, ABF92, GP92]. It is interesting to notice that the now classical "duel" paradigm [Vis85] allows a drastic improvement on string searching average efficiency [Gal92] while a refined analysis of possible repetitions and periods in a word allowed to derive the worst case complexity, $(1 + O(\frac{1}{m})).n$, and to achieve it [CH92].

It seems reasonable to expect that a classification of repetitions in d-dimensional patterns will allow to derive the theoretical complexity bound and achieve it. To support this strong assertion, we point out that, as extensively studied for string searching, lower bounds on the worst-case complexity depend on the maximal number of occurrences of a given pattern that can be found in a text. That is, in a d-dimensional space, possible space coverings. Also, if such a compact representation of p is not a tiling, certainly p overlaps with himself: some subpattern is repeated inside p. Notice that preliminary results, as well as algorithmic issues, are presented in [NR92].

---

* This work was partially supported by the ESPRIT II Basic Research Actions Program of the EC under contract No. 3075 (project ALCOM).

Let us turn now to the state of the art and a definition of repetitions or periods. In dimension 1, the problem has been extensively studied by various mathematicians, whose results are grouped in [Lot83]. Notably, it is proved that a word w is selfoverlapping iff exists a word x such that

$$w = s(x).x^k = x^k p(x) \tag{1}$$

where $p(x)$ and $s(x)$ are a prefix and a suffix of x. Equivalently, w is a truncation of some word of $x^* = \cup_m \{x^m\}$. When $w \neq x$, w is called *periodic*, and x is a *period* of w. Additionnally, a unicity property holds. A *primitive* word w being a word that is not the power of any other word, we have:

Theorem 1. *Given a word w, let x_1, x_2 be two primitive words such that:*

$$w = x_1^{k_1}.p(x_1) = x_2^{k_2}p(x_2), k_1 > 1, k_2 > 1 . \tag{2}$$

Then $x_1 = x_2$ and $k_1 = k_2$.

In that case, the word w satisfies the (more restrictive) periodicity definition: a self-overlap accross the middle point. We prefer to maintain the definition of period of [Lot83], and distinguish non-degenerated periodicities ($|x| < |w|/2$) and degenerated periodicities. Notice that degenerated periodicities have consequences on the average case of string searching algorithms. A study can be found for Knuth-Morris-Pratt in [Rég89]. Also, they provide the variations on the second order term, $O(\frac{1}{m})$ of 1D complexity [CH92] (although this is not stated explicitly in the paper). A hint on their importance in d-dimension. As a matter of fact, they appear in the first order [NR92]. In the following, we use equivalently the terms of self-overlapping patterns and periodic patterns. A recent work [AB92] started the study of 2-dimensional periods. In that work, a pattern is periodic iff it selfoverlaps **and** this overlapping includes the center. This is a natural extension of 1D non-degeneracy notion, and will be called in the following non-degenerated periodic patterns. [AB92] introduces the notion of *sources*, the locations where an overlap can originate. The main result is a classification of pattern periodicities according to such locations: non-periodic, line periodic, radiant periodic and lattice periodic. A geometric regularity of this pattern of "candidates to be a source" is claimed, but actual sources only form a subset of that location pattern (the difference occurs essentially for radiant periodicity). That classification is incomplete, as the characterization of such subsets is not provided. We will give it. Also, it is claimed that the 1D property in Theorem 1: "a periodic pattern has a (unique) generator, the period", does not generalize to 2 dimensions. We invalidate that claim, showing that a periodic pattern actually is the repetition of a smaller pattern. This leads to provide the lacking criterium cited above, and to fully describe the characters repetitions in all periodic patterns. This result also provides a (natural) classification of space coverings in 3 classes (3 stands for $2^{d-1} + 1$ for a dimension d equal to 2). We discuss in 5 the pertinence of this new classification, where degenerated periodicities are essential, and the relationship to the previous classification. Notably, we discuss the algorithmic consequences and the extension to d-dimensional spaces.

2 Formalism

A d-dimensional pattern p is a d-dimensional array $p[[1 \ldots l_1], \ldots, [1 \ldots l_d]]$ where l_i is some integer, called the i-th dimension. We note $\bar{j} = j \bmod d$ and call $(\mathbf{e_1}, \ldots, \mathbf{e_d})$ an euclidean basis of the d-dimensional space. Most results on p will be proved on the vectorial subset P defined below:

Definition 2. Given a pattern p, let P be the subset:

$$\{\mathbf{v}; \mathbf{v} = \sum_i v_i \mathbf{e_{\bar{i}}}, (|v_1|, \ldots, |v_d|) \in \prod_1^d [1 \ldots l_i]\} . \tag{3}$$

Let c be the application $P \leftarrow P$ defined by:

$$c(\mathbf{v}) = \sum_{u_i \geq 0} u_i \mathbf{e_{\bar{i}}} + \sum_{u_i < 0} (l_i - u_i) \mathbf{e_{\bar{i}}} . \tag{4}$$

Given a vector $\mathbf{u}$, one notes:

$$P_\mathbf{u} = \{\mathbf{v}; \mathbf{v} = \mathbf{w} + \mathbf{u}, \mathbf{w} \in P\} . \tag{5}$$

A vector $\mathbf{u}$ is a *translation vector* iff $P_\mathbf{u} \cap P \neq \emptyset$. Let T be the set of translation vectors.

In the following, we note, for $\mathbf{u} \in c(P)$, $p(\mathbf{u})$ the element $p[u_1, \ldots, u_d]$ of p. For other $\mathbf{u}$, we define $p(\mathbf{u}) = p(c(\mathbf{u}))$. We are interested in shifts such that the two copies are consistent in the overlapping area. That is:

Definition 3. A translation vector $\mathbf{u}$ is an *invariance* vector for p iff:

$$\mathbf{v} \in P, \mathbf{v} + \mathbf{u} \in P \Rightarrow p(\mathbf{u} + \mathbf{v}) = p(\mathbf{v}) . \tag{6}$$

An invariance vector $\mathbf{u}$ is *non-degenerated* if exists an invariance vector $\mathbf{v}$ (possibly equal) such that $\mathbf{u} + \mathbf{v} \in T$. We note I and $\tilde{I}$ the set of invariance vectors and the set of non-degenerated invariance vectors. It generates

$$\bar{I} = \{\lambda \mathbf{u} + \mu \mathbf{v}; \mathbf{u} \in \tilde{I}, \mathbf{v} \in \tilde{I}\} . \tag{7}$$

Note 4. An invariance vector is associated to a "source" in [AB92] terminology. In [AB92], a source must also satisfy the condition: $2\mathbf{u} \in T$, which is one of our cases of degeneracies.

To express our results, we need the fundamental notion of direction:

Definition 5. A *direction* in a d-dimensional space is a sequence:

$$(\epsilon_1, \ldots, \epsilon_d) \in [0, 1]^{d-1} \tag{8}$$

The 2^{d-1} directions so-defined can be numbered from 1 to 2^{d-1} by the application:

$$(\epsilon_1, \ldots, \epsilon_d) \rightarrow n = 1 + \sum_{i=2}^d \epsilon_i 2^{i-1} \tag{9}$$

For any $\mathbf{u}$, we note $dir(\mathbf{u})$ the number associated to his direction, determined by the set of equations:

$$\text{if } u_1 > 0 \text{ then } \epsilon_i = \begin{cases} 0 \text{ if } u_{i+1} \geq 0 \\ 1 \text{ otherwise} \end{cases} \tag{10}$$
$$\text{else } dir(\mathbf{u}) = dir(-\mathbf{u})$$

We can also define a canonical numbering on the corners of p.

Definition 6. Given two opposite corners in direction i, we number C_i (respectively C_{d+i}) the one with smaller (respectively greater) 1-coordinate. We note $C_l^i, 1 \leq l \leq d$ its coordinates in p.

Definition 7. A pattern is k-self overlapping if it may overlap with himself on exactly k directions.

As k ranges in $\{0, 2^{d-1}\}$, this determines $2^{d-1} + 1$ classes. In dimension 2, more precisely studied in this paper, we get:

Definition 8. We classify 2-dimensional patterns in 3 disjoint classes:

(1) A pattern is *non-overlapping* iff it has no invariance vector.
(2) A pattern is *mono-overlapping* if all its invariance vectors lie in the same direction.
(3) A pattern is *bi-overlapping* iff exist invariance vectors in both directions.

Definition 9. Given a pattern p, and $\mathbf{u}$ a vector, the $\mathbf{u}$-strip for that pattern is:

$$S_{\mathbf{u}} = S_{\mathbf{u}}^+ \cup S_{\mathbf{u}}^- , \tag{11}$$

where $S_{\mathbf{u}}^+ = \{\mathbf{v} \in P; \mathbf{u} + \mathbf{v} \notin P\}$ and $S_{\mathbf{u}}^- = \{\mathbf{v} \in P; \mathbf{v} - \mathbf{u} \notin P\}$.
The $\mathbf{u}$-dead zone is:

$$DZ_{\mathbf{u}} = \{\mathbf{v} \in P; \mathbf{v} + \mathbf{u} \notin P \text{ and } \mathbf{v} - \mathbf{u} \notin P\} = S_{\mathbf{u}}^+ \cap S_{\mathbf{u}}^- . \tag{12}$$

Let us define:

$$F_{\mathbf{u}}^+ = \{\mathbf{w} \in P_{-\mathbf{u}}; p(\mathbf{w} + \mathbf{u}) \neq p(\mathbf{w}))\}, F_{\mathbf{u}}^- = \{\mathbf{w} \in P_{\mathbf{u}}; p(\mathbf{w} - \mathbf{u}) \neq p(\mathbf{w}))\} . \tag{13}$$

A pattern is $\mathbf{u}$-selfoverlapping iff $F_{\mathbf{u}}^+ \cup F_{\mathbf{u}}^-$ is empty. We define the $\mathbf{u}$-free zone as:

$$F_{\mathbf{u}} = (F_{\mathbf{u}}^+ \cap F_{\mathbf{u}}^-) \cup (F_{\mathbf{u}}^+ - P_{\mathbf{u}}) \cup (F_{\mathbf{u}}^- - P_{-\mathbf{u}}) , \tag{14}$$

Intuitively, $DZ_{\mathbf{u}}$ is the subset of the pattern that is not constrained by a $\mathbf{u}$-period. One always has: $DZ_{\mathbf{u}} \cap F_{\mathbf{u}} = \emptyset$. $F_{\mathbf{u}}^+ \cup F_{\mathbf{u}}^-$ is the zone of non-invariance. The free zone is its subset; one excludes points $p(\mathbf{t} + \mathbf{u}) = p(\mathbf{t}) \neq p(\mathbf{t} - \mathbf{u})$ or $p(\mathbf{t} + \mathbf{u}) \neq p(\mathbf{t}) = p(\mathbf{t} - \mathbf{u})$.

In order to derive properties of patterns with several periods and to define a minimal representation of the set of periods, we define an **order** on invariance vectors:

Definition 10. Let $\leq$ be the order on the d-dimensional vectorial space:

$$\mathbf{u} \leq \mathbf{v} \Leftrightarrow dir(\mathbf{u}) = dir(\mathbf{v}) \ and \ \mathbf{v} \in P_{\mathbf{u}} \ . \tag{15}$$

Analytically, this is equivalent to:

$$|u_i| \leq |v_i|, u_i.v_i > 0, \ 1 \leq i \leq d \ . \tag{16}$$

Remark. This notion coincides with the "monotone ordering" of sources in [AB92].

Finally, we recall the basic definitions for lattices, as defined for elliptic functions (see for instance [Lan87]).

Definition 11. Given two non-colinear vectors ω_1 and ω_2, we note L_{ω_1, ω_2} the sub-lattice:

$$\{\lambda\omega_1 + \mu\omega_2; \lambda \in Z, \mu \in Z\} \cap T \ . \tag{17}$$

(ω_1, ω_2) is said a basis of the lattice. The *vectorial fundamental parallelogram* with respect to this basis is:

$$VFP_{\omega_1, \omega_2} = \{\alpha\omega_1 + \beta\omega_2 \in P; 0 \leq \alpha < 1, 0 \leq \beta < 1\} \ . \tag{18}$$

The *fundamental parallelogram* located in C_i is:

$$FP_{\omega_1, \omega_2} = \{p(C_i + \mathbf{w}); \mathbf{w} \in VFP_{\omega_1, \omega_2}\} \ . \tag{19}$$

Definition 12. Given a set S of vectors, a S-path is a sequence $(\mathbf{w}_i)_{i \in N}$ such that:

$$\mathbf{w}_{i+1} - \mathbf{w}_i \in S \cup (-S) \ . \tag{20}$$

A S-path from a vector $\mathbf{w}$ to an other vector $\mathbf{w}'$ is a sequence $(\mathbf{w}_i)_{0\ldots n}$ such that:

$$\mathbf{w}_0 = \mathbf{w}, \mathbf{w}_n = \mathbf{w}', \mathbf{w}_{i+1} - \mathbf{w}_i \in S \cup (-S) \ . \tag{21}$$

It is *monotonic* with the restriction:

$$\mathbf{w}_{i+1} - \mathbf{w}_i \in S \ . \tag{22}$$

Given a pattern p, a S-path is said p-valid iff all its elements lie in P.

Let $\phi_{\omega_1, \omega_2}$ be the application $P \rightarrow VFP_{\omega_1, \omega_2}$ that associates to any vector in P it's image in VFP_{ω_1, ω_2} by a $\{\omega_1, \omega_2\}$-path.

3 Combining periods

This section is devoted to the study of patterns having several invariance vectors. One would expect the sum of invariance vectors is an invariance vector. Unfortunately, it is not always the case. Nevertheless, basic equality $p((u + v) + w) = p(w)$ is satisfied *almost everywhere*, and we may characterize this zone of non-invariance.

We exhibit first constraints on the patterns, that will be extended in the following section by a study of possible generating subpatterns. Our first lemma states general conditions under which the sum of invariance vectors is an invariance vector, formalizing analytically and generalizing the result in [AB92].

Lemma 13. *Let p be a d-dimensional "rectangular" array. If u is an invariance vector, then: $ku \in T, k \in Z$ is an invariance vector. Let u, v be two invariance vectors such that u, v and $u + v$ lie in the same direction and way. Then $k_1 u + k_2 v, k_1, k_2 \in N$, is an invariance vector if it lies in T.*

Proof. The property for P is a direct consequence of the definition (see [AB92] for dimension 2). It is enough to remark that, for a (convex) rectangular array, whenever $t \in P$ and $t + (u + v) \in P$, either $t + u \in P$ or $t + v \in P$.

Our next theorems will rely on a very simple but useful lemma:

Lemma 14. *Let u, v be two vectors in different directions. Assume $t = u + v$ lies in the same direction as u. Then, for any vector $w \in P$, exists a valid (u, v)-path to the fundamental parallelogram $FP_{u,t}$.*

Proof. We include $FP_{u,t}$ in two strips parallel to u and v, of widths $|v|$ and $2|u|$. Any point has a v-path to the first strip, then a u-path to some point α of the intersection. If this point does not lie in $FP_{u,t}$, either one of $\alpha + u$ or $\alpha - u$ lies in it. Clearly, this path is p-valid.

We get as an immediate corollary the fundamental result, also stated in [AB92]:

Theorem 15. *Assume $d = 2$. Let u, v be two invariance vectors in different directions. Then, $u + v$ and $u - v$ are invariance vectors when they lie in T.*

Remark. It is interesting to see that the implication: $dir(u) \neq dir(v) \Rightarrow u - v$ *is an invariance vector* does not hold in higher dimensions.

Proof. Let w and w' be two vectors in P such that $w - w' \in \{t, -t\}$. w and w' has valid (u, v)-path to two points of the parallelogram. As t is a border of that parallelogram, these points must coincide. Hence, $p(w) = p(w')$.

Finally, the next proposition deals with the remaining case:

Proposition 16. *Let u, v be two invariance vectors in the same direction. Let $t = u + v$. Then,*

$$F_t^+ \subseteq S_u^+ \cap S_v^+ \cap P_{-t}, F_t^- \subseteq S_u^- \cap S_v^- \cap P_t . \tag{23}$$

That is, t is an invariance vector in $P - (F_t^+ \cup F_t^-)$.

Proof. Let $\mathbf{w}$ be a vector in $P_{\mathbf{t}}^-$. Then:

$$\mathbf{w} \in P_{-\mathbf{u}} \Rightarrow \left\{ \begin{array}{l} p(\mathbf{w}+\mathbf{u}) =_{\mathbf{u}} p(\mathbf{w}) \\ p(\mathbf{w}+\mathbf{t}) =_{\mathbf{v}} p(\mathbf{w}+\mathbf{u}) \end{array} \right\} \Rightarrow p(\mathbf{w}) = p(\mathbf{w}+\mathbf{t}) \Rightarrow \mathbf{w} \notin F_{\mathbf{t}}^+ \ . \quad (24)$$

Hence, $F_{\mathbf{t}}^+ \subseteq S_{\mathbf{u}}^+ = P - S_{\mathbf{u}}^+$; similarly, $F_{\mathbf{t}}^+ \subseteq S_{\mathbf{v}}^+$, which yields our inclusion. The same reasoning applies for $F_{\mathbf{t}}^-$.

This proposition shows that only some borders of the pattern are not constrained, while the $\mathbf{u} + \mathbf{v}$ invariance holds in the interior. These borders will be precised in the next section, where a canonical decomposition of periodicities is proposed.

4 Canonical Decomposition

The aim of this section is twofold. First, we define a basic set of generators for I. Also, we characterize a subpattern that generates P. Second, we characterize I as a subset of $L_{\mathbf{u},\mathbf{v}}$ plus a set of degenerated invariance vectors.

4.1 Period Characterization and Periodicities Classification

From the definition of an invariance vector, it appears that $S_{\mathbf{u}}^+$ (or $S_{\mathbf{u}}^-$) is a generator of the pattern, as:

$$\mathbf{w} \in P \Leftrightarrow \exists \mathbf{t} \in S_{\mathbf{u}}^+, \exists k \in Z : \mathbf{w} = \mathbf{t} + k\mathbf{u} \ . \quad (25)$$

Then, $p(\mathbf{u}) = p(\mathbf{t})$. A similar property holds for $S_{\mathbf{u}}^-$. We want a minimal representation of generators. In Theorem 18, we derive characteristic parameters for a set of colinear invariance vectors. Then, we address the case when two non-colinear invariance vectors exist, and show that the generating strip reduces to a (linear) word. We state below the associated definition:

Definition 17. A primitive word s is said a linear i-pseudoperiod of a bidimensional pattern p iff:

$$s\mathbf{e_i} \in \bar{I}, \bar{I} \not\subset \{\mu\mathbf{e_i}; \mu \in Z\} \ . \quad (26)$$

We define:

$$L = min\{x; -x\mathbf{e_i} + \mathbf{e_{i+1}^-} \in \bar{I}, x \in N\} \ ,$$
$$E_{s,L} = (s \bmod L)\mathbf{e_i} + (s/L+1)\mathbf{e_{i+1}^-} \ ,$$
$$F_{s,L} = (L + s \bmod L)\mathbf{e_i} + (s/L)\mathbf{e_{i+1}^-} \ ,$$

and we call $(E_{s,L}, F_{s,L})$ the canonical i-basis. If

$$p[\phi_{(E_{s,L},F_{s,L})}(\mathbf{w})] = p[\mathbf{w}] \quad (27)$$

holds in P, s is said a linear i-period. If it holds out of a subpattern A, then s is said a A-linear i-pseudoperiod.

Remark. It is easy to see that $(E_{s,L}, F_{s,L})$ and $(se_i, -Le_i + e_{i+1}^-)$ generate the same lattice. $(E_{s,L}, F_{s,L})$ is canonical in the usual sense of elliptic functions. Properties of such a basis may be found in [Lan87]. Intuitively, $(E_{s,L}, F_{s,L})$ represents the "smallest" invariance vectors, and s repeats indefinitely in the pattern.

We first consider sets of colinear invariance vectors.

Theorem 18. *Let Δ be a set of colinear invariance vectors. Exists $\mathbf{u}, \lambda_{\mathbf{u}}$ such that:*

$$\wedge \lambda_i \neq 1 \qquad \wedge \lambda_i \wedge \lambda_{\mathbf{u}} = 1 \tag{28}$$

$$\Delta = \{\lambda \mathbf{u}; \lambda \geq \lambda_{\mathbf{u}}\} \cup \{\lambda_i \mathbf{u}\} . \tag{29}$$

$\mathbf{u}$ *is called a pseudo-period.*

If I contains a non-colinear invariance vector $\mathbf{v}$, then. $\Delta_{\mathbf{u}}$ reduces to the segment:

$$\{\lambda \mathbf{u}; \lambda \geq \lambda_{\mathbf{u}}\} . \tag{30}$$

Remark. As in 1D, all non degenerated periods are multiple of the pseudo-period, *but* this pseudo-period is not always a period.

Proof. By definition, $\Delta = \{k\mathbf{v}, k \in Z\} \cap T$. Let $\mu = \wedge_i \{\mu_i; \mu_i \mathbf{v} \in I\}$ and note $\mathbf{u} = \mu \mathbf{v}$. Ordering Δ yields $\lambda_1 < \ldots \lambda_k$ satisfying $\wedge_{i=1}^{k-1} \lambda_i \neq 1, \wedge_{i=1}^{k} \lambda_i = 1$. Now, repeatedly applying Proposition 14 shows that $\mathbf{u}$ is an invariance vector out of $DZ_{\lambda_1 \mathbf{u}} \cup \ldots \cup DZ_{\lambda_k \mathbf{u}} = DZ_{\lambda_k \mathbf{u}}$. Hence, $\lambda_k \mathbf{u} + \alpha \mathbf{u}$ is an invariance vector out of this zone, strictly included in $DZ_{(\lambda_k + \alpha)\mathbf{u}}$. This establishes that $(\lambda_k + \alpha)\mathbf{u} \in I$. We note $\lambda_k = \lambda_{\mathbf{u}}$.

Assume now the existence of a non-colinear invariance vector $\mathbf{v}$. Note $\lambda \mathbf{u}$ the minimal element of $\Delta_{\mathbf{u}}$. If $2\lambda \mathbf{u} \notin I$, then Δ reduces to $\lambda \mathbf{u}$, and the property trivially holds. Assume now that exists $\nu, \lambda < \nu < \lambda_{\mathbf{u}}$ such that $\nu \mathbf{u} \notin I$. If $\lambda \mathbf{u}$ and $\mathbf{v}$ are ordered, then $F_{\mathbf{v} - \lambda \mathbf{u}} \subseteq DZ_{(min(\mathbf{v}, \lambda \mathbf{u}))} \subseteq DZ_{\nu \mathbf{u}}$. If they are not ordered, then the coordinates of $\mathbf{v} - \lambda \mathbf{u}$ are upper bounded by $\lambda \mathbf{u}$ coordinates, hence $\nu \mathbf{u}$ coordinates, then the inclusion also holds. As s repeats regularly out of $DZ_{\nu \mathbf{u}}$, which is a set irrelevant to $\nu \mathbf{u}$ invariance, we get $\nu \mathbf{u} \in I$, a contradiction.

We turn now to the case where non-colinear multiplicities occur. Our canonical decomposition is based on the following theorem:

Theorem 19. *Let p be a 2-dimensional patterns with two invariance vectors $\mathbf{u}$ and $\mathbf{v}$ in the same direction i such that $\mathbf{u} + \mathbf{v} \in T$. Let x_{max} and y_{max} (respectively x_{min} and y_{min}) be the x and y coordinates of largest (respectively smallest) absolute value. Let α_j be the j-th row of the fundamental parallelogram. Let σ be the canonical permutation:*

$$j \leftarrow (j - y_{min}) \bmod y_{max} . \tag{31}$$

Then, exists a unique primitive word s and an integer l such that:

$$\alpha_1 \alpha_{\sigma(1)} \ldots \alpha_{\sigma^{y_{max}-1}(1)} = (s)^l . \tag{32}$$

Note $L = |\alpha_1 \ldots \alpha_{\sigma^{k-1}(1)}| \pmod{|s|}$, with $\sigma^k(1) = 2$. Then s is a $F_{\mathbf{v}-\mathbf{u}}$-linear i-pseudoperiod of p.

Proof. We show that, for any location of the fundamental parallelogram in the pattern, the j-th row is followed in direction i by row $\sigma(j)$. Let $\mathbf{w}$ point to its last element. As $\mathbf{w} + \mathbf{e_i} \notin VFP$, either $\mathbf{w} + \mathbf{e_i} - \mathbf{u}$ or $\mathbf{w} + \mathbf{e_i} + \mathbf{v}$ points to one element of the parallelogram. Also, neither $\mathbf{w} - \mathbf{u}$ nor $\mathbf{w} + \mathbf{v}$ can point to the parallelogram. Hence, $\mathbf{t}$ points to the first element of a sequence, whose index is:

$$\begin{cases} j - u_{i\mp 1} \text{ if } j > |u_{i\mp 1}| (i) \\ j + v_{i\mp 1} \text{ otherwise } (ii) \end{cases} \tag{33}$$

which is precisely $\sigma(j)$. As $\mathbf{u}, \mathbf{v}$ are invariance vectors, $\mathbf{w} + \mathbf{e_i}$ points to the same sequence. As fundamental parallelograms realize a tiling of the plane, sequence $\alpha_1 \alpha_{\sigma(1)} \ldots \alpha_{\sigma^{y_{max}-1}(1)}$ repeats indefinitely. If it rewrites w^k, the same repetition property holds for w, and notably for the associated primitive word s'. To get the formula for L, it is enough to remark that row 2 starts with $\alpha_{\sigma^{-1}(2)}$.

This theorem allows for defining a canonical basis of periodicities. Basically, we show that all invariance vectors are generated by a 2-set. Namely, we have the 2D-generalization of 1D unicity property 1 :

Theorem 20. *A pattern admits at most one i-pseudoperiod.*

Proof. Assume two i-pseudoperiods s and s' exist. Assume, say, that $|s| \leq |s'|$. Our construction implies that

$$s' = sa = as \tag{34}$$

which implies by the conjugacy theorem [Lot83] that s and s' are powers of a same word, a contradiction with the primitivity property.

We are now ready to give our classification of periodicities:

Definition 21. We classify periodicities in three classes. A pattern is:

(1) **non-periodic:** all invariance vectors are degenerated.
(2) **monoperiodic:** all non-degenerated invariance vectors are colinear. Let $\mathbf{u}$ be the pseudoperiod.
(3) **biperiodic:** exist two non-colinear non-degenerated invariance vectors. It divides into two subclasses:
   (a) **lattice periodic:** the linear pseudo-periods are periods.
   (b) **radiant periodic:** the linear pseudoperiods are not periods.

We will discuss in 5 the relationship to the definition 8 and to the covering classification. We only state the relationship to [AB92] classification. *Lattice, radiant and line periodic* patterns as defined in [AB92] are *lattice biperiodic , radiant biperiodic and monoperiodic* according to our classification. Our definitions of *non-periodic* patterns coincide. But our proposed classification takes into account all periodicities (e.g includes degenerated periodicities), which is essential to a space covering classification. Also, it allows a full characterization of the relative position of sources, as detailed in the next subsection.

4.2 Invariance Vectors Distribution

From the results above, we get:

Theorem 22. *Let p be a periodic pattern. If p is:*

(1) monoperiodic with pseudoperiod $\mathbf{u}$:

$$\begin{cases} \tilde{I} & \subseteq \{\lambda\mathbf{u}; |\lambda| \geq \lambda_{\mathbf{u}}\} + \cup_i \{\lambda_i \mathbf{u}\} \\ I - \tilde{I} \subseteq S^+_{\lambda_{\mathbf{u}}\mathbf{u}} \cup S^-_{\lambda_{\mathbf{u}}\mathbf{u}} \end{cases} \tag{35}$$

where (λ_i) *satisfies condition 28.*

(2) biperiodic with linear pseudoperiod s and canonical basis $(E_{s,L}, F_{s,L})$:

$$\begin{cases} \tilde{I} & \subseteq L_{(E_{s,L}, F_{s,L})} \\ I - \tilde{I} \subseteq \{\mathbf{v}; \phi_{(E_{s,L}, F_{s,L})}(P \cap P_{\mathbf{v}}) \nsubseteq V F_{(E_{s,L}, F_{s,L})}\} \end{cases} \tag{36}$$

Proof. Assertion for monoperiodic patterns is a straightforward consequence of Theorem 18 and of degeneracy and monoperiodicity definitions. Now, when p is biperiodic, let $\mathbf{v} \in I$ be some non-degenerated vector. Assume it is interior to some fundamental parallelogram of the canonical lattice. Mapping the fundamental parallelogram located in $\mathbf{v}$ into it yields the equation: $s = s_1 s_2 = s_2 s_1$, a contradiction as explained above.

We now characterize the subset of the canonical lattice that actually is in $\tilde{I}$, according to the type of biperiodicity.

Theorem 23. *Let p be a biperiodic pattern with linear i-pseudoperiod s, and i-canonical basis* $(E_{s,L}, F_{s,L})$. *Then:*

$$p \text{ is lattice periodic} \Leftrightarrow s \text{ is a period} . \tag{37}$$

If p is radiant periodic, exists $(\alpha_1, \alpha_2) \in N^{*2}$ *such that:*

$$\tilde{I} = \{\mathbf{u} \in L_{(E_{s,L}, F_{s,L})}; |u_1| \geq \alpha_1, |u_2| \geq \alpha_2\} . \tag{38}$$

Note 24. As stated in [AB92], all basic sources "radiate" from the same corner. The equidistribution appearing on the figures of that work is now precised. Only points falling in a canonical segment (bolded characters in Figure 1) define actual periodicities. The other ones (italic characters) do not, due to the intersection of the shifted copies so-defined with the dead zone. In our Figure, all points on that segment define periodicities, but this is not general).

```
b g c d e f g h a b c d e
c d b c d e f g h a b c d
g h a b c d e f g h a b c
f g h a b c d e f g h a b
e f g h a b c d e f g h a
d e f g h a b c d e f g h
c d e f g h a b c d e f g
b c d e f g h a b c a b g
a b c d e f g h a b a c d
```

Figure 1.

5 Pattern and Covering Classification

We classify the possible coverings as a function of the number of directions where the pattern p may overlap. Also, all periods, including the "degenerated" ones are to be taken into account. This relies on the fundamental notion of *mutual consistency* of overlapping copies of p.

Definition 25. Given a pattern p and two vectors $\mathbf{u}, \mathbf{v}$ in P, two sets $P_\mathbf{u}$ and $P_\mathbf{v}$ are said mutually consistent iff:

$$\mathbf{w} \in P_\mathbf{u} \cap P_\mathbf{v} \Rightarrow p(\mathbf{w} - \mathbf{u}) = p(\mathbf{w} - \mathbf{v}) \tag{39}$$

Intuitively, this means that whenever a copy of p is shifted by $\mathbf{u}$ and $\mathbf{v}$, the two copies can coincide in the text. Remark that, whenever $P_\mathbf{u} \cap P_\mathbf{v} = \emptyset$, $P_\mathbf{u}$ and $P_\mathbf{v}$ are mutually consistent. We get the following lemma, easy to prove, but fundamental for algorithmic issues [BYR90, NR92]:

Lemma 26. *Given a pattern p, let $\mathbf{u}$ and $\mathbf{v}$ be two non-ordered invariance vectors. $P_\mathbf{u}$ and $P_\mathbf{v}$ overlap iff $\mathbf{u} - \mathbf{v} \in T$. Also, two overlapping sets $P_\mathbf{u}$ and $P_\mathbf{v}$ are mutually consistent iff $\mathbf{u} - \mathbf{v}$ is an invariance vector.*

Proof. It is enough to remark that two copies of p shifted by $\mathbf{u}$ and $\mathbf{v}$ overlap between themselves as p and a copy shifted by $\mathbf{u} - \mathbf{v}$. That is, mutual consistency is equivalent to have $\mathbf{u} - \mathbf{v}$ as an invariance vector, or to have $P_\mathbf{u} \cap P_\mathbf{v} = \emptyset$.

The following theorem steadily follows:

Theorem 27. *Let $P, P_\mathbf{u}, P_\mathbf{v}$ be three overlapping copies of a pattern. They are mutually consistent in either one of the two cases (and no other):*

(a) p is lattice periodic. Then $\mathbf{u}$ and $\mathbf{v}$ lie on the canonical lattice with its origin in a corner of p.

(b) p is monoperiodic or radiant periodic. Then, $\mathbf{u}, \mathbf{v} \in \{k\lambda_\mathbf{G}\mathbf{G}; k \in Z\}$, where $\mathbf{G}$ is a pseudoperiod of p. .

We first state the definitions:

Definition 28. A $\mathbf{u}$-overlapping sequence is a set of copies $(p_i)_{i \in I}$ of p, such that two copies p_i and p_{i+1} are shifted by $\mathbf{u}$.

A $\mathbf{u}$-diagonal covering is a tiling of $\mathbf{u}$-overlapping sequences.

A $(\mathbf{u}, \mathbf{v})$-lattice covering is a set of interleaved $\mathbf{u}$-overlapping sequences where two neighbouring sequences are shifted by $\mathbf{v}$. Given a pattern p, it is *regular* if $\mathbf{u} + \mathbf{v} \in T$, else it is said *extended*.

It follows from Theorem 27:

Theorem 29. *A maximal covering of the 2-dimensional space by a pattern p is either of the three following.*

(1) tiling,

(2) u-diagonal covering, *where* u *is an invariance vector.*

(3) (u, v)-*lattice coverings, where* (u, v) *is a basis of the canonical lattice.*

Additionnally, p has a maximal covering in class I iff p is in overlapping class I. Also, only lattice periodic patterns have regular lattice coverings.

Remark that extended lattice coverings are an extension of the covering notion, where some "holes" appear in the representation. Nevertheless, this is pertinent for algorithmic issues as it allows to determine the **maximum** number of occurrences of a given pattern. (This number may not be given by the maximal covering, when degeneracies appear). This parameter is clearly related to the worst-case complexity. This is extensively studied in a companion paper [NR92]. Notably, one proves a complexity $1.n$ for non-overlapping patterns and a complexity $\alpha.n, \alpha > 1$ for bi-overlapping patterns. We believe that the refinements of periodicity classification presented here will allow to extend the results to mono-overlapping patterns.

Finally, note this classification follows from our overlapping classification in Definition 8, while it is "orthogonal" to the one in [AB92], as well as to our periodicities classification. Class 3 contains patterns with periods in two directions. A radiant periodic pattern (periodicity class (3)) with its pseudoperiods in some direction i lies either in class (2) or in class (3)(b). This depends of the existence (or not) of a degenerated invariance vector in the other direction. In any case, its maximal covering is a diagonal covering (class (2)).

6 Conclusion

We exhibited here the relationship between the periods of a pattern and the possible space coverings by the same pattern. This is relevant both to the derivation of the theoretical complexity of d-dimensional pattern matching and to algorithmic issues. Notably, combining the duel paradigm with knowledge on periods should improve average and worst-case complexity of pattern matching. This is treated in a companion paper. We proved here that a periodic pattern is generated by a subpattern, and exhibit the subpattern as well as the generating law. This considerably refines and achieves the previous classification by [AB92] and allows for a classification of space coverings. Additionnally, it provides tools for a generalization to any dimension. Notably, the number of periodicity classes appear linear in the dimension. Finally, it provides knowledge to derive efficient pattern perprocessing.

Acknowledgements

We wish to thank Pierre Nicodème for many fruitful discussions.

References

[AB92]   A. Amir and G. Benson. Two-dimensional periodicity and its application. In *SODA '92*, 1992. Proc. 3-rd Symposium on Discrete Algorithms, Orlando,FL.

[ABF92] A. Amir, G. Benson, and M. Farach. Alphabet independent two dimensional matching. In *STOC'92*, pages 59–67, 1992. Victoria,BC.

[BYR90] R. Baeza-Yates and M. Régnier. Fast algorithms for two dimensional and multiple pattern matching. In *SWAT'90*, volume 447 of *Lecture Notes in Computer Science*, pages 332–347. Springer-Verlag, 1990. Preliminary draft in Proc. Swedish Workshop on Algorithm Theory, Bergen, Norway. To appear in IPL.

[CH92] R. Cole and R. Hariharan. Tighter Bounds on the Exact Complexity of String Matching. In *FOCS'92*. IEEE, 1992. Proc. 33-rd IEEE Conference on Foundations of Computer Science, Pittsburgh, USA.

[Gal92] Z. Galil. Hunting Lions in the Desert Optimally or a Constant-Time Optimal Parallel String-Matching Algorithm. In *STOCS'92*, 1992. Victoria,BC.

[GP92] Z. Galil and K. Park. Truly Alphabet Independent Two-Dimensional Pattern Matching. In *FOCS'92*. IEEE, 1992. Proc. 33-rd IEEE Conference on Foundations of Computer Science, Pittsburgh, USA.

[Lan87] S. Lang. *Elliptic Functions*. Springer-Verlag, New-York, 1987.

[Lot83] Lothaire. *Combinatorics on Words*. Addison-Wesley, Reading, Mass., 1983.

[NR92] P. Nicodème and M. Régnier. Towards 2D Pattern Matching Complexity, 1992. in preparation.

[Rég89] M. Régnier. Knuth-Morris-Pratt algorithm: an analysis. In *MFCS'89*, volume 379 of *Lecture Notes in Computer Science*, pages 431–444. Springer-Verlag, 1989. Proc. Mathematical Foundations for Computer Science 89, Porubka, Poland.

[Vis85] U. Vishkin. Optimal Parallel Pattern Matching in Strings. *Information and Control*, 67:91–113, 1985.

[ZT89] R.F. Zhu and T. Takaoka. A technique for two-dimensional pattern matching. *CACM.*, 32(9):1110–1120, 1989.

Approximate String-Matching over Suffix Trees *

Esko Ukkonen

Department of Computer Science, University of Helsinki
P. O. Box 26, SF–00014 University of Helsinki, Finland
email: ukkonen@cs.Helsinki.FI

Abstract. The classical approximate string–matching problem of finding the locations of approximate occurrences P' of pattern string P in text string T such that the edit distance between P and P' is $\leq k$ is considered. We concentrate on the special case in which T is available for preprocessing before the searches with varying P and k. It is shown how the searches can be done fast using the suffix tree of T augmented with the suffix links as the preprocessed form of T and applying dynamic programming over the tree. Three variations of the search algorithm are developed with running times $O(mq + n)$, $O(mq \log q + \text{size of the output})$, and $O(m^2q + \text{size of the output})$. Here $n = |T|$, $m = |P|$, and q varies depending on the problem instance between 0 and n. In the case of the unit cost edit distance it is shown that $q = O(\min(n, m^{k+1}|\Sigma|^k))$ where Σ is the alphabet.

1 Introduction

The *approximate string–matching problem* is to find the approximate occurrences of a pattern in a text. We will consider the problem in the following form: Given *text* $T = t_1t_2 \cdots t_n$ and *pattern* $P = p_1p_2 \cdots p_m$ in alphabet Σ, and a number $k \geq 0$, find the end locations j of all substrings P' of T such that the edit distance between P and P' is $\leq k$.

The edit distance between P and P' is the minimum possible total cost of a sequence of editing steps that convert P to P'. Each editing step applies a rewriting rule of the forms $a \rightarrow \epsilon$ (deletion), $\epsilon \rightarrow b$ (insertion), or $a \rightarrow b$ (change) where $a, b \in \Sigma$, $a \neq b$.

The problem has the following four subcases:

1. $k = 0$, no preprocessing of T (*exact on–line string–matching*).
2. $k = 0$, with preprocessing of T (*exact off–line string–matching*).
3. $k > 0$, no preprocessing of T (*approximate on–line string–matching*).
4. $k > 0$, with preprocessing of T (*approximate off–line string–matching*).

Case 1 leads to the well–known Boyer–Moore and Knuth–Morris–Pratt algorithms. Case 2 has optimal solutions based on suffix trees [16, 25] or on suffix automata ('DAWG') [3, 6, 7]. Case 3 has recently received lot of attention [8, 9, 26]. The simplest solution is by dynamic programming in time $O(mn)$ where $m = |P|$ and $n = |T|$. For the *k–differences problem* (each edit operation has cost 1) fast special methods are possible, including $O(kn)$ time algorithms, see e.g. [14, 10, 23, 19, 5, 21].

* This work was supported by the Academy of Finland and by the Alexander von Humboldt Foundation (Germany).

This paper deals with Case 4, which also could be called the problem of approximate string searches over indexed files. The problem is to find a suitable preprocessing for T and an associated search algorithm that finds the approximate occurrences of P using the preprocessed T for varying P and k. We show how this can be solved fast using the suffix tree (for simplicity, the algorithms will be formulated for the suffix–trie) of T augmented with the suffix links, and applying dynamic programming over the tree. Recall that a suffix tree for T is, basically, a trie representing all the suffixes of T. It can be constructed in time $O(n)$. Therefore the preprocessing phase of our algorithms will be linear.

Perhaps the most natural way of applying dynamic programming over a suffix tree is to make a depth–first traversal that finds all substrings P' of T at a distance $\leq k$ from P. (Note that this is not exactly our problem; we want only the end points of such strings P'.) The search is easy to organize because all possible substrings of T can be found along some path starting from the root of the tree. Each path is followed until the edit distance between the corresponding substring and all prefixes of P becomes $> k$. The backtracking point can be found using the column of edit distances that is evaluated at each node visited during the traversal. This type of method is described and analyzed by Baeza–Yates & Gonnet [2] (see also Remark 2 of [13]). The method is further applied in [2, 11] for finding significant alignments between all pairs of substrings of T.

In the worst case, the above method evaluates $\Theta(mn)$ columns of edit distances which is more than the n columns evaluated by the simple on–line algorithm with no preprocessing of T. In this paper we show how to apply dynamic programming over the suffix tree such that in the worst case the number of evaluated columns stays $\leq n$ and can in a good case be much smaller.

To explain the idea, let $T = \mathbf{aaaaaaaabbbbbbbb}$, $P = \mathbf{abbb}$, and $k = 1$. In this case there is lot of repetition in the on–line dynamic programming algorithm. It evaluates a table which has a column of $m + 1$ entries for each symbol t_j of T. We call an entry *essential* if its value is $\leq k$. The occurrences of P can be found using only the essential entries: if the last entry of a column is essential then there is an approximate occurrence whose edit distance from P is $\leq k$ ending at that column. A column and its essential part in particular can depend only on a substring of T of length $O(m)$. We call this substring a *viable k–approximate prefix* of P in T. If two columns have same viable prefix then their essential part must be identical. In our example, the eight columns corresponding to the eight a's at the beginning of T will have the same viable prefix and hence the same essential part of the column.

To avoid evaluating a column whose viable prefix has occurred earlier we store columns into the suffix tree. A column with viable prefix Q is stored with the state that can be reached along the Q–path from the root. The search algorithm performs a traversal over the tree that spells out string T. The traversal can follow both the normal trie transitions and the suffix transitions. During the traversal, new columns are evaluated for each t_j except if we can conclude that the viable prefix at t_j will be the same as some older prefix. In this case the evaluation can be skipped; we have already stored a column with the same essential part.

The number of columns evaluated by the method is $\leq n$ and proportional to q where q is the total number of different viable prefixes in T. For small k, q can be considerably smaller than n.

We elaborate the above idea into three algorithms of different degree of sophistication. The introductory Algorithm A (Section 4) runs in time $O(mq + n)$ and always needs time $\Omega(n)$. This undesirable dependency on n is eliminated by using more complicated data structures in Algorithm B (Section 5) which has running time $O(mq \log q)$ + size of the output). Algorithm C (Section 6) is finally an easy-to-implement simplification of Algorithms A and B. It can evaluate more than n columns and has running time $O(m^2q$ + size of the output). We also show that $q \leq \min(n, \frac{12}{5}(m+1)^{k+1}(|\Sigma|+1)^k) = O(\min(n, m^{k+1}|\Sigma|^k))$.

The exponential growth of q as a function of k suggests that while our methods can be very fast for small k, their running time rapidly approaches the time of the on-line algorithm when k grows. In an interesting paper [17] (see also [1]), Myers points out that this inherent difficulty in our problem can be relieved by dividing P into smaller subpatterns and performing the search with a reduced error level for each subpattern. This filters out the interesting regions of T where one then attempts to expand the approximate occurrences of the subpatterns into k-approximate occurrences of the whole P. A simpler 'q-gram' method along similar lines is described in [13].

2   The approximate string matching problem

An *edit operation* is given by any rewriting rule of the form $a \to \epsilon$ (a *deletion*), $\epsilon \to a$ (an *insertion*), or $a \to b$ (a *change*), where a, b are any symbols in alphabet Σ, $a \neq b$, and ϵ is the empty string. Each operation $x \to y$ has a cost $c(x \to y) > 0$.

Operation $a \to a$ is called the *identity operation* for all $a \in \Sigma$. It has cost $c(a \to a) = 0$.

Let $A = a_1 a_2 \cdots a_m$ and $B = b_1 b_2 \cdots b_n$ be strings over Σ. A *trace* from A to B is any sequence $\tau = (x_1 \to y_1, x_2 \to y_2, \ldots, x_h \to y_h)$ of edit operations and identity operations such that $A = x_1 x_2 \cdots x_h$ and $B = y_1 y_2 \cdots y_h$. The cost of a trace τ is $c(\tau) = \sum_{i=1}^{h} c(x_i \to y_i)$. The *edit distance* $E(A, B)$ between A and B is the minimum possible cost of a trace from A to B [24]. The *unit cost edit distance* which means that each edit operation has cost $= 1$ is denoted as $E_1(A, B)$.

The intuition behind this definition is that $E(A, B)$ will be the minimum possible total cost of a sequence of editing steps that convert A into B such that each symbol is rewritten at most once. Distance $E(A, B)$ can be evaluated in time $O(mn)$ by a very simple form of dynamic programming [24]. The method evaluates an $(m+1) \times (n+1)$ table e such that $e(i, j) = E(a_1 \cdots a_i, b_1 \cdots b_j)$. Hence $E(A, B) = e(m, n)$.

If $E(A, B) \leq k$ we say that B is a k-*approximation* of A.

Definition. Let $P = p_1 p_2 \cdots p_m$ be a *pattern string* and $T = t_1 t_2 \cdots t_n$ a *text string* over Σ, and let k be a number ≥ 0. The *approximate string matching problem with threshold* k is to find all j such that the edit distance $E(P, P')$ between P and some substring $P' = t_{j'} \cdots t_j$ of T ending at t_j is $\leq k$. Then P has a k-*approximate occurrence* P' at position j of T.

The approximate string matching problem can be solved on-line, without preprocessing of T, with a very slightly modified form of the dynamic programming for the the edit distance [18]: Let $D(i, j)$ be the minimum edit distance between the prefix $P_i = p_1 \cdots p_i$ of P and the substrings of T ending at t_j. The $(m+1) \times (n+1)$ table $D(i,j)$, $0 \le i \le m$, $0 \le j \le n$, of such values can be evaluated from

$$D(0,j) = 0, \quad 0 \le j \le n; \tag{1}$$

$$D(i,j) = \min \begin{cases} D(i-1,j) + c(p_i \to \epsilon) \\ D(i-1,j-1) + (\text{if } p_i = t_j \text{ then } 0 \text{ else } c(p_i \to t_j)) \\ D(i,j-1) + c(\epsilon \to t_j) \end{cases} \tag{2}$$

for $1 \le i \le m$, $0 \le j \le n$. It should be emphasized that all entries $D(0,j)$ on row 0 of this table have value 0 while in the corresponding table for the edit distance between P and T only the $(0,0)$–entry gets value 0.

The solution to the problem can be read from the last row of table D: there is a k–approximate occurrence of P in T at position j if and only if $D(m,j) \le k$.

In the sequel, an important technical tool will be the length $L(i,j)$ of the *shortest* substring of T ending at t_j whose edit distance from P_i equals $D(i,j)$. Value $L(i,j)$ obviously satisfies

$$L(0,j) = 0, \quad 0 \le j \le n; \tag{3}$$

$$L(i,j) = \text{if } D(i,j) = D(i-1,j) + c(p_i \to \epsilon) \text{ then } L(i-1,j) \tag{4}$$

$$\text{elsif } D(i,j) = D(i-1,j-1) + (\text{if } p_i = t_j \text{ then } 0 \text{ else } c(p_i \to t_j))$$

$$\text{then } L(i-1,j-1) + 1$$

$$\text{else } L(i,j-1) + 1$$

for $1 \le i \le m$, $0 \le j \le n$.

Tables D and L can be conveniently evaluated, column–by–column, in an on–line, left–to–right scan over T. Columns $D(*,j)$ and $L(*,j)$ can be produced from $D(*,j-1)$, $L(*,j-1)$, and symbol t_j of T. The evaluation can be organized as function dp, given below, which will return $(D(*,j), L(*,j))$ as $dp(D(*,j-1), L(*,j-1), t_j)$:

function $dp(d'(0 \ldots m), l'(0 \ldots m), t)$:
    $d(0) \leftarrow l(0) \leftarrow 0$;
    for $j \leftarrow 1$ **to** m **do**
        $d(i) \leftarrow d(i-1) + c(p_i \to \epsilon)$
        $l(i) \leftarrow l(i-1)$
        if $d'(i-1) + (\text{if } p_i = t_j \text{ then } 0 \text{ else } c(p_i \to t_j)) < d(i)$ **then**
            $d(i) \leftarrow d'(i-1) + (\text{if } p_i = t_j \text{ then } 0 \text{ else } c(p_i \leftarrow t_j))$
            $l(i) \leftarrow l'(i-1) + 1$;
        if $d'(i) + c(\epsilon \to t_j) < d(i)$ **then**
            $d(i) \leftarrow d'(i) + c(\epsilon \to t_j)$
            $l(i) \leftarrow l'(i) + 1$
    return(d, l).

This takes time $O(m)$ and the evaluation of D and L therefore takes total time of $O(mn)$. Other on–line algorithms running in $O(kn)$ expected time [20, 4] (these methods can easily be incorporated into procedure dp) or in $O(kn)$ worst–case time (for the unit cost edit distance) [10, 23] are also known.

In the next sections we develop algorithms that are off–line with respect to T. We assume that T has been preprocessed into a suffix tree and study how the evaluation of D can be organized in a more efficient way.

3 k–approximate prefixes of P

The on–line solution to our problem in Section 2 has the drawback that dynamic programming is explicitly repeated over identical repeated substrings of T. This may create unnecessary work because the content of each column $D(*, j)$ of D depends only on a relatively short substring of T. If such a substring occurs again in T, the dynamic programming would give a column that is equal to an old column. Our new algorithms avoid the repetition of such identical calculations.

To make this precise we first define the *essential entries* of D. The approximate string matching problem can be solved using only entries $D(i, j) \leq k$ of D. Therefore we call each entry $D(i, j) \leq k$ an *essential* entry. By (1), (2), an essential entry depends only on other essential entries in the sense that the inessential entries of D could be replaced by default value ∞ without affecting the content of the essential part.

Let $D(*, i)$ and $D(*, j)$ be any two columns of D and let $L(*, i)$ and $L(*, j)$ be the corresponding columns of L. Then pairs $(D(*, i), L(*, i))$ and $(D(*, j), L(*, j))$ are called *equivalent*, denoted $(D(*, i), L(*, i)) \equiv (D(*, j), L(*, j))$, if the essential entries of $D(*, i)$ and $D(*, j)$ have identical contents and the corresponding entries of $L(*, i)$ and $L(*, j)$ have identical contents. In other words, if $D(h, i) \leq k$ or $D(h, j) \leq k$ for some $0 \leq h \leq m$, then $D(h, i) = D(h, j)$ and $L(h, i) = L(h, j)$.

Next we define the substring Q_j of T that determines the essential part of $D(*, j)$. Recall here that the Knuth–Morris–Pratt algorithm of exact string matching has the property that it finds at each text location j the *longest* prefix $p_1 \cdots p_i$ of pattern P that occurs at j, i.e., $p_1 \cdots p_i = t_{j-i+1} \cdots t_j$ is a 0–approximation of $p_1 \cdots p_i$ that occurs at j. The use of Q_j can be seen as a generalization of this to the approximate case: Q_j will be a k–approximation of $p_1 \cdots p_i$ that occurs at j in T.

Let $T_j = t_1 \cdots t_j$ be the prefix of T ending at j, and let $\lambda(T_j) = L(i, j)$ where i is the largest index such that $D(i, j)$ is essential. Obviously, $P_i = p_1 \cdots p_i$ is the longest prefix of P that has a k–approximation at the end of T_j. String $t_{j-\lambda(T_j)+1} \cdots t_j$ is such an approximation, in fact, the shortest one.

Definition. String $Q_j = t_{j-\lambda(T_j)+1} \cdots t_j$ is called the *viable k–approximate prefix* of P at j (*viable prefix* at j, for short). If $\lambda(T_j) = 0$ then $Q_j = \epsilon$.

String Q_j is 'viable' in the sense that it can be a prefix of a k–approximate occurrence of the whole P.

Viable prefix Q_i determines the essential part of column $D(*, i)$:

Theorem 1. *If $Q_i = Q_j$ then $(D(*, i), L(*, i)) \equiv (D(*, j), L(*, j))$.*

Proof. It is helpful to consider table D as a solution to a shortest path problem in the *edit graph* associated with our pattern matching problem.

Such a graph consists of nodes $G(i, j)$, $0 \leq i \leq m$, $0 \leq j \leq n$, and of weighted directed arcs that form a regular grid as follows: There is an arc $(G(i-1, j), G(i, j))$ with weight $c(p_i \rightarrow \epsilon)$ for all $1 \leq i \leq m$, $0 \leq j \leq n$; an arc $G(i-1, j-1), G(i, j))$ with weight 0 if $p_i = t_j$ and with weight $c(p_i \rightarrow t_j)$ otherwise for all $1 \leq i \leq m$, $1 \leq j \leq n$; and an arc $(G(i, j-1), G(i, j))$ with weight $c(\epsilon \rightarrow t_j)$ for all $1 \leq i \leq m$, $1 \leq j \leq n$. Then $D(i, j)$ gives the length of a shortest path in this graph among all paths that lead from any node $G(0, j')$ on the row 0 to node $G(i, j)$. Value $L(i, j)$

indicates the start node of a steepest path: $L(i,j)$ is the smallest value such that a shortest path to $G(i,j)$ starts from $G(0, j - L(i,j))$.

Let now i and j be as in the theorem and let h be the largest index such that $D(h,i) \leq k$. Hence $|Q_i| = L(h,i)$. Then for each $r \leq h$, there is a shortest path to $G(r,i)$ that starts from some node $G(0, i - |Q_i|), G(0, i - |Q_i| + 1), \ldots, G(0,i)$. To evaluate the essential entries of column $D(*,i)$ correctly it therefore suffices to consider only subgraph G_i of the edit graph spanned by nodes $G(r,s)$, $0 \leq r \leq m$, $i - |Q_i| \leq s \leq i$. Similarly, to evaluate the essential entries of column $D(*,j)$ correctly it suffices to consider only subgraph G_j spanned by nodes $G(r,s)$, $0 \leq r \leq m$, $j - |Q_j| \leq s \leq j$. Graphs G_i and G_j have identical topology and weights because $Q_i = Q_j$. Hence their shortest path problems have identical solutions, in particular, the essential entries of $D(*,i)$ and $D(*,j)$ have to be identical as well as the corresponding entries of $L(*,i)$ and $L(*,j)$. $\square$

Example. Let $T = \text{aaaaaaaabbbbbbbb}$, $P = \text{abbb}$, and $k = 1$. Assume the unit cost model of the edit distance (each edit operation has cost $= 1$). Then table D is

		a	a	a	a	a	a	a	a	b	b	b	b	b	b	b	b
	0	0	0	0	0	0	0	0	0	0	0	0	0	0	0	0	0
a	1	0	0	0	0	0	0	0	0	1	1	1	1	1	1	1	1
b	2	1	1	1	1	1	1	1	1	0	1	1	1	1	1	1	1
b	3	2	2	2	2	2	2	2	2	1	0	1	1	1	1	1	1
b	4	3	3	3	3	3	3	3	3	2	1	0	1	1	1	1	1

and table L is

0	0	0	0	0	0	0	0	0	0	0	0	0	0	0	0	0
0	1	1	1	1	1	1	1	1	0	0	0	0	0	0	0	0
0	1	1	1	1	1	1	1	1	2	1	1	1	1	1	1	1
0	1	1	1	1	1	1	1	1	2	3	2	2	2	2	2	2
0	1	1	1	1	1	1	1	1	2	3	4	3	3	3	3	3

The viable prefixes are $Q_1 = \text{a}$ (because $D(2,1)$ is the last essential entry of $D(*,1)$, and $L(2,1) = 1$), $Q_2 = Q_3 = \ldots = Q_8 = \text{a}$, $Q_9 = \text{ab}$, $Q_{10} = \text{abb}$, $Q_{11} = \text{abbb}$, $Q_{12} = \ldots = Q_{16} = \text{bbb}$. There are five different viable prefixes. $\square$

For each j we let $Q'_j = Q_{j-1}t_j$. The following theorem says that viable prefix Q_j can not start properly before Q_{j-1}.

Theorem 2. Q_j is a suffix of Q'_j.

Proof. Using the interpretation of D as a solution to the shortest path problem (see the proof of Theorem 1), one first notices that values $L(h,j)$ are non–decreasing when h grows: If $h < h'$ then $L(h,j) \leq L(h',j)$. The rest of the proof is a simple case analysis of how $L(h,j)$ where h is the largest index such that $D(h,j) \leq k$ can depend on the entries of $L(*, j - 1)$. $\square$

4 Dynamic programming over suffix trees

We will evaluate table D using T represented as a suffix tree. First we recall the alternative forms of such trees.

Suffix tree of T. The suffix tree of T is a data structure representing all the suffixes $T^i = t_i \cdots t_n$, $1 \leq i \leq n+1$, of T. We distinguish three versions of such a structure.

The uncompacted version of a suffix tree is called a *suffix trie* of T, denoted $STrie(T)$. It is the unique deterministic finite–state automaton that recognizes the suffixes T^i of T and nothing else, and has tree–shaped transition graph. The transition graph is the trie representing strings T^i.

Let *root* denote the initial state and g the transition function of $STrie(T)$. We say that there is a *goto–transition* from state r to state s on input $a \in \Sigma$ if $s = g(r,a)$. If there is a goto–transition path from r to s on input symbols whose catenation is string x we write $s = g(r,x)$.

We augment $STrie(T)$ with the *suffix function* f, defined for each state s, $s \neq root$, as follows: As $s \neq root$, there is a symbol a and a string x in Σ^* such that $g(root, ax) = s$. We set $f(s) = r$ where r is the state such that $g(root, x) = r$. We say that there is a *suffix transition* from s to r. A suffix transition does not consume any input.

The size of $STrie(T)$ is $O(|T|^2)$. $STrie(T)$ is easy to construct (see e.g. [23, 22]) but its quadratic size makes it impractical. Fortunately, $STrie(T)$ has linear size representations that can be constructed in linear time, namely the (compact) *suffix tree* [25, 16, 22] and the *suffix automaton* (DAWG) [3, 6, 7].

For simplicity, the suffix trie $STrie(T)$ consisting of functions g and f will be used in the description of our algorithms. However, the actual implementation will be done using the standard linear–size suffix tree or suffix automaton for T. This does not change the complexity bounds derived here for $STrie(T)$.

Algorithm A. The algorithm will traverse in $STrie(T)$ a path of goto and suffix transitions that starts from *root* and spells out in its goto–transitions string T. Combined with this the columns of D that correspond to different viable prefixes Q_i will be evaluated. Each such column $D(*, i)$ together with column $L(*, i)$ will be stored with state $r_i = g(root, Q_i)$ as $d(r_i) \leftarrow D(*, i)$, $l(r_i) \leftarrow L(*, i)$.

The traversal goes through states $r_0, s_1, \ldots, r_1, s_2, \ldots, r_{n-1}, s_n, \ldots, r_n$ where $r_0 = root$, $r_i = g(root, Q_i)$, and $s_i = g(root, Q_i')$. The transition from r_{i-1} to s_i is a goto–transition for t_i because $s_i = g(root, Q_i') = g(root, Q_{i-1}t_i) = g(r_{i-1}, t_i)$. The transition path from s_i to r_i consists of zero or more suffix transitions; such a path exists by Theorem 2.

Consider the subpath from r_{j-1} to r_j. The goto–transition $g(r_{j-1}, t_j) = s_j$ is taken first. After that there are two cases:

Case 1. If s_j has already been visited during the traversal, then follow the suffix transition path until the first state r is encountered such that $d(r)$ and $l(r)$ have non–empty values. Then $r = r_j$.

Case 2. If s_j has not been visited yet, then evaluate a pair (d, l) of columns as $(d, l) \leftarrow dp(d(r_{j-1}), l(r_{j-1}), t_j)$. Then (see Lemma 4 below) $(d, l) \equiv (D(*, j), L(*, j))$. This equivalence implies that $d(h) = D(h, j)$ and $l(h) = L(h, j) = |Q_j|$, where h is such that $d(h)$ is the last essential entry of d. The algorithm then follows the suffix link path from s_j to the state r whose depth (distance from *root*) is $|Q_j|$. Then $r = r_j$ and the algorithm saves columns (d, l) as $d(r) \leftarrow d$, $l(r) \leftarrow l$.

To make the whole traversal the above is repeated for $j = 1, \ldots, n$. As an initialization we set $d(root) \leftarrow D(*, 0)$, $l(root) \leftarrow L(*, 0)$. By (2), entry $D(h, 0)$ of $D(*, 0)$

is given as $D(h, 0) = \sum_{i=1}^{h} c(p_i \to \epsilon)$, and by (4), entry $L(h, 0)$ of $L(*, 0)$ is given as $L(h, 0) = 0$.

The algorithm has to output j whenever $D(m, j) \leq k$. This is implemented such that Algorithm A outputs j whenever $d(r_j)(m) \leq k$ during the traversal.

Consider then the correctness of Algorithm A. We need a notation: If x is a suffix of y, we write $y|x$, and if, moreover, y is a suffix of z, we write $z|y|x$.

The crucial point where Algorithm A saves compared to the on–line algorithm is Case 1. Assume that $s_j = g(root, Q'_j)$ has been visited earlier. This means that s_j has to belong to the suffix link path between s_i and r_i for some $i < j$, that is, $Q'_i|Q'_j|Q_i$. On the other hand we have:

Lemma 3. If $Q'_i|Q'_j|Q_i$ for some $i < j$, then $Q_j = Q_i$.

Proof. This is immediate when D is viewed as a solution to the shortest path problem (see the proof of Theorem 1). □

This implies, noting Theorem 1, that a pair of columns equivalent to $(D(*, j), L(*,$ has already been stored as $(d(r_i), l(r_i))$. The dynamic programming can be skipped; the algorithm just follows the suffix transition path from s_j to $r_i = r_j$. Hence Case 1 is correct.

It is correct to use in Algorithm A columns that are only equivalent to the actual columns of D and L. The essential entries of a new column of D are determined by the essential entries of the previous column. Therefore we have the following lemma.

Lemma 4. If $(d', l') \equiv (D(*, j-1), L(*, j-1))$ and $(d, l) = dp(d', l', t_j)$, then $(d, l) \equiv (D(*, j), L(*, j))$.

Hence Algorithm A correctly outputs all j such that $D(m, j) \leq k$.

Analysis. Let $\mathbf{Q} = \{Q_i \mid 1 \leq i \leq n\}$, and let $q = |\mathbf{Q}|$ be the size of $\mathbf{Q}$, i.e. the number of different viable prefixes. Moreover, let $\mathbf{Q}' = \{Q'_i \mid 1 \leq i \leq n\}$ and $q' = |\mathbf{Q}'|$.

Algorithm A evaluates $\leq q'$ pairs of columns of D and L, and stores q of them. As the evaluation of each pair of columns takes time and space $O(m)$, and the time consumption for the rest is proportional to n (note that the traversal takes n goto–transitions and at most n suffix transitions), we obtain:

Theorem 5. *Algorithm A runs in time $O(mq' + n)$ and needs working space $O(mq)$ for storing the columns of the tables.*

Next we analyze the growth of q in more detail in the special case of the unit cost edit distance. Let $U_k(P) = \{x \in \Sigma^* \mid E_1(P, x) \leq k\}$ be the set of strings whose unit cost edit distance from P is $\leq k$. The size of $U_k(P)$ has the following bound, c.f. Lemma 3 of [17].

Theorem 6. $|U_k(P)| \leq \frac{12}{5}(m + 1)^k(|\Sigma| + 1)^k$.

Proof. The size of $U_k(P)$ is $\leq$ the number of different traces (edit scripts) of length $\leq k$ that can be applied on P. Each trace consists of $\leq k$ actual editing steps and of zero or more identity steps $a \to a$. The number of traces equals the number of different possibilities to select the actual steps. This can be estimated by bounding the number of different ways of applying exactly k steps that can include both actual steps *and* identity steps.

The k steps are divided into two groups: The steps of the form $a \to x$ where $a \in \Sigma$, $x \in \Sigma \cup \{\epsilon\}$ (= group A; this contains the possible identity operations), and the steps of the form $\epsilon \to a$ where $a \in \Sigma$ (= group B).

In group A, each step $a \to x$ has a unique p_i such that $a = p_i$. Moreover, x can be selected in $|\Sigma| + 1$ different ways. Hence a group A consisting of t steps can be selected in $\leq \binom{m}{t}(|\Sigma| + 1)^t$ different ways.

In group B, each step $\epsilon \to a$ can be selected in $(m+1)|\Sigma|$ different ways because ϵ refers to any of the $m+1$ intervals between the m letters of P, and because a can be selected independently of ϵ in $|\Sigma|$ different ways. Each interval can be selected arbitrarily many times. Hence a group B consisting of t steps can be selected in $\leq (m+1)^t|\Sigma|^t$ different ways.

This gives

$$|U_k(P)| \leq \sum_{t=0}^{k} [\binom{m}{t}(|\Sigma| + 1)^t + (m+1)^{k-t}|\Sigma|^{k-t}]$$

$$= \sum_{t=0}^{k} [\binom{m}{t}(|\Sigma| + 1)^t + (m+1)^t|\Sigma|^t]$$

$$\leq 2\sum_{t=0}^{k}(m+1)^t(|\Sigma| + 1)^t \leq \frac{12}{5}(m+1)^k(|\Sigma| + 1)^k$$

where we have assumed that $m \geq 1$ and $|\Sigma| \geq 2$. $\square$

As $q \leq \sum_{i=k}^{m} U_k(p_1 \cdots p_i) \leq m \cdot |U_k(P)|$, we have by Theorem 6

$$q \leq \frac{12}{5}(m+1)^{k+1}(|\Sigma| + 1)^k = O(m^{k+1}|\Sigma|^k). \tag{5}$$

As $q' \leq |\Sigma|q$, we further obtain

$$q' \leq \frac{12}{5}(m+1)^{k+1}(|\Sigma| + 1)^{k+1} = O(m^{k+1}|\Sigma|^{k+1}). \tag{6}$$

Noting that $q \leq q' \leq n$, Theorem 5 with (5) and (6) gives:

Theorem 7. *Algorithm A runs for the k-differences problem in time $O(m \cdot \min(n, m^{k+1}|\Sigma|^{k+1}) + n)$ and needs working space of $O(m \cdot \min(n, m^{k+1}|\Sigma|^k))$.*

5   Finding the next viable prefix fast

The method of this section can be understood as an advanced implementation of Algorithm A. Algorithm A always needs time $\Omega(n)$ because it scans symbol by symbol over the whole text T. In Algorithm B to be developed next this dependency on n will be eliminated. Columns of D for different viable prefixes will be found using dictionary operations implemented with balanced search trees. The method is based on Lemma 3 and its implementation heavily depends on the special properties of $STrie(T)$.

Assume that Algorithm A has performed the dynamic programming at t_i, has obtained (d, l) equivalent to $(D(*, i), L(*, i))$, and has stored them as $d(r_i) \leftarrow d$, $l(r_i) \leftarrow l$ where $r_i = g(root, Q_i)$. Algorithm A will next examine the state $s_{i+1} = g(r_i, t_{i+1})$. If s_{i+1} has already been visited, Algorithm A knowns by Lemma 3 that dynamic programming can be skipped because Q_{i+1} has to be equal to Q_h for some $h \leq i$. State $r_{i+1} = g(root, Q_h) = g(root, Q_{i+1})$ is found by following the suffix link path from s_{i+1}. Then Algorithm A will examine $s_{i+2} = g(r_{i+1}, t_{i+2})$, and so on. Finally an unvisited state s_j will be found, and dynamic programming is resumed.

To find s_j *directly* after s_i, we first observe:

• the set of different viable prefixes can grow at s_i and again at s_j, but it remains unchanged between them;

• the set of the visited states remains unchanged between s_i and s_j;

• the string on the path from *root* to any state $s_{i+1}, \ldots, s_j$ is of the form $Q_h a$ for some $a \in \Sigma$, $h \leq i$.

Hence states $s_{i+1}, \ldots, s_j$ belong to the set

$$S_i = \{s \mid s = g(root, Q_h a) \text{ for some } h \leq i, a \in \Sigma\}$$

of states that are at the distance of one goto–transition from some state that can be reached from *root* along some viable prefix Q_h.

Algorithm B. For any state s of $STrie(T)$, let $Key(s)$ denote the string such that $g(root, Key(s)) = s$, and for a set S of states, let $Keys(S)$ be the set of strings $Key(s)$, where $s \in S$. We will associate with each state s in S_i value $loc(s)$ (to be defined precisely below) that gives the smallest index $h > i$ such that $Key(s)$ 'could be' equal to Q'_h. During Algorithm B the *uneliminated* states s in S_i will be kept in dictionary H. The records in the dictionary are of the form $(s, loc(s))$ where $loc(s)$ is used as the search–key for s. The dictionary has to support insertions, deletions, and minimum extractions. By extracting the minimum element from H we get the state s with the smallest $loc(s)$. This state s will be s_j and $j = loc(s_j)$. Then new columns have to be evaluated by dynamic programming from $d(r)$ and $l(r)$, where $r = father(s_j)$, and from symbol a such that $g(r, a) = s_j$.

For a precise definition of $loc(s)$ we need the concepts of *elimination* and *covering*. To introduce the latter, consider strings Q'_v, $i+1 \leq v \leq j$, in more detail. As already mentioned, each $Q'_v = Q_{v-1} t_v$ has to be equal to $Q_h a$ for some Q_h, $h \leq i$. Hence we have $Q_{v-1} = Q_h$. Moreover, viable prefix Q_{v-1} is the longest among *all* viable prefixes of T that are suffixes of $T_{v-1} = t_1 t_2 \cdots t_{v-1}$:

Lemma 8. *If $T_{v-1} | Q_e$ then $Q_{v-1} | Q_e$.*

Proof. Use the interpretation of D as a solution to the shortest path problem as presented in the proof of Theorem 1. $\square$

This implies that each Q'_v, $i+1 \leq v \leq j$, has to be the *longest* string in $Keys(S_i)$ that is a suffix of T_v. If more than one string in $Keys(S_i)$ is a suffix of T_v, then these strings have to be suffixes of the longest one. With this in mind we make the following definition.

Definition. String X *covers* an occurrence of string Y at v if $T_v | X | Y$.

String $Key(s)$ is the longest element of $Keys(S_i)$ at v if and only if $T_v|Key(s)$ and no other string in $Keys(S_i)$ covers $Key(s)$ at v.

We still need the concept of elimination. Its purpose is to incorporate Lemma 3 into our algorithm.

Definition. Strings Q'_h and Q_h *eliminate* a state s and string $Key(s)$ if $Q'_h|Key(s)|Q_h$.

Note that the states visited by Algorithm A and the eliminated states defined here are same. By Lemma 3, dynamic programming need not be performed when entering an eliminated state.

We now define

$$loc(s) = \begin{cases} \infty, & \text{if } Key(s) \text{ is eliminated by some } Q'_h, Q_h \text{ where } h \le i; \\ v, & \text{otherwise,} \end{cases}$$

where $v > i$ is the first occurrence of $Key(s)$ after location i in T that is not covered by some other string in $Keys(S_i)$. Note that $loc(s)$ is defined for all states s, not only for members of S_i. The algorithm also maintains these values for all s.

The algorithm selects $j \leftarrow \min_{s \in S_i} loc(s)$ using dictionary H that contains $(s, loc(s))$ for states s in S_i. The dynamic programming is performed next at s_j such that $loc(s_j) = j$.

After this some loc-values have to be changed and H must be updated such that it represents S_j instead of S_i. The algorithm follows the suffix link path from s_j to $r_j = g(root, Q_j)$. All states s on this path become now eliminated if they are not eliminated earlier (this can be the case for all $s \ne s_j$). Hence $loc(s) \leftarrow \infty$; this is implemented simply by removing s from H.

We have still to add into H new elements corresponding to $S_j - S_i$ and to make the updates on loc-values due to covering. This happens only if r_j is a new state not visited earlier. Then $(s, loc(s))$ is inserted into H for all uneliminated s such that $s = g(r_j, a)$ for some symbol a. Moreover, the appropriate changes to $loc(w)$ have to be done for all w such that $Key(w)$ is covered by some $Key(s)$.

Here, again, the suffix transitions can be used. We call a state w *primary* if $Key(w) = t_1 \cdots t_h$ for some h. (Note that the suffix transitions constitute a tree, with primary states as the leaves and *root* as the root.) The next lemma follows from the definition of loc and gives a method for updating; recall that f denotes the suffix function.

Lemma 9. *If w is an eliminated state then $loc(w) = \infty$; if w is primary but not eliminated then $loc(w) = depth(w)$; otherwise*

$$loc(w) = \min loc(w') \tag{7}$$

where the minimum is over all w' such that $f(w') = w$ and w' is not in S_i.

This means, each $loc(w)$ that needs updating can be found by traversing the suffix link path from each new state $s \in S_j - S_i$. At each uneliminates state w, $w \ne s$, on such path the updated $loc(w)$ is evaluated from (7). As there are at most $|\Sigma|$ different w' such that $f(w') = w$, the minimization in (7) can be done in time $O(\log|\Sigma|)$. If $(w, loc(w))$ is in H, the update is performed in H, too.

In summary, Algorithm B starts by inserting $(root, loc(root) = 0)$ into an initially empty dictionary H. Then $(s_j, j) \leftarrow extract\text{-}min(H)$ is performed, H and the loc–values are updated, and this is repeated until H becomes empty. Whenever a column $d(r)$ is stored such that $d(r)(m) \leq k$, state r is marked for output. The final output phase lists all occurrences of $Key(r)$ in T, for all states r marked for output. These occurrences can be found from $STrie(T)$ by standard methods.

The preprocessing phase creates $STrie(T)$ and initializes values $loc(s)$ using the method of Lemma 9 with $S_i = \emptyset$.

Theorem 10. *Algorithm B runs in time $O(mq \log q + size\ of\ the\ output)$ and needs working space of $O(mq)$ for dictionary H and the columns of dynamic programming tables.*

Proof. Algorithm B evaluates q' columns of D and L. Dictionary H is implemented as a balanced search tree which takes $O(\log |H|)$ time per dictionary operation. The algorithm performs the following q' times: selection of next s_j from H in time $O(\log |H|)$; evaluation of new columns in time $O(m)$; traversal from s_j to r_j, removal of the eliminated states from H in time $O(m \log |H|)$; insertion of states $s = g(r_j, a)$ into H in time $O(|\Sigma| \log |H|)$. Moreover, for each new state s inserted into H during the algorithm, $loc(w)$ has to be updated for states w on the suffix link path from s to $root$ and the corresponding changes have to be done in H. The length of each such path is $O(m)$, hence the updates take total time of $O(|H| m (\log |\Sigma| + \log |H|))$.

This gives total time bound $O(q'(\log |H| + m + m \log |H|) + |H| m (\log |\Sigma| + \log |H|))$ which is $O(mq \log q)$ because $q' \leq |\Sigma| q$, $|H| \leq |\Sigma| q$, and $|\Sigma|$ is assumed constant.

The output time can be made linear in the size of the output if some care is devoted to the elimination of duplicated output.

The space requirement is $O(mq)$ for the columns and $O(|\Sigma| q)$ for H, hence $O(mq)$. □

Theorem 10 together with upper bound (6) of q shows that for small k and large n Algorithm B can be faster than Algorithm A.

6   Simple algorithm

Dictionary H and the other mechanisms of Algorithm B for maintaining values $loc(s)$ create relatively large overhead. We describe next Algorithm C, a simplified version of Algorithm B that uses only elimination of states but does not use loc–values. Algorithm C is easy to implement and has low overhead.

Algorithm C makes a depth–first–search over the uneliminated states. All states with a saved pair (d, l) of columns are now kept in a stack. When there is a transition $g(r, a) = s$ from the top state r of the stack to an uneliminated state s, new columns are evaluated as $(d, l) \leftarrow dp(d(r), l(r), a)$. Columns (d, l) and state r' are saved in the stack; state r' is the state on the suffix link path from s such that its distance from $root$, $depth(r')$, equals the length of the viable prefix associated with (d, l).

The resulting algorithm is given below. Function $viable\text{-}prefix\text{-}length(d, l)$ gives the length of the viable prefix represented by columns (d, l), i.e., the value of $l(h)$ where h is the largest index such that $d(h) \leq k$. Function $output\text{-}mark(r)$ adds state

r to the list of states that represent the locations of the k–approximate occurrences of P in T.

Algorithm C.

1. $eliminated(root) \leftarrow$ **true**
2. $search(root, D(*, 0), L(*, 0))$.
3.   **procedure** $search(r, d'(0 \ldots m), l'(0 \ldots m))$:
4.   **for each state** $s = g(r, a)$ for some $a \in \Sigma$ **do**
5.   **if not**($eliminated(s)$) **then**
6. $(d, l) \leftarrow dp(d', l', a)$
7. $length \leftarrow viable\text{-}prefix\text{-}length(d, l)$
8.   **if** $depth(s) > length$ **do**
9. $eliminated(s) \leftarrow$ **true**; $s \leftarrow f(s)$
10.   **until** $depth(s) = length$ **or** $eliminated(s)$
11.   **if** $depth(s) = length$ **and not**($eliminated(s)$) **then**
12.   **if** $d(m) \leq k$ **then** $output\text{-}mark(s)$
13. $eliminated(s) \leftarrow$ **true**
14. $w \leftarrow s$
15.   **while** $f(w) \neq root$ **and**
                 $eliminated(f(w')) =$ **true for all** w' such that $f(w') = f(w)$ **do**
16. $w \leftarrow f(w)$; $eliminated(w) \leftarrow$ **true**
17. $search(s, d, l)$.

In Algorithm C the selection order of the next state s is not based on $loc(s)$. Therefore Algorithm C can select a state s that would have never been selected by Algorithm B; the optimal selection order implemented in Algorithm B can result into total covering of s and therefore into an elimination of s before it would come selected.

Fortunately, it is not a fatal error to select such an s. It only means that the algorithm first finds a too short viable prefix for some locations of T but will find the correct, long–enough prefix later. All different essential parts of columns of D will ultimately be evaluated.

Each viable prefix is of length $O(m)$. Before finding the correct prefix Algorithm C may find one or more of its proper suffixes. Therefore the total number of extra columns evaluated is $O(mq)$. In any case, the algorithm evaluates the same q' columns as Algorithm B. Thus the total number of columns is $O(mq + q') = O(mq)$ and we have the following theorem.

Theorem 11. *Algorithm C runs in time* $O(m^2q + \text{size of the output})$ *and needs working space of* $O(m^2q)$.

7   Concluding remarks

Several relevant questions concerning the new algorithms remained unanswered. Most notably, these include theoretical analysis of the expected running times and experimental comparison of these and related algorithms from [2, 13, 17].

For modestly long T it is feasible to implement our algorithms using the (compact) suffix tree of T. Adapting the methods for suffix automata seems simple, too.

However, for very long texts it is better to use the more space economical suffix array [15, 12] instead. The details and a practical fine–tuning of such an implementation are a subject for further study.

References

1. Altschul, S., Gish, W., Miller, W., Myers, E. & Lipman, D. (1990): A basic local alignment search tool. *J. of Molecular Biology 215*, 403–410.
2. Baeza–Yates, R. A. & Gonnet, G. H.: All–against–all sequence matching (Extended Abstract).
3. Blumer,A., Blumer,J., Haussler, D., Ehrenfeucht, A., Chen, M.T. and Seiferas, J. (1985): The smallest automaton recognizing the subwords of a text. *Theor. Comp. Sci. 40*, 31-55.
4. Chang, W. & Lampe, J. (1992): Theoretical and empirical comparisons of approximate string matching algorithms. *Proc. Combinatorial Pattern Matching 1992*, (Tucson, April 1992), Lect. Notes in Computer Science 644 (Springer–Verlag 1992), pp. 175–184.
5. Chang, W. & Lawler, E (1990): Approximate string matching in sublinear expected time. *Proc. IEEE 1990 Ann. Symp. on Foundations of Computer Science*, pp. 116-124.
6. Crochemore, M. (1986): Transducers and repetitions. *Theor. Comp. Sci. 45*, 63-86.
7. Crochemore, M. (1988): String matching with constraints. *Proc. MFCS'88 Symposium.* Lect. Notes in Computer Science 324 (Springer–Verlag 1988), pp. 44–58.
8. Dowling, G. R. & Hall, P. (1980): Approximate string matching. *ACM Comput. Surv. 12*, 381-402.
9. Galil, Z. & Giancarlo, R. (1988): Data structures and algorithms for approximate string matching. *J. Complexity 4*, 33–72.
10. Galil, Z. & Park, K. (1989): An improved algorithm for approximate string matching. *SIAM J. on Computing 19*, 989–999.
11. Gonnet, G. H. (1992): *A tutorial introduction to Computational Biochemistry using Darwin.* Informatik E. T. H. Zuerich, Switzerland.
12. Gonnet,G.H., Baeza-Yates,R.A. & Snider,T. (1991): Lexicographical indices for text: Inverted files vs. PAT trees. Report OED-91-01, UW Centre for the New Oxford English Dictionary and Text Research, 1991.
13. Jokinen, P. & Ukkonen, E. (1991): Two algorithms for approximate string matching in static texts. *Proc. MFCS'91*, Lect. Notes in Computer Science 520 (Springer–Verlag 1991), pp. 240-248.
14. Landau, G. & Vishkin, U. (1988): Fast string matching with k differences. *J. Comp. Syst. Sci. 37*, 63-78.
15. Manber, U. & Myers, G. (1990): Suffix arrays: A new method for on–line string searches. In: *SODA-90*, pp. 319–327.
16. McCreight, E. M. (1976): A space economical suffix tree construction algorithm. *J. ACM 23*, 262-272.
17. Myers, E. W.: A sublinear algorithm for approximate keyword searching. TR 90–25, Department of Computer Science, The Univ. of Arizona, Tucson (to appear in *Algorithmica*).
18. Sellers, P. H. (1980): The theory and computation of evolutionary distances: Pattern recognition. *J. Algorithms* 1, 359–373.
19. Tarhio, J. & Ukkonen, E. (1990): Boyer-Moore approach to approximate string matching. *2nd Scand. Workshop on Algorithm Theory*, Lect. Notes in Computer Science 447 (Springer–Verlag 1990), pp. 348-359. Full version is to appear in *SIAM J. Comput. 22*.

20. Ukkonen, E. (1985): Finding approximate patterns in strings. *J. Algorithms 6*, 132–137.
21. Ukkonen, E. (1992): Approximate string–matching with q-grams and maximal matches. *Theoretical Computer Science 92*, 191–211.
22. Ukkonen, E. (1992): Constructing suffix trees on–line in linear time. In: J. van Leeuwen (ed.), *Algorithms, Software, Architecture. Information Processing 92*, vol. I, pp. 484–492. Elsevier.
23. Ukkonen, E. & Wood, D.: Approximate string matching with suffix automata. *Algorithmica* (to appear in 1993).
24. Wagner, R. A. & Fischer, M. J. (1974): The string-to-string correction problem. *J. ACM 21*, 168-173.
25. Weiner, P. (1973): Linear pattern matching algorithms. *Proc. 14th IEEE Symp. Switching and Automata Theory*, pp. 1-11.
26. Wu, S. & Manber, U. (1992): Fast text searching allowing errors. *Comm. ACM 35*, 83–91.

Multiple Sequence Comparison and
n-Dimensional Image Reconstruction [*]

Martin Vingron[1] and Pavel A. Pevzner[2]

[1] Department of Mathematics, University of Southern California
Los Angeles, CA 90089-1113
[2] Computer Science Department, The Pennsylvania State University
333 Whitmore Laboratory, University Park, PA 16802

Abstract. Calculation of dot-matrices is a widespread tool in pairwise sequence comparison. In recent studies the usefulness of dot-matrices for multiple sequence alignment has been proved. Viewing dot-matrices as projections of unknown n-dimensional points, we consider the multiple alignment problem (for n sequences) as an n-dimensional image reconstruction problem with noise. From this perspective we introduce and develop the filtering method due to Vingron and Argos (*J. Mol. Biol.* (1991), **218**, pp. 33-43). We discuss a conjecture of theirs regarding the number of iterations their algorithm requires and demonstrate that this number may be large. An improved version of the original algorithm is introduced that avoids costly dot-matrix multiplications and runs in $O(n^3 \cdot L^3)$ time (L is the length of the longest sequence). This is equivalent to only one iteration of the original algorithm. We also discuss applications to DNA/protein sequence comparisons.

1 Introduction

As a consequence of the human genome project vast amounts of biological sequence data have been determined biochemically during the last years and stored in various sequence data bases. In the context of evaluating this information there arose the task of identifying similarities among biological sequences. We refer to this task as the *sequence comparison* problem. Biologists usually consider two variants of the sequence comparison problem: *pairwise alignment* (see [Wat84] for review) and *multiple alignment* (see [CWC92]). Even though these problems are clearly formulated the second one still does not lend itself to an efficient solution. In fact, exact methods ([MRS85], [Got86], [CaL88], [AlL89], [Pev92]) are costly in terms of computer time and memory and therefore not widely used by biologists.

Finding similarities among sequences is in a certain sense more general than aligning two sequences. When similar regions between two sequences are known, an alignment roughly corresponds to an ordered subset of the similarities. This viewpoint gave rise to many heuristic procedures for multiple sequence alignment which

* This research was supported by grants from the National Science Foundation (DMS 90-05833, DMS 90-05833) and the National Institute of Health (GM-36230). This paper was written when P.A.P. was at the Department of Mathematics, University of Southern California, Los Angeles.

first screen for similarities among sequences. Subsequently a subset of the similarities is extracted that is laid out to give an alignment. Those methods will in general find not an *optimal*, but a biologically plausible (*suboptimal*) solution. Motivated by such considerations many authors base their methods on the search for *words* which occur either identically or similarly in all or most all sequences. Such words are frequently called *consensus* elements. Algorithms that base alignments on the search for consensus elements have been developed by Waterman et al. [WAG84], Sobel and Martinez [SoM86], Karlin et al. [KMG88] and Vingron and Argos [ViA89]. More recently methods have been proposed which instead of directly searching for consensus elements among the sequences compute pairwise similarities and then assemble the overall multiple alignment from pairwise comparisons (Vihinen [Vih88], Vingron and Argos [ViA91], Schuler et al. [SAL91], Roytberg [Roy92], Miller [Mil93]).

In molecular biology similarities between sequences are frequently depicted in the form of *dot-matrices* ([MaL81], [Arg87]). Formally we will view a dot-matrix simply as a matrix with all entries either 0 or 1, where a 1 at position (i, j) denotes the existence of similarity between the i-th position of the first sequence and the j-th position of the second one. The criteria used to decide whether a position from the first sequence and a position from the second one are similar vary from being purely combinatorial (e.g. a match of length m with at most k mismatches starting at positions i of the first sequence and j of the second one) to using correlation coefficients between physical parameters of the sequence residues [Arg87]. However, no such criterion is perfect in its ability to distinguish "real" (biologically relevant) similarities from such which are due to chance. Consequently any collection of similarities between two sequences is likely to contain *false positives* in addition to the real matching regions. In many biological applications, eliminating noise from dot-plots is highly valuable, possibly even more useful than an optimal multiple alignment.

The availability of several sequences sharing biologically relevant similarities may provide a remedy in this situation. When n sequences are given one can calculate $\binom{n}{2}$ pairwise dot-matrices. Any false positives are unlikely to occur (in a sense which has to be defined) "consistently" among all the comparisons. On the other hand if there were a word which is shared by all sequences this should be visible in all those comparisons. The practical problem is to reverse this observation: Given the $\binom{n}{2}$ dot-matrices find any pairwise similarities that occur consistently among all pairs and thus point towards regions shared by all sequences.

The above problem has already attracted the attention of several researchers. Vihinen [Vih88] proposed to "superimpose" pairwise dot-matrices by choosing one reference sequence and relating all others to this one. This approach has been taken significantly farther by Roytberg [Roy92]. Gotoh [Got90] has defined a notion of consistency for three sequences. Schuler et al. [SAL91] use diagonals instead of dots as the basic entity and devise a heuristic to assemble those into a multiple alignment. An exact algorithm for assembling an n-dimensional dot-matrix from 2-dimensional dot-matrices should check consistency between all pairwise comparisons. This is done in Vingron and Argos [ViA91].

2 n-Dimensional image reconstruction

We want to represent the problem of finding consistent similarities in a simple geometric framework. Consider M integer points in n-dimensional space

$$(i_1^1, \ldots i_n^1), \ldots, (i_1^M, \ldots i_n^M)$$

for which we do not know the coordinates . Suppose we observe the projection of these points onto each pair of dimensions s and t, $1 \leq s < t \leq n$:

$$(i_s^1, i_t^1), \ldots, (i_s^M, i_t^M)$$

as well as some additional points (*noise*). Suppose also that we cannot distinguish a point representing a *real* projection (i_s^m, i_t^m) from one representing noise. In this context the *n-dimensional image reconstruction problem* is to reconstruct M n-dimensional points $(i_1^m, \ldots, i_n^m)$, $1 \leq m \leq M$ from $\frac{n(n-1)}{2}$ projections (with noise) onto coordinates s and t for $1 \leq s < t \leq n$.

In this construction each consensus shared by n biological sequences corresponds to an integer point $(i_1, \ldots i_n)$ in n-dimensional space where i_s is the coordinate of the first position of the consensus in the s-th sequence, $1 \leq s \leq n$. In practice, it is hard to find the integer points $(i_1, \ldots, i_n)$ corresponding to consensuses. On the other hand, it is relatively easy to find (with considerable noise though) the projections (i_s, i_t) of all consensuses $(i_1, \ldots i_n)$ onto every pair of coordinates s and t . This observation establishes the link between the multiple alignment problem and the n-dimensional image reconstruction problem.

3 Consistency and filtering

From the given dots in the side-planes we propose to keep only those that fulfill the following criterion of consistency: The point (i, j) in projection s, t is called *consistent* if for every other dimension u there exists an integer k such that

(i) the point (i, k) belongs to the projection s, u,
(ii) the point (j, k) belongs to the projection t, u.

Obviously each "real" point, i.e. one that was generated as a projection of an n-dimensional point, is consistent. On the other hand random points representing noise are expected to be inconsistent. This observation allows one to filter out most (though possibly not all) of the noise from the side-planes. It leads to an algorithm that multiplies and compares dot-matrices, which was first described in [ViA91].

The suggested dot-matrix multiplication algorithm [ViA91] is an iterative procedure which requires $O(n^3 \cdot L^3)$ time per iteration (L is the length of the longest sequences). It will be shown that the number of iterations may be very large. We will construct a case (based on one proposed by A. Dress, personal communication) for which the number of iterations is of the order of the length of the sequences. This motivates an improved algorithm running in time $O(n^3 \cdot L^3)$ overall, which is equivalent to the run-time of only one iteration of the matrix-multiplication algorithm. In practical applications the input data for the algorithm are sparse matrices which makes the algorithm even faster. The running time becomes $O(L^2 n^3)$ if the number of dots in every dot-matrix is $O(L)$. Expressed in the overall number M of dots in all $\binom{n}{2}$ dot-matrices, the running time is $O(nLM)$.

4 Consistency and dot-matrix multiplication

We model the collection of dot-matrices as an *n-partite graph* $G(V_1 \cup V_2 \cup \ldots \cup V_n, E)$, where V_i is the set of positions in the *i*-th sequence. We join the vertex $i \in V_s$ with $j \in V_t$ by an (undirected) *edge e* if and only if there exists a dot at position (i, j) of the dot-matrix comparing sequences s and t. An edge $e \in E$ will be written as $e = (s, i|t, j)$ to indicate that it joins vertices $i \in V_s$ and $j \in V_t$. We denote a *triangle* formed by three edges $(s, i|t, j), (t, j|u, k)$ and $(s, i|u, k)$ as $(s, i|t, j|u, k)$. A triangle *meets* a set V_s if one of the three vertices is an element of V_s. We now define an edge $(s, i|t, j)$ to be *consistent* if for every V_u, $u \neq s, t$, $1 \leq u \leq n$ there exists a triangle $(s, i|t, j|u, k)$ for some k. A subset $E' \subseteq E$ is called *consistent* if for all edges $(s, i|t, j) \in E'$ there exists a triangle $(s, i|t, j|u, k)$ all edges of which are in E'. The *n-partite* graph G is defined to be *consistent* if its edge-set is consistent. Clearly, if $G'(V_1 \cup V_2 \cup \ldots \cup V_n, E')$ and $G''(V_1 \cup V_2 \cup \ldots \cup V_n, E'')$ are consistent graphs then their *union* $G' \cup G''(V_1 \cup V_2 \cup \ldots \cup V_n, E' \cup E'')$ is consistent. Therefore we can associate to any *n-partite* graph a unique maximal consistent subgraph. Our main problem is the following *consistency problem:* Given $G = (V_1 \cup V_2 \cup \ldots \cup V_n, E)$, find the maximal consistent subset $E' \subseteq E$.

The $|V_s| \times |V_t|$ *adjacency* matrix A_{st} between V_s and V_t defined as

$$(A_{st})_{ij} = \begin{cases} 1, & \text{if } (s, i|t, j) \in E \\ 0, & \text{otherwise} \end{cases}$$

is exactly the dot-matrix for the sequences s and t. Each such matrix corresponds to a subset of E and we will apply the operations $\cup, \cap, \subset$ to the matrices A_{st}. We reformulate the above definition of consistency in the terms of *boolean multiplication* (denoted by "$\circ$") of the adjacency matrices [ViA91]. An *n*-partite graph is consistent exactly if

$$A_{st} \subseteq A_{su} \circ A_{ut} \quad \forall s, t, u : 1 \leq s, t, u \leq n, u \neq s \neq t \tag{1}$$

5 An algorithm for the consistency problem

The characterization (1) suggests the following simple procecedure to solve the consistency problem: Keep only those 1's in the adjacency matrix which are present both in the matrix itself and in all products $A_{su} \circ A_{ut}$: $A_{st} \leftarrow A_{st} \cap (\bigcap_{u \neq s,t} A_{su} \circ A_{ut})$. Doing this once for all adjacency matrices will also change the matrices used for the products when the computation commenced. This leads to the following iterative procedure:

Algorithm 1 ("matrix-multiplication algorithm"):

To distinguish between the different generations of modified adjacency matrices we introduce a superscript. We start with the adjacency matrices $A_{st}^{(0)} := A_{st}$ of the given *n*-partite graph. The iteration is defined by:

$$A_{st}^{(l+1)} := A_{st}^{(l)} \cap (\bigcap_{u \neq s,t} A_{su}^{(l)} \circ A_{ut}^{(l)}).$$

Once this is done for all indices s and t the process is repeated until at some iteration m we have $A_{st}^{(m+1)} = A_{st}^{(m)}$ for all $1 \leq s, t \leq n$. □

The following theorem has been proven in [ViA91].

Theorem 1. *The matrix-multiplication algorithm converges and the result is the maximal consistent subgraph of G.*

6 The number of iterations in the matrix-multiplication algorithm may be large

For an n-partite graph $G(V, E)$ where each component contains L elements each iteration of the matrix-multiplication algorithm takes $O(n^3 L^3)$ operations (for the sake of simplicity we ignore fast matrix-multiplication algorithms, [AHU74]). Vingron and Argos [ViA91] noticed that in the examples they studied the number of iterations the matrix-multiplication algorithm required was less than $n-1$. However, A. Dress (personal communication) constructed a graph requiring a larger number of iterations. Based on his construction we will describe a graph for which $O(\log L)$ iterations are needed before the algorithm terminates. We also construct a more complicated example for which the number of iterations is $O(L)$. At each iteration of the matrix-multiplication algorithm we reduce the overall number of vertices in the dot-matrices; therefore the number of iterations is bounded by $O(L^2 \cdot n^2)$. The question if the number of iterations might be larger than $O(L)$ is still open.

We will now present an iterative rule of constructing a certain n-partite graph. We start with a small set of edges which are not consistent. Then successively more and more edges are added to make the prior ones appear consistent. This will not make the graph consistent but will make it increasingly harder for the matrix-multiplication algorithm to uncover this fact.

Consider a 4-partite graph $G = (V_1 \cup V_2 \cup V_3 \cup V_4, E)$ in which each edge belongs to a triangle and let $e_1, \ldots e_d$ be inconsistent edges of G. An example of such a graph $G^\Delta = (V_1 \cup V_2 \cup V_3 \cup V_4, \{e_1, e_2, e_3\})$ is given in Figure 1. The 3 edges of G^Δ are not consistent because none of them belongs to a triangle meeting the 4th component. We can make any $e_i = (s, i|t, j)$ consistent by adding two additional edges and an additional vertex which establish a triangle meeting the needed component (Fig.1). We can do this for every inconsistent edge in a given edge-set. We call this operation *completion* and denote the resulting graph as $c(G)$. Every such operation adds new inconsistent edges such that it is reasonable to apply the same procedure again. Repeated application of k. completions to a graph will be denoted $c^{(k)}(G)$.

Lemma 2. *Let G be a graph in which each edge belongs to a triangle. Edges in $c^{(k-1)}(G)$ are consistent in $c^{(k)}(G)$. Edges in $c^{(k)}(G) \setminus c^{(k-1)}(G)$ are not consistent in $c^{(k)}(G)$.*

Lemma 3. *Applying one iteration of the matrix-multiplication algorithm to $c^{(k)}(G)$ results in $c^{(k-1)}(G)$.*

Denote the average number of vertices in the components of the graph $c^{(k)}(G^\Delta)$ as L_k. The following theorem demonstrates that the graph $c^{(k)}(G^\Delta)$ is a "hard case" for the matrix-multiplication algorithm

Theorem 4. *The number of iterations of the matrix-multiplication algorithm for the graph $c^{(k)}(G^\Delta)$ is $O(\log L_k)$.*

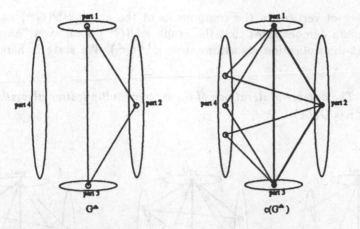

Fig. 1. Completion of graph G^Δ.

Proof. With one iteration of the algorithm being the inverse operation to the completion (Lemma 3) it follows that after having applied k iterations to $c^{(k)}(G^\Delta)$ we are left with G^Δ. This in itself is not consistent and after one further iteration the edge-set will be empty and the algorithm will terminate. From the definition of completion and Lemma 2 it follows that the number of edges in $c^{(k)}(G) \setminus c^{(k-1)}(G)$ is doubled in comparison to $c^{(k-1)}(G) \setminus c^{(k-2)}(G)$. Starting with 3 edges of G^Δ we would have $3 + 3 \cdot \sum_{i=1}^{k} 2^k$ edges and $3 + 3 \cdot \sum_{i=1}^{k} 2^{k-1} = 3 \cdot 2^k$ vertices in $c^{(k)}(G^\Delta)$. This implies that the average number of vertices in the components of $c^{(k)}(G^\Delta)$ is $L_k = \frac{3}{4} \cdot 2^k$. Therefore, the number of iterations of the algorithm for the graph $c^{(k)}(G^\Delta)$ is $k + 1 = \log_2 L_k + \log_2 \frac{4}{3} + 1$. □

The above construction freely introduces new vertices into the graph. Better exploitation of the vertices allows us to construct a graph for which the matrix-multiplication algorithm requires even $O(L)$ iterations. Once again consider a 4-partite graph G in which each edge belongs to a triangle and let $\{e_1, \ldots e_d\}$ be inconsistent edges of G. With every such inconsistent edge $e = (s, i|t, j)$ we associate the component u that is met by the triangle(s) containing the edge. We will call the set $\{\{s,t\}, u\}$ the *class* of e. For every class $\{\{s,t\}, u\}$ (there are 12 of them) define 1 new vertex in the component w that is not part of the class. Join this vertex with the ends of all inconsistent edges belonging to the class. This certainly makes all inconsistent edges between the components V_s and V_t of the class $\{\{s,t\}, u\}$ consistent. On the other hand there will not be an edge linking this new vertex in the component w to any vertex in component u and thus none of the newly introduced edges can be consistent. Doing this procedure for each class we get a graph $b(G)$ with at most 12 new vertices and at most $2d$ new edges. The difference between $b(G)$ and $c(G)$ is that in $c(G)$ we added a new vertex for *each inconsistent edge* while in $b(G)$ we add a new vertex for *each of 12 classes* $\{\{s,t\}, u\}$. Repeated application of k operations $b(G)$ to a graph will be denoted $b^{(k)}(G)$. Fig.2 illustrates the difference between $c(G)$ and $b(G)$. Analogs of Lemmas 2 and 3 hold for $b^{(k)}(G)$. Denote the average number of vertices in the components of the graph $b^{(k)}(G^\Delta)$ as L_k. The following theorem demonstrates that the graph $b^{(k)}(G^\Delta)$ is an even "harder case" for the matrix-multiplication algorithm than $c^{(k)}(G^\Delta)$. We state it here without proof.

Theorem 5. *The number of iterations of the matrix-multiplication algorithm for the graph* $b^{(k)}(G^\Delta)$ *is* $O(L_k)$.

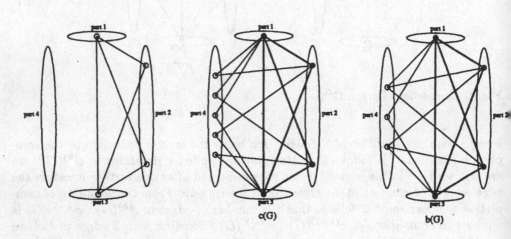

Fig. 2. Comparison of the graphs $c(G)$ and $b(G)$.

7 An efficient algorithm for the consistency problem

Denote $V = V_1 \cup \ldots \cup V_n$ and for the sake of simplicity assume that $|V_1| = \ldots = |V_n| = L$. Each iteration of the dot-matrix multiplication algorithm for the n-partite graph $G(V, E)$ needs $O(n^3 L^3)$ time. In this section we give an algorithm for the calculation of the maximal consistent subset of $G(V, E)$ running in $O(n^3 L^3)$ time overall.

Let T be the set of triangles in G. The algorithm uses a bipartite graph $B(E \cup T, A)$ relating edges in E to the triangles in T they are part of. The arc-set A is defined by the rule: $(s, i|t, j) \in E$ is joined with all triangles $(s, i|t, j|u, k) \in T$ containing $(s, i|t, j)$.

For a given edge $e = (s, i|t, j)\epsilon E$ we denote the set of all triangles $(s, i|t, j|u, k)$ meeting a component V_u of the n-partite graph G as $D(e, V_u)$. The number of triangles in $D(e, V_u)$ is the *degree* $d(e, V_u) = |D(e, V_u)|$ of edge e with respect to component V_u. This leads to a reformulation of the consistency notion: An edge $e = (s, i|t, j)$ is consistent if and only if $d(e, V_u) > 0$ for all u distinct from s and t, $1 \leq u \leq n$. For every edge e we calculate the vector $\mathbf{d}(e) = (d(e, V_1), ..., d(e, V_n))$ with the intent of keeping it up to date throughout the computation. We define $D((s, i|t, j), V_s)$ and $D((s, i|t, j), V_t)$ to be empty sets; therefore at least two components of $\mathbf{d}(e)$ equal zero. An inconsistent edge e is now revealed by having at least three components of $d(e)$ equaling zero.

Algorithm 2

Edges are divided into "potentially consistent" ones E^+ and inconsistent ones E^-. The idea is to keep track of the vectors $d(e)$ while deleting inconsistent edges.

Preprocessing

At the preprocessing stage we construct the bipartite graph B and determine the sets $D(e, V_u)$ for all e and u.

For each edge $e = (s, i|t, j)\epsilon E$ we compute the degrees $d(e)$ and check $n - 2$ conditions

$$d(e, V_u) > 0$$

for $u \neq s, t$. If all these conditions hold assign $(s, i|t, j)$ to E^+, otherwise assign $(s, i|t, j)$ to E^-.

Iteration of algorithm.

During the iteration of the algorithm we exclude the edges belonging to E^- from $G(V, E)$ (or in other words we exclude some vertices from the part E of the bipartite graph B) and transfer some edges from E^+ into E^-. The following pseudocode results in the graph $G(V, E^+)$ which is the maximal consistent subgraph of $G(V, E)$:

```
while (E⁻ is non-empty)
  { for (each edge e = (s, i|t, j) in E⁻)
    { /* delete edge e from G */
       for (each triangle (s, i|t, j|u, k) containing e)
       { /* modify d((s, i|u, k), Vₜ) */
          d((s, i|u, k), Vₜ) = d((s, i|u, k), Vₜ) − 1
          if ( d((s, i|u, k), Vₜ) < 1)
             if ((s, i|u, k)ϵE⁺)
                transfer (s, i|u, k) from E⁺ to E⁻
          /* modify d((t, j|u, k), Vₛ) */
          d((t, j|u, k), Vₛ) = d((t, j|u, k), Vₛ) − 1
          if ( d((t, j|u, k), Vₛ) < 1)
             if ((t, j|u, k)ϵE⁺)
                transfer (t, j|u, k) from E⁺ to E⁻
          delete triangle (s, i|t, j|u, k) from the part T of B
       }
       delete edge e from the part E of the bipartite graph B
    }
  }
```

Theorem 6. *Algorithm 2 computes the maximal consistent subgraph of $G(V, E)$ in $O(|V| \cdot |E|) \leq O(n^3 L^3)$ time.*

Proof. Each triangle $(s, i|t, j|u, k)$ in G is uniquely defined by the vertex $(s, i)\epsilon V$ and the edge $(t, j|u, k)\epsilon E$. Therefore the bipartite graph $B = (E \cup T, A)$ can be computed in $O(|T|) \leq O(|V| \cdot |E|)$ time (see Bern and Eppstein, [BeE92], for a $O(|E|^{\frac{3}{2}})$ estimate of $|T|$). All sets $D(e, V_u)$, all degrees $d(e, V_u)$ and the partition E into E^+ and E^- can be computed in $O(|A|)$ time, where $|A| \leq 3 \cdot |T|$ (each triangle is adjacent to at most 3 arcs in B). Therefore the complexity of preprocessing is $O(|V| \cdot |E|)$. The complexity of iteration is the number of triangles excluded on the iteration. Therefore the complexity of all iterations does not exceed the number of all triangles $O(|V| \cdot |E|)$ and the overall complexity of the algorithm is $O(|V| \cdot |E|) \leq O(n^3 L^3)$. $\qquad\qquad$ □

8　Example

Figure 3 illustrates how filtering of the raw data coming from pairwise sequence comparisons allows one to distinguish between significant and insignificant sequence similarities. Figure 3a shows a dotplot comparing two amino acid sequences. The dots shown represent residue pairs which are contained in the 20 best non-intersecting local suboptimal alignments [WaE87]. The sequences used are from the protein *Elongation Factor Tu* from human and fruit fly.

20 suboptimal alignments provide a very noisy version of this sequence comparison. We then used 4 more sequences of the same protein family to obtain $\binom{6}{2}$ comparisons. Calculating the maximal consistent subset for the given set of comparisons yields considerably clearer plots which have maintained most of the correct alignment. Figure 3b shows the comparison between the same two sequences as in Fig. 3a but after filtering. Finally we used a set of 12 sequences instead of only 6 and obtained the comparison between human and fruit fly *Elongation Factor Tu* shown in Fig. 3c. It is obvious that most of the noise though not all of it has disappeared. There are hardly any points left outside the main diagonal which represents the correct alignment. The main diagonal is interrupted at various positions. This is due to other comparisons containing gaps which are placed at slightly different positions in every comparison. As a result in the surrounding of those gaps there are no consistent dots. This might indicate "important" (corresponding, e.g, to structurally and/or functionally conserved fragments) and "non-important" regions of an alignment.

9　Conclusion

In conclusion we would like to formulate three open problems.

Characterization of consistent graphs: It would be interesting to find a characterization of consistent graphs not in *local* but in *global* terms, possibly in terms of cliques. At a first glance, consistent graphs look similar to *partition intersection graphs* which are the unions of n-cliques (McMorris and Meacham, [McM83]) and thus have to be consistent. However Sampath Kannan (personal communication) constructed a consistent graph which is not a partition intersection graph.

Expected number of consistent points: Given m random "real" n-dimensional points (they correspond to at most m real points in each projection) and t random "noise" points in each projection, find the expected number of consistent points after application of the filtering procedure.

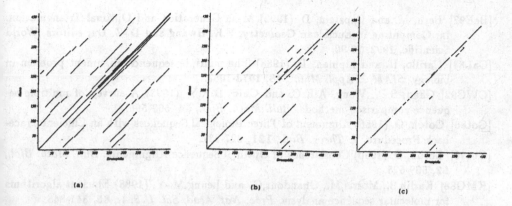

Fig. 3. (a) Dot-plot containing 20 non-intersecting local suboptimal alignments ([WaE87]) of the protein *Elongation factor Tu* from *Drosophila melanogaster* and *Homo sapiens*. Dots along a diagonal are linked to form a straight line. (b) After comparing the same two sequences and four more of the same family (*Elongation factor Tu* from *Mus musculus, Xenopus laevis, Saccharomyces cerevisiae, Rhizomucor racemosus*) all resulting comparisons are filtered and the remaining dots in the pair from (a) shown. (c) A set of 12 sequences comprising the ones from (a) and (b) as well as 6 others (*Dictyostelium discoideum, Sulfolobus acidocaldarius, Halobacterium marismortui, Methanococcus vannielii, Thermococcus celer, Thermoplasma acidophilum*) are compared and filtered and the remaining dots in the pair from (a) shown.

Noise elimination based on diagonals: Using entire matching regions (diagonals, see [SAL91], [Mil93]) as the basic entity in the search for commonalities among sequences promises a reduction in the length of the input word. Reconciling our notion of consistency with such a more efficient representation of pairwise similarities might lead to new, fast and practical algorithms.

Acknowledgment

M.V. wishes to thank Andreas Dress for several helpful discussions. Thanks are also due to David Eppstein, Sampath Kannan, Webb Miller, Michael Roytberg and Tandy Warnow for useful suggestions and criticism.

References

[AHU74] Aho, A.V., Hopcroft, J.E. and Ullman: The Design and Analysis of Computer Algorithms. Addison-Wesley, 1974.

[AiL89] Altschul, S.F and Lipman, D.J. (1989) Trees, stars, and multiple biological sequence alignment. *SIAM J. Appl. Math.*, 49, 179-209.

[Arg87] Argos, P. (1987) A Sensitive Procedure to Compare Amino Acid Sequences. *J. Mol. Biol.*, 193, 385-396.

[BeE92] Bern, M. and Eppstein, D. (1992) Mesh Generation and Optimal Triangulation. In: Computing in Euclidean Geometry, F.K. Hwang and D.-Z. Du, editors, World Scientific, 1992, 23-90.

[CaL88] Carillo, H. and Lipman, D. (1988) The multiple sequence alignment problem in biology. *SIAM J. Appl. Math.*, 48, 1073-1082.

[CWC92] Chan, S.C., Wong, A.K.C. and Chiu, D.K.Y. (1992) A survey of multiple sequence comparison methods. *Bull. Math. Biol.*, 54, 563-598.

[Got86] Gotoh, O. (1986) Alignment of Three Biological Sequences with an Efficient Traceback Procedure. *J. Theor. Biol.*, 121, 327-337.

[Got90] Gotoh, O. (1990) Consistency of Optimal Sequence Alignments. *Bull. Math. Biol.*, 52, 509-525.

[KMG88] Karlin,S., Morris,M., Ghandour,G. and Leung,M.-Y. (1988) Efficient algorithms for molecular sequence analysis. *Proc. Nat. Acad. Sci. U.S.A.*, 85, 841-845.

[MaL81] Maizel, J.V. and Lenk R.P. (1981) Enhanced graphic matrix analysis of nucleic acid and protein sequences. *Proc. Nat. Acad. Sci. USA*, 78, 7665-7669.

[McM83] McMorris, F.R. and Meacham, C.F. (1983) Partition intersection graphs. *Ars Combin.*, 16-B, 133-138.

[Mil93] Miller, W. (1993) Building multiple alignments from pairwise alignments. *Comp. Appl. Biosci.*, to appear.

[MRS85] Murata, M., Richardson, J.S. and Sussman, J.L. (1985) Simultaneous comparison of three protein sequences. *Proc. Nat. Acad. Sci. U.S.A.*, 82, 3073-3077.

[Pev92] Pevzner P.A. Multiple alignment, communication cost, and graph matchings. *SIAM J. Appl. Math.*, 52, 1992, 1763-1779.

[Roy92] Roytberg, M.A. (1992) A search for common patterns in many sequences. *Comp. Appl. Biosci.*, 8, 57-64.

[SAL91] Schuler, G.D., Altschul S.F. and Lipman, D.J. (1991) A Workbench for Multiple Sequence Alignment Construction and Analysis. *PROTEINS: Structure, Function and Genetics*, 9, 180-190.

[SoM86] Sobel, E. and Martinez, H. (1986) A multiple sequence alignment program. *Nucleic Acids Res.*, 14, 363-374.

[Vih88] Vihinen, M. (1988) An algorithm for simultaneous comparison of several sequences. *Comp. Appl. Biosci.*, 4, 89-92.

[ViA89] Vingron, M. and Argos, P. (1989) A fast and sensitive multiple sequence alignment algorithm. *Comp. Appl. Biosci.*, 5, 115-121.

[ViA91] Vingron and Argos (1991) Motif recognition and alignment for many sequences by comparison of dot-matrices *J. Mol. Biol.* 218, pp. 33-43.

[Wat84] Waterman, M.S. (1984) General methods of sequence comparison. *Bull. Math. Biol.*, 46, 473-500.

[WAG84] Waterman, M.S., Arratia, R. and Galas, D.J. (1984) Pattern recognition in several sequences: Consensus and Alignment. *Bull. Math. Biol.*, 46, 515-527.

[WaE87] Waterman, M.S. and Eggert, M. (1987) A new algorithm for best subsequence alignments with application to tRNA-rRNA comparisons. *J. Mol. Biol.*, 197, 723-725.

A New Editing based Distance between Unordered Labeled Trees *

Kaizhong Zhang

Department of Computer Science
University of Western Ontario
London, Ont. N6A 5B7
CANADA
e-mail: kzhang@csd.uwo.ca

Abstract. This paper considers the problem of computing a new editing based distance between unordered labeled trees. The problem of approximate unordered tree matching is also considered. We present algorithms solving these problems in sequential time $O(|T_1| \times |T_2| \times \max\{deg(T_1), deg(T_2)\} \times \log_2(\max\{deg(T_1), deg(T_2)\}))$. Our previous result shows that computing the editing distance between unordered labeled trees is NP-complete.

1 Introduction

Unordered labeled trees are trees whose nodes are labeled and in which only ancestor relationships are significant (the left-to-right order among siblings is not significant). Such trees arise naturally in genealogical studies, for example, the genetic study of the tracking of diseases. For many such applications, it would be useful to compare unordered labeled trees by some meaningful distance metric. The editing distance metric, used with some success for ordered labeled trees [4], is a natural such metric.

In section 2 we review the previous NP-completeness results on editing distance between unordered trees. In section 3 we introduce the new distance metric between trees, which we call constrained editing distance. In section 4 we will investigate the properties of the new distance metric. In section 5 we present an algorithm to compute the new distance metric and analyse the complexity. In section 6 we consider the approximate unordered tree matching problem.

2 Preliminaries

In this section we will first introduce some basic definitions and then review the previous results on unordered trees. Unless otherwise stated, all trees we consider in the paper are rooted, labeled, and unordered.

---

* Research supported by the Natural Sciences and Engineering Research Council of Canada under Grant No. OGP0046373.

2.1 Editing Operations

We consider three kinds of operations. Changing a node n means changing the label on n. Deleting a node n means making the children of n become the children of the parent of n and then removing n. Inserting is the complement of deleting. This means that inserting n as the child of m will make n the parent of a *subset* (as opposed to a consecutive subsequence [11]) of the current children of m.

Following [5, 11, 12], we represent an edit operation as $a \rightarrow b$, where a is either λ or a label of a node in tree T_1 and b is either λ or a label of a node in tree T_2. We call $a \rightarrow b$ a change operation if $a \neq \lambda$ and $b \neq \lambda$; a delete operation if $b = \lambda$; and an insert operation if $a = \lambda$.

Let S be a sequence $s_1, ..., s_k$ of edit operations. An S-derivation from tree A to tree B is a sequence of trees $A_0, ..., A_k$ such that $A = A_0$, $B = A_k$, and $A_{i-1} \rightarrow A_i$ via s_i for $1 \leq i \leq k$. Let γ be a cost function which assigns to each edit operation $a \rightarrow b$ a nonnegative real number $\gamma(a \rightarrow b)$.

We constrain γ to be a distance metric. That is, i) $\gamma(a \rightarrow b) \geq 0$, $\gamma(a \rightarrow a) = 0$; ii) $\gamma(a \rightarrow b) = \gamma(b \rightarrow a)$; and iii) $\gamma(a \rightarrow c) \leq \gamma(a \rightarrow b) + \gamma(b \rightarrow c)$.

We extend γ to the sequence of editing operations S by letting $\gamma(S) = \sum_{i=1}^{|S|} \gamma(s_i)$.

2.2 Editing Distance and Editing Distance Mapping

The results in this subsection are from [12]. We will omit the proofs.

Editing Distance. [12] defined the *editing distance* between two trees by considering the minimum cost editing operations sequence that transforms one tree to the other. Formally the distance between T_1 and T_2 is defined as:

$$D_e(T_1, T_2) = \min_S \{\gamma(S) \mid S \text{ is an edit operation sequence taking } T_1 \text{ to } T_2\}.$$

Editing Distance Mappings. The edit operations give rise to a mapping which is a graphical specification of what edit operations apply to each node in the two unordered labeled trees.

Suppose that we have a numbering for each tree. Let $t[i]$ be the ith node of tree T in the given numbering. Formally we define a triple (M_e, T_1, T_2) to be an editing distance mapping from T_1 to T_2, where M_e is any set of pair of integers (i, j) satisfying:

(1) $1 \leq i \leq |T_1|$, $1 \leq j \leq |T_2|$;
(2) For any pair of (i_1, j_1) and (i_2, j_2) in M_e,
  (a) $i_1 = i_2$ iff $j_1 = j_2$ (one-to-one)
  (b) $t_1[i_1]$ is an ancestor of $t_1[i_2]$ iff $t_2[j_1]$ is an ancestor of $t_2[j_2]$ (ancestor order preserved)

We will use M_e instead of (M_e, T_1, T_2) if there is no confusion. Let M_e be an editing distance mapping from T_1 to T_2. Then we can define the cost of M_e:

$$\gamma(M_e) = \sum_{(i,j) \in M_e} \gamma(t_1[i] \rightarrow t_2[j]) + \sum_{i \notin M_e} \gamma(t_1[i] \rightarrow \lambda) + \sum_{j \notin M_e} \gamma(\lambda \rightarrow t_2[j])$$

The relation between an editing distance mapping and a sequence of editing operations is as follows:

Lemma 1. *Given S, a sequence $s_1, \ldots, s_k$ of edit operations from T_1 to T_2, there exists a editing distance mapping M_e from T_1 to T_2 such that $\gamma(M_e) \leq \gamma(S)$. Conversely, for any mapping M_e, there exists a sequence of editing operations such that $\gamma(S) = \gamma(M_e)$.*

Based on the lemma, the following theorem states the relation between the editing distance and the editing distance mappings. This is why we call this kind mapping editing distance mapping.

Theorem 2. $D_e(T_1, T_2) = \min_{M_e}\{\gamma(M_e) \mid M_e$ *is a mapping from T_1 to $T_2\}$*

One of the results in [12] is that finding $D_e(T_1, T_2)$ is NP-complete even if the trees are binary trees with a label alphabet of size two. Kilpelainen and Mannila [2] showed that even the inclusion problem for unordered trees is NP-complete. In fact we recently proved a stronger result that the problem of finding the minimum cost mapping (edit distance) between two unordered trees and the problem of finding the largest common subtree of two unordered trees are both MAX SNP-hard which means that there is no polynomial time approximation scheme (PTAS) for these problems unless P=NP. Since unordered trees are important in some applications, one would like to find distances that can be efficiently computed.

3 A New Editing based Distance Metric between Unordered Trees

Our new distance metric is based on a restriction of the mappings allowed between two trees. The intuitive idea is that two separate subtrees of T_1 should be mapped to two separate subtrees in T_2. This idea was proposed by Tanaka and Tanaka [7] in their definition for a structure preserving mapping between two ordered labeled trees although the definition in [7] dose not capture the idea precisely. Tanaka and Tanaka [7] also showed that in some applications (e.g., classification tree comparison) the structural preserving mapping is more meaningful than editing distance mapping. We refined the definition for ordered trees [10] and in this paper extend the definition from ordered trees to unordered trees

3.1 New Mappings

Suppose again that we have a numbering for each tree. Let $t[i]$ be the ith node of tree T in the given numbering. Let $T[i]$ be the subtree rooted at $t[i]$ and $F[i]$ be the unordered forest obtained by deleting $t[i]$ from $T[i]$.

Formally we define a triple (M, T_1, T_2) to be a mapping from T_1 to T_2, where M is any set of pairs of integers (i, j) satisfying:

(1) $1 \leq i \leq |T_1|, 1 \leq j \leq |T_2|$;
(2) For any pair of (i_1, j_1) and (i_2, j_2) in M_e,
  (a) $i_1 = i_2$ iff $j_1 = j_2$ (one-to-one)
  (b) $t_1[i_1]$ is an ancestor of $t_1[i_2]$ iff $t_2[j_1]$ is an ancestor of $t_2[j_2]$ (ancestor order preserved);

(3) For any triple (i_1, j_1), (i_2, j_2) and (i_3, j_3) in M, let $t_1[I]$ be the $lca(t_1[i_1], t_1[i_2])$ and $t_2[J]$ be the $lca(t_2[j_1], t_2[j_2])$, where lca represents least common ancestor. $t_1[I]$ is not an ancestor or a descendant of $t_1[i_3]$ iff $t_2[J]$ is not an ancestor or a descendant of $t_2[j_3]$.

We will use M instead of (M, T_1, T_2) if there is no confusion. Let M be a mapping from T_1 to T_2. Then we can similarly define the cost of M:

$$\gamma(M) = \sum_{(i,j) \in M} \gamma(t_1[i] \to t_2[j]) + \sum_{i \notin M} \gamma(t_1[i] \to \lambda) + \sum_{j \notin M} \gamma(\lambda \to t_2[j])$$

Mappings can be composed. Let M_1 be a mapping from T_1 to T_2 and M_2 be a mapping from T_2 to T_3. Define

$$M_1 \circ M_2 = \{(i, j) \mid \exists k \ s.t. \ (i, k) \in M_1 \ and \ (k, j) \in M_2\}.$$

Lemma 3. *1) $M_1 \circ M_2$ is a mapping between T_1 and T_3. 2) $\gamma(M_1 \circ M_2) \leq \gamma(M_1) + \gamma(M_2)$.*

Proof: (1) follows from the definition of mapping. Let us check condition (3) only. Let (i_1, j_1), (i_2, j_2) and (i_3, j_3) be in $M_1 \circ M_2$. By the definition of $M_1 \circ M_2$, there are k_1, k_2 and k_3 such that (i_1, k_1), (i_2, k_2) and (i_3, k_3) are in M_1 and (k_1, j_1), (k_2, j_2) and (k_3, j_3) are in M_2. Let I be the $lca(i_1, i_2)$, K be the $lca(k_1, k_2)$ and J be the $lca(j_1, j_2)$. By the definition of M_1 and M_2, I is not an ancestor or descendant of i_3 iff K is not an ancestor or descendant of k_3. Moreover K is not an ancestor or descendant of k_3 iff J is not an ancestor or descendant of j_3. Therefore I is not an ancestor or descendant of i_3 iff J is not an ancestor or descendant of j_3.

  (2) Let M_1 be the mapping from T_1 to T_2. Let M_2 be the mapping from T_2 to T_3. Let $M_1 \circ M_2$ be the composed mapping from T_1 to T_3 and let I and J be the corresponding deletion and insertion sets. Three general situations occur. $(i, j) \in M_1 \circ M_2$, $i \notin M_1$, or $j \notin M_2$. In each case this corresponds to an editing operation $\gamma(x \to y)$ where x and y may be nodes or may be λ. In all such cases, the triangle inequality on the distance metric γ ensures that $\gamma(x \to y) \leq \gamma(x \to z) + \gamma(z \to y)$. $\square$

3.2   A New Editing Based Distance between Trees

We can now define a dissimilarity measure between T_1 and T_2 as:

$$D(T_1, T_2) = \min_M \{\gamma(M) \mid M \ is \ a \ mapping \ from \ T_1 \ to \ T_2\}$$

In fact this dissimilarity measure is a distance metric.

Theorem 4. *1) $D(T, T) = 0$;*
2) $D(T_1, T_2) = D(T_2, T_1)$;
3) $D(T_1, T_3) \leq D(T_1, T_2) + D(T_2, T_3)$.

Proof: 1) and 2) follow directly from the definition of the mapping. For 3), consider the minimum cost mappings M_1 between T_1 and T_2 and M_2 between T_2 and T_3. It is easy to see the following:

$$D(T_1, T_3) \leq \gamma(M_1 \circ M_2) \leq \gamma(M_1) + \gamma(M_2) = D(T_1, T_2) + D(T_2, T_3).$$

$\square$

The relation between our new distance metric D and the editing distance metric D_e is: $D_e(T_1, T_2) \leq D(T_1, T_2)$. The reason is that any new mapping we defined is always a editing distance mapping.

4 Properties of the New Distance

In this section we will present several lemmas which will be the basis for the algorithm in the next section.

Lemma 5. *Let $t_1[i_1], t_1[i_2], ...t_1[i_{n_i}]$ be the children of $t_1[i]$ and $t_2[j_1], t_2[j_2], ...t_2[j_{n_j}]$ be the children of $t_2[j]$, then $D(\theta, \theta) = 0$;*
$D(F_1[i], \theta) = \sum_{k=1}^{n_i} D(T_1[i_k], \theta);$ $D(T_1[i], \theta) = D(F_1[i], \theta) + \gamma(t_1[i] \to \lambda);$
$D(\theta, F_2[j]) = \sum_{k=1}^{n_j} D(\theta, T_2[j_k]);$ $D(\theta, T_2[j]) = D(\theta, F_2[j]) + \gamma(\lambda \to t_2[j]).$

Lemma 6. *Let $t_1[i_1], t_1[i_2], ...t_1[i_{n_i}]$ be the children of $t_1[i]$ and $t_2[j_1], t_2[j_2], ...t_2[j_{n_j}]$ be the children of $t_2[j]$, then*

$$D(T_1[i], T_2[j]) = \min \begin{cases} D(T_1[i], \theta) + \min_{1 \leq s \leq n_i}\{D(T_1[i_s], T_2[j]) - D(T_1[i_s], \theta)\} \\ D(\theta, T_2[j]) + \min_{1 \leq t \leq n_j}\{D(T_1[i], T_2[j_t]) - D(\theta, T_2[j_t])\} \\ D(F_1[i], F_2[j]) + \gamma(t_1[i] \to t_2[j]) \end{cases}$$

Proof: Consider the minimum-cost mapping M between $T_1[i]$ and $T_2[j]$. There are four cases: (1) $i \in M$ and $j \notin M$, (2) $i \notin M$ and $j \in M$, (3) $i \in M$ and $j \in M$, (4) $i \notin M$ and $j \notin M$.

Case 1: let (i, t) in M. Since $j \notin M$, t must be a node in $F_2[j]$. Let $t_2[j_t]$ be the child of $t[j]$ on the path from $t_2[t]$ to $t_2[j]$. Thus $D(T_1[i], T_2[j]) = D(T_1[i], T_2[j_t]) + D(\theta, T_2[j_1]) + ... + D(\theta, T_2[j_{t-1}]) + D(\theta, T_2[j_{t+1}]) + ... + D(\theta, T_2[j_{n_j}]) + \gamma(\lambda, t_2[j])$. Since $D(\theta, T_2[j]) = \gamma(\lambda, t_2[j]) + \sum_{k=1}^{n_j} D(\theta, T_2[j_k])$, we can rewrite the right hand side of the formula as $D(\theta, T_2[j]) + D(T_1[i], T_2[j_t]) - D(\theta, T_2[j_t])$. Since the range of k is from 1 to n_j, we take the minimum of these corresponding costs.

Case 2 is similar to case 1.

Case 3: since $i \in M$ and $j \in M$, by the condition of mapping, (i, j) must be in M. Since $M - (i, j)$ is a mapping between $F_1[i]$ and $F_2[j]$, and for any mapping M' between $F_1[i]$ and $F_2[j]$, $(i, j) \cup M'$ is a mapping between $T_1[i]$ and $T_2[j]$, we know $D(T_1[i], T_2[j]) = D(F_1[i], F_2[j]) + \gamma(t_1[i], t_2[j])$.

Case 4 is similar to Case 3. The formula would be $D(T_1[i], T_2[j]) = D(F_1[i], F_2[j]) + \gamma(t_1[i], \lambda) + \gamma(\lambda, t_2[j])$. Since $\gamma(t_1[i], t_2[j]) \leq \gamma(t_1[i], \lambda) + \gamma(\lambda, t_2[j])$, we do not have to include this case in our final formula. $\square$

Before we proceed to the next lemma we need the following definition.

Define a restricted mapping $RM(i,j)$ between $F_1[i]$ and $F_2[j]$ as follows:

1) $RM(i,j)$ is a mapping between $F_1[i]$ and $F_2[j]$,

2) if (l,k) is in $RM(i,j)$ and $t_1[l]$ is in $T_1[i_s]$ and $t_2[k]$ is in $T_2[j_t]$, then for any (l_1,k_1) in $RM(i,j)$ $t_1[l_1]$ is in $T_1[i_s]$ if and only if $t_2[k_1]$ is in $T_2[j_t]$.

Since a restricted mapping is a mapping, the cost of a restricted mapping is well defined.

Lemma 7. *Let* $t_1[i_1], t_1[i_2], ... t_1[i_{n_i}]$ *be the children of* $t_1[i]$ *and* $t_2[j_1], t_2[j_2], ... t_2[j_{n_j}]$ *be the children of* $t_2[j]$, *then*

$$D(F_1[i], F_2[j]) = \min \begin{cases} D(F_1[i], \theta) + \min_{1 \le s \le n_i} \{D(F_1[i_s], F_2[j]) - D(F_1[i_s], \theta)\} \\ D(\theta, F_2[j]) + \min_{1 \le t \le n_j} \{D(F_1[i], F_2[j_t]) - D(\theta, F_2[j_t])\} \\ \min_{RM(i,j)} \gamma(RM(i,j)) \end{cases}$$

Proof: We consider the minimum-cost mapping M between $F_1[i]$ and $F_2[j]$. There are four cases.

Case 1: there is a $1 \le s \le n_i$ such that if $(k,l) \in M$, then $t_1[k]$ is a node in subtree $T_1[i_s]$; and there are (k_1, l_1) and (k_2, l_2) in M such that $t_2[l_1]$ is a node in $T_2[i_{t_1}]$ and $t_2[l_2]$ is a node in $T_2[i_{t_2}]$, where $1 \le t_1 \ne t_2 \le n_j$. Note that in this case i_s cannot be in M. This is similar to case 1 in Lemma 6, and hence we have following formula: $D(F_1[i], F_2[j]) = D(F_1[i], \theta) + \min_{1 \le s \le n_i} D(F_1[i_s], F_2[j]) - D(F_1[i_s], \theta)$.

Case 2: there is a $1 \le t \le n_j$ such that if $(k,l) \in M$, then $t_2[l]$ is a node in subtree $T_2[i_t]$; and there are (k_1, l_1) and (k_2, l_2) in M such that $t_1[k_1]$ is a node in $T_2[i_{s_1}]$ and $t_1[k_2]$ is a node in $T_2[i_{s_2}]$, where $1 \le s_1 \ne s_2 \le n_i$. This is similar to case 1.

Case 3: there are s and t such that if $(k,l) \in M$ then $t_1[k]$ is node in $T_1[i_s]$ and $t_2[l]$ is node in $T_2[i_t]$. In this case M is a restricted mapping and therefore $D(F_1[i], F_2[j]) = \min_{RM(i,j)} \gamma(RM(i,j))$.

Case 4: there are $(k_1, l_1), (k_2, l_2), (x_1, y_1), (x_2, y_2)$ in M such that $t_1[k_1]$ is a node in $T_1[i_{s_1}]$, $t_1[k_2]$ is a node in $T_1[i_{s_2}]$, $t_2[y_1]$ is a node in $T_2[i_{t_1}]$, and $t_2[y_2]$ is a node in $T_2[i_{t_2}]$, where $1 \le s_1 \ne s_2 \le n_i$ and $1 \le t_1 \ne t_2 \le n_j$. We will show that in this case the mapping M is a restricted mapping between $F_1[i]$ and $F_2[j]$.

Suppose this is not true. Then, w.l.o.g., we assume that there are (a_1, b_1) and (a_2, b_2) in M such that $t_1[a_1]$ and $t_1[a_2]$ belong to the same subtree and $t_2[b_1]$ and $t_2[b_2]$ belong to different subtrees. Let $(a_3, b_3) \in M$ such that $t_1[a_3]$ and $t_1[a_1]$ belong to different subtrees of $F_1[i]$. Now consider $lca(t_1[a_1], t_1[a_2])$ and $t_1[a_3]$. Since $t_1[a_1]$ and $t_1[a_2]$ belong to the same subtree which is different from the subtree $t_1[a_3]$ belongs to, we know that $lca(t_1[a_1], t_1[a_2])$ and $t_1[a_3]$ are not in ancestor or descendant relationship. However if we consider $lca(t_2[b_1], t_2[b_2])$ and $t_2[b_3]$, it is easy to see that $lca(t_2[b_1], t_2[b_2]) = t_2[j]$ which is an ancestor of $t_2[b_3]$. This means that M is not a valid mapping. Contradiction. $\square$

From lemma 7, in order to compute $D(F_1[i], F_2[j])$, $\min_{RM(i,j)} \gamma(RM(i,j))$ have to be computed first. The next two lemmas will establish the relationship between $\min_{RM(i,j)} \gamma(RM(i,j))$ and $D(T_1[i_s], T_2[j_t])$, where $1 \le s \le n_i$ and $1 \le n_j$. We need the following definition in the next two lemmas.

Given $I = \{i_1, i_2, ... i_{n_i}\}$ and $J = \{j_1, j_2, ... j_{n_j}\}$, We define a partial function between I and J $PF(i,j)$ as follows:

1) $PF(i,j)$ is a set of pairs (s,t) such that $1 \le s \le n_i$ and $1 \le t \le n_j$,
2) let (s,t) and (x,y) be in $PF(i,j)$, $s = x$ if and only if $t = y$.
The cost of a partial function, $gamma(FP(i,j))$, is defined as follows:
$\sum_{(s,t)\in PF(i,j)} D(T_1[i_s], T_2[j_t]) + \sum_{s\notin PF(i,j)} D(T_1[i_s], \theta) + \sum_{t\notin PF(i,j)} D(\theta, T_2[j_t])$.

Lemma 8. $\min_{RM(i,j)} \gamma(RM(i,j)) = \min_{PF(i,j)} \gamma(PF(i,j))$

Proof: Given a restricted mapping $RM(i,j)$, we define a partial function $PF(i,j)$ as follows.

$$PF(i,j) = \{(s,t)| \text{ there is } (k,l) \text{ in } RM \text{ s.t. } t_1[k] \ (t_2[l]) \text{ is a node in } T_1[i_s] \ (T_2[j_t])\}$$

By the definition of restricted mapping this is indeed a partial function. Furthermore it is easy to see that $\gamma(PF(i,j)) \le \gamma(RM(i,j))$.

On the other hand, given a partial function $PF(i,j)$, we can construct a restricted mapping $RM(i,j)$ as follows.

$$RM(i,j) = \left\{(k,l)| \begin{matrix} \text{there exists } (s,t) \in PF(i,j) \text{ s.t. } (k,l) \text{ is in} \\ \text{the minimum cost mapping between } T_1[i_s] \text{ and } T_2[j_t] \end{matrix} \right\}$$

It is clear that this is a restricted mapping. Therefore $\gamma(RM(i,j)) \le \gamma(PF(i,j))$.

Hence $\min_{RM(i,j)} \gamma(RM(i,j)) = \min_{PF(i,j)} \gamma(PF(i,j))$. $\square$

In fact we can add one more condition in the definition of partial function, namely $|PF(i,j)| = \min\{n_i, n_j\}$. The reason is that if $|PF(i,j)| < \min\{n_i, n_j\}$ then there are s and t such that $s \notin PF(i,j)$ and $t \notin PF(i,j)$. Because D is a distance metric, $D(T_1[i_s], T_2[j_t]) \le D(T_1[i_s], \theta) + D(\theta, T_2[j_t])$. Therefore $\gamma(PF'(i,j) = \{(s,t)\cup PF(i,j)\}) \le \gamma(PF(i,j))$.

Combining lemma 7, 8 and the above observation, we have proved the following lemma.

Lemma 9. Let $t_1[i_1], t_1[i_2], ...t_1[i_{n_i}]$ be the children of $t_1[i]$ and $t_2[j_1], t_2[j_2], ...t_2[j_{n_j}]$ be the children of $t_2[j]$, then

$$D(F_1[i], F_2[j]) = \min \begin{cases} D(F_1[i], \theta) + \min_{1 \le s \le n_i} D(F_1[i_s], F_2[j]) - D(F_1[i_s], \theta) \\ D(\theta, F_2[j]) + \min_{1 \le t \le n_j} D(F_1[i], F_2[j_t]) - D(\theta, F_2[j_t]) \\ \min_{|PF(i,j)|=\min\{n_i,n_j\}} \gamma(PF(i,j)) \end{cases}$$

5 Algorithm and Complexity

We first consider how to compute $\min\limits_{|PF(i,j)|=\min\{n_i,n_j\}} \gamma(PF(i,j))$ and then present our simple algorithm.

5.1 Algorithm

From the definition of $PF(i,j)$ and $\gamma(PF(i,j))$, it is clear that this problem is related to the minimum cost bipartite matching problem. If $n_i = n_j$, then this is exactly the minimum cost bipartite matching problem. If $n_i \neq n_j$, then we have to consider those extra trees in one of the forests. Suppose that $n_i > n_j$. One way to solve this problem is to add $n_i - n_j$ empty trees to $F[j]$ and then use bipartite matching. However this will result in redundant computation. We will reduce this problem directly to the minimum cost maximum flow problem by adding only one empty tree to $F[j]$.

Given $F_1[i]$ and $F_2[j]$, w.l.o.g., we assume that $n_i > n_j$. Let $I = \{i_1, i_2, ...i_{n_i}\}$ and $J = \{j_1, j_2, ...j_{n_j}\}$, where i_k, $1 \le k \le n_i$, represents tree $T_1[i_k]$, and j_l, $1 \le l \le n_j$, represents tree $T_2[j_l]$.

We construct a graph $G = (V, E)$ as follows:

vertex set: $V = \{s, t, e\} \cup I \cup J$, where s is the source, t is the sink and e represents an empty tree;

edge set: $[s, i_k], [j_l, t], [e, t]$ with cost zero, $[i_k, j_l]$ with cost $D(T_1[i_k], T_2[j_l])$, and $[i_k, e]$ with cost $D(T_1[i_k], \theta)$. All the edges have capacity one except $[e, t]$ whose capacity is $n_i - n_j$.

G is a network with integer capacities, nonnegative costs, and the maximum flow $f^* = n_i = \max\{n_i, n_j\}$, see figure 1.

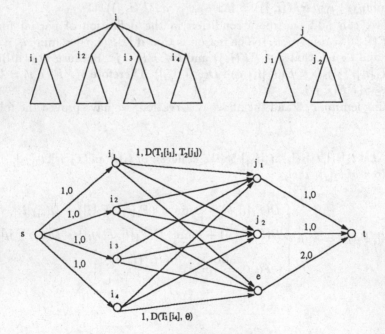

Fig. 1. Reduction to the minimum cost flow problem.

Now let us examine the meaning of $\gamma(PF(i,j))$. It is easy to see that given a $PF(i,j)$, $\gamma(PF(i,j))$ represents the cost of the following maximum flow on G: for any $(k,l) \in PF(i,j)$ the flow on edge $[i_k, j_l]$ is one; for any $k \notin PF(i,j)$ the flow on edge $[i_k, e]$ is one; the flow on edge $[s, i_k]$ and edge $[j_l, t]$ are one; the flow on edge $[e, t]$ is $n_i - n_j$; and all the flows on the other edges are zero.

Therefore $\min\limits_{|PF(i,j)|=\min\{n_i,n_j\}} \gamma(PF(i,j))$ is exactly the cost of the minimum cost maximum flow of G. Hence we can use minimum cost maximum flow algorithm to compute $\min\limits_{|PF(i,j)|=\min\{n_i,n_j\}} \gamma(PF(i,j))$.

We are now ready to give our algorithm.

Input: T_1 and T_2
Output: $D(T_1[i], T_2[j])$, where $1 \le i \le |T_1|$ and $1 \le j \le |T_2|$

$D(\theta, \theta) = 0;$
for $i = 1$ to $|T_1|$
$\quad D(F_1[i], \theta) = \sum_{k=1}^{n_i} D(T_1[i_k], \theta)$
$\quad D(T_1[i], \theta) = D(F_1[i], \theta) + \gamma(t_1[i] \to \lambda)$

for $j = 1$ to $|T_2|$
$\quad D(\theta, F_2[j]) = \sum_{k=1}^{n_j} D(\theta, T_2[j_k])$
$\quad D(\theta, T_2[j]) = D(\theta, F_2[j]) + \gamma(\lambda \to t_2[j])$

for $i = 1$ to $|T_1|$
$\quad$ for $j = 1$ to $|T_2|$

$$D(F_1[i], F_2[j]) = \min \begin{cases} D(F_1[i], \theta) + \min\limits_{1 \le s \le n_i} \{D(F_1[i_s], F_2[j]) - D(F_1[i_s], \theta)\} . \\ D(\theta, F_2[j]) + \min\limits_{1 \le t \le n_j} \{D(F_1[i], F_2[j_t]) - D(\theta, F_2[j_t])\} \\ \min\limits_{|PF(i,j)|=\min\{n_i,n_j\}} \gamma(PF(i,j)) \end{cases}$$

$$D(T_1[i], T_2[j]) = \min \begin{cases} D(T_1[i], \theta) + \min_{1 \le s \le n_i} \{D(T_1[i_s], T_2[j]) - D(T_1[i_s], \theta)\} \\ D(\theta, T_2[j]) + \min_{1 \le t \le n_j} \{D(T_1[i], T_2[j_t]) - D(\theta, T_2[j_t])\} \\ D(F_1[i], F_2[j]) + \gamma(t_1[i] \to t_2[j]) \end{cases}$$

5.2 Complexity

The complexity of computing $D(T_1[i], T_2[j])$ is, by lemma 6, bounded by $O(n_i + n_j)$. The complexity of computing $D(F_1[i], F_2[j])$ is bounded by the $O(n_i + n_j)$ plus the complexity of the minimum cost maximum flow computation.

Our graph is a graph with integer capacities, nonnegative edge costs, and maximum flow $f^* = max\{n_i, n_j\}$. The complexity of finding minimum cost maximum flow for such a graph with n vertices and m edges is $O(m|f^*| \log_{(2+m/n)} n) \le O(m|f^*| \log_2 n)$ [6]. For our graph, $n = n_i + n_j + 3$ and $m = n_i * n_j + 2n_i + n_j$; therefore the complexity is bounded by $O(n_i * n_j * \max\{n_i, n_j\} * \log_2(max\{n_i, n_j\}))$.

Hence for any pair i and j, the complexity of computing $D(T_1[i], T_2[j])$ and $D(F_1[i], F_2[j])$ is bounded by $O(n_i * n_j * \max\{n_i, n_j\} * \log_2(max\{n_i, n_j\}))$. Therefore the complexity of our algorithm is

$$\sum_{i=1}^{|T_1|} \sum_{j=1}^{|T_2|} O(n_i \times n_j \times \max\{n_i, n_j\} \times \log_2(max\{n_i, n_j\}))$$

$$\leq \sum_{i=1}^{|T_1|} \sum_{j=1}^{|T_2|} O(n_i \times n_j \times \max\{deg(T_1), deg(T_2)\} \times \log_2(max\{deg(T_1), deg(T_2)\}))$$

$$\leq O(\max\{deg(T_1), deg(T_2)\} \times \log_2(max\{deg(T_1), deg(T_2)\}) \times \sum_{i=1}^{|T_1|} n_i \times \sum_{j=1}^{|T_2|} n_j)$$

$$\leq O(|T_1| \times |T_2| \times \max\{deg(T_1), deg(T_2)\} \times \log_2(max\{deg(T_1), deg(T_2)\}))$$

6 Approximate Unordered Tree Matching

Approximate unordered tree matching is a natural extension of approximate string matching [3, 8, 1] and approximate ordered tree matching [11, 13]. We omit details in this section.

In this section, we use $D[i]$ to represent subtree rooted at $d[i]$ and use $D_f[i]$ to represent the forest obtained by removing $d[i]$ from $D[i]$.

We first define the operation of removing at a node.
Removing at node $d[i]$ means removing the subtree rooted at $d[i]$.

Define a subtree set $S(D)$ as follows: $S(D)$ is a set of numbers satisfying:

(1) $i \in S(D)$ implies that $1 \leq i \leq |D|$
(2) $i, j \in S(D)$ implies that neither is an ancestor of the other.

Define $R(D, S(D))$ to be the tree D with removing at all nodes in $S(D)$.

Now we can give the definition of approximate unordered tree matching. Given tree D and P, for each i, we want to compute

$$D_r(D[i], P) = \min_S\{D(R(D[i], S(D[i])), P)\}.$$

The minimum here is over all possible subtree sets $S(D[i])$.

In the following we give an algorithm for approximate unordered tree matching with time complexity $O(|P| \times |D| \times deg(P) \times \log_2(max\{deg(P), deg(D)\}))$. The algorithm needs a subroutine to compute $\min_{PF(i,j)} \gamma_r(PF(i,j))$. We can similarly reduce this problem to the minimum cost flow problem. However we have to modify the definition of the cost of a partial function:

$$\gamma_r(PF(i,j)) = \sum_{(s,t)\in PF(i,j)} D_r(D[i_s], P[j_t]) + \sum_{s\notin PF(i,j)} D_r(\theta, P[i_s]).$$

Input: D and P
Output: $D_r(D[i], P)$, where $1 \le i \le |D|$

$D(\theta, \theta) = 0;$
for $i = 1$ to $|D|$
$\quad D_r(D_f[i], \theta) = 0$
$\quad D_r(D[i], \theta) = 0$

for $j = 1$ to $|P|$
$\quad D_r(\theta, P_f[j]) = \sum_{k=1}^{n_j} D_r(\theta, P[j_k])$
$\quad D_r(\theta, P[j]) = D_r(\theta, P_f[j]) + \gamma(\lambda \to p[j])$

for $i = 1$ to $|D|$
$\quad$ for $j = 1$ to $|P|$

$$D_r(D_f[i], P_f[j]) = \min \begin{cases} \min\limits_{1 \le s \le n_i} D_r(D_f[i_s], P_f[j]) + \gamma(d[i_s] \to \lambda) \\ D_r(\theta, P_f[j]) + \min\limits_{1 \le t \le n_j} D_r(D_f[i], P_f[j_t]) - D_r(\theta, P_f[j_t]) \\ \min\limits_{PF(i,j)} \gamma_r(PF(i,j)) \end{cases}$$

$$D_r(D[i], P[j]) = \min \begin{cases} D_r(\theta, P[j]) \\ \gamma(d[i] \to \lambda) + \min_{1 \le s \le n_i} D_r(D[i_s], P[j]) \\ D_r(\theta, P[j]) + \min_{1 \le t \le n_j} D_r(D[i], P[j_t]) - D_r(\theta, P[j_t]) \\ D_r(D_f[i], P_f[j]) + \gamma(d[i] \to p[j]) \end{cases}$$

7 Conclusion

Motivated by the NP-complete results in [2, 12], we have defined a constrained editing distance metric between unordered labeled trees. We present an algorithm for computing this distance metric based on a reduction to the minimum cost maximum flow problem. Our algorithm is generalizable with the same complexity to the approximate unordered tree matching problem.

The work presented is part of a project to develop a comprehensive tool for approximate tree pattern matching [9]. The proposed algorithms and their implementation will be integrated into this tool.

References

1. G. M. Landau and U. Vishkin, 'Fast parallel and serial approximate string matching', *J. Algorithms*, vol. 10, pp.157-169, 1989
2. Pekka Kilpelainen and Heikki Mannila, 'Ordered and unordered tree inclusion', To appear *SIAM J. on Computing*
3. P. H. Sellers, 'The theory and computation of evolutionary distances' *J. Algorithms* vol. 1, pp.359-373, 1980
4. Bruce Shapiro and Kaizhong Zhang, 'Comparing multiple RNA secondary structures using tree comparisons' *Comput. Appl. Biosci.* vol. 6, no. 4, pp.309-318, 1990
5. K. C. Tai, 'The tree-to-tree correction problem', *J. ACM*, vol. 26, pp.422-433, 1979
6. Robert E. Tarjan, 'Data structures and network algorithms', CBMS-NSF Regional Conference Series in Applied Mathematics, 1983

7. E. Tanaka and K. Tanaka, 'The tree-to-tree editing problem', *International Journal of Pattern Recognition and Artificial Intelligence*', vol. 2, no. 2, pp.221-240 1988

8. E. Ukkonen, 'Finding approximate patterns in strings', *J. Algorithm*, vol. 6, pp.132-137, 1985

9. Jason T.L. Wang, Kaizhong Zhang, Karpjoo Jeong and Dennis Shasha 'ATBE: A system for approximate tree matching', To appear *IEEE Trans. on Knowledge and Data Engineering*

10. Kaizhong Zhang, 'An editing based distance between ordered labeled trees', In preparation.

11. Kaizhong Zhang and Dennis Shasha, 'Simple fast algorithms for the editing distance between trees and related problems', *SIAM J. Computing* vol. 18, no. 6, pp.1245-1262, 1989

12. Kaizhong Zhang, Rick Statman and Dennis Shasha, 'On the editing distance between unordered labeled trees' *Information Processing Letters* no. 42, pp.133-139, 1992

13. Kaizhong Zhang, Dennis Shasha, and Jason Wang, 'Approximate tree matching in the presence of variable length don't cares', To appear *J. Algorithms*

Lecture Notes in Computer Science

For information about Vols. 1–610
please contact your bookseller or Springer-Verlag

Vol. 611: M. P. Papazoglou, J. Zeleznikow (Eds.), The Next Generation of Information Systems: From Data to Knowledge. VIII, 310 pages. 1992. (Subseries LNAI).

Vol. 612: M. Tokoro, O. Nierstrasz, P. Wegner (Eds.), Object-Based Concurrent Computing. Proceedings, 1991. X, 265 pages. 1992.

Vol. 613: J. P. Myers, Jr., M. J. O'Donnell (Eds.), Constructivity in Computer Science. Proceedings, 1991. X, 247 pages. 1992.

Vol. 614: R. G. Herrtwich (Ed.), Network and Operating System Support for Digital Audio and Video. Proceedings, 1991. XII, 403 pages. 1992.

Vol. 615: O. Lehrmann Madsen (Ed.), ECOOP '92. European Conference on Object Oriented Programming. Proceedings. X, 426 pages. 1992.

Vol. 616: K. Jensen (Ed.), Application and Theory of Petri Nets 1992. Proceedings, 1992. VIII, 398 pages. 1992.

Vol. 617: V. Mařík, O. Štěpánková, R. Trappl (Eds.), Advanced Topics in Artificial Intelligence. Proceedings, 1992. IX, 484 pages. 1992. (Subseries LNAI).

Vol. 618: P. M. D. Gray, R. J. Lucas (Eds.), Advanced Database Systems. Proceedings, 1992. X, 260 pages. 1992.

Vol. 619: D. Pearce, H. Wansing (Eds.), Nonclassical Logics and Information Proceedings. Proceedings, 1990. VII, 171 pages. 1992. (Subseries LNAI).

Vol. 620: A. Nerode, M. Taitslin (Eds.), Logical Foundations of Computer Science – Tver '92. Proceedings. IX, 514 pages. 1992.

Vol. 621: O. Nurmi, E. Ukkonen (Eds.), Algorithm Theory – SWAT '92. Proceedings. VIII, 434 pages. 1992.

Vol. 622: F. Schmalhofer, G. Strube, Th. Wetter (Eds.), Contemporary Knowledge Engineering and Cognition. Proceedings, 1991. XII, 258 pages. 1992. (Subseries LNAI).

Vol. 623: W. Kuich (Ed.), Automata, Languages and Programming. Proceedings, 1992. XII, 721 pages. 1992.

Vol. 624: A. Voronkov (Ed.), Logic Programming and Automated Reasoning. Proceedings, 1992. XIV, 509 pages. 1992. (Subseries LNAI).

Vol. 625: W. Vogler, Modular Construction and Partial Order Semantics of Petri Nets. IX, 252 pages. 1992.

Vol. 626: E. Börger, G. Jäger, H. Kleine Büning, M. M. Richter (Eds.), Computer Science Logic. Proceedings, 1991. VIII, 428 pages. 1992.

Vol. 628: G. Vosselman, Relational Matching. IX, 190 pages. 1992.

Vol. 629: I. M. Havel, V. Koubek (Eds.), Mathematical Foundations of Computer Science 1992. Proceedings. IX, 521 pages. 1992.

Vol. 630: W. R. Cleaveland (Ed.), CONCUR '92. Proceedings. X, 580 pages. 1992.

Vol. 631: M. Bruynooghe, M. Wirsing (Eds.), Programming Language Implementation and Logic Programming. Proceedings, 1992. XI, 492 pages. 1992.

Vol. 632: H. Kirchner, G. Levi (Eds.), Algebraic and Logic Programming. Proceedings, 1992. IX, 457 pages. 1992.

Vol. 633: D. Pearce, G. Wagner (Eds.), Logics in AI. Proceedings. VIII, 410 pages. 1992. (Subseries LNAI).

Vol. 634: L. Bougé, M. Cosnard, Y. Robert, D. Trystram (Eds.), Parallel Processing: CONPAR 92 – VAPP V. Proceedings. XVII, 853 pages. 1992.

Vol. 635: J. C. Derniame (Ed.), Software Process Technology. Proceedings, 1992. VIII, 253 pages. 1992.

Vol. 636: G. Comyn, N. E. Fuchs, M. J. Ratcliffe (Eds.), Logic Programming in Action. Proceedings, 1992. X, 324 pages. 1992. (Subseries LNAI).

Vol. 637: Y. Bekkers, J. Cohen (Eds.), Memory Management. Proceedings, 1992. XI, 525 pages. 1992.

Vol. 639: A. U. Frank, I. Campari, U. Formentini (Eds.), Theories and Methods of Spatio-Temporal Reasoning in Geographic Space. Proceedings, 1992. XI, 431 pages. 1992.

Vol. 640: C. Sledge (Ed.), Software Engineering Education. Proceedings, 1992. X, 451 pages. 1992.

Vol. 641: U. Kastens, P. Pfahler (Eds.), Compiler Construction. Proceedings, 1992. VIII, 320 pages. 1992.

Vol. 642: K. P. Jantke (Ed.), Analogical and Inductive Inference. Proceedings, 1992. VIII, 319 pages. 1992. (Subseries LNAI).

Vol. 643: A. Habel, Hyperedge Replacement: Grammars and Languages. X, 214 pages. 1992.

Vol. 644: A. Apostolico, M. Crochemore, Z. Galil, U. Manber (Eds.), Combinatorial Pattern Matching. Proceedings, 1992. X, 287 pages. 1992.

Vol. 645: G. Pernul, A M. Tjoa (Eds.), Entity-Relationship Approach – ER '92. Proceedings, 1992. XI, 439 pages, 1992.

Vol. 646: J. Biskup, R. Hull (Eds.), Database Theory – ICDT '92. Proceedings, 1992. IX, 449 pages. 1992.

Vol. 647: A. Segall, S. Zaks (Eds.), Distributed Algorithms. X, 380 pages. 1992.

Vol. 648: Y. Deswarte, G. Eizenberg, J.-J. Quisquater (Eds.), Computer Security – ESORICS 92. Proceedings. XI, 451 pages. 1992.

Vol. 649: A. Pettorossi (Ed.), Meta-Programming in Logic. Proceedings, 1992. XII, 535 pages. 1992.

Vol. 650: T. Ibaraki, Y. Inagaki, K. Iwama, T. Nishizeki, M. Yamashita (Eds.), Algorithms and Computation. Proceedings, 1992. XI, 510 pages. 1992.

Vol. 651: R. Koymans, Specifying Message Passing and Time-Critical Systems with Temporal Logic. IX, 164 pages. 1992.

Vol. 652: R. Shyamasundar (Ed.), Foundations of Software Technology and Theoretical Computer Science. Proceedings, 1992. XIII, 405 pages. 1992.

Vol. 653: A. Bensoussan, J.-P. Verjus (Eds.), Future Tendencies in Computer Science, Control and Applied Mathematics. Proceedings, 1992. XV, 371 pages. 1992.

Vol. 654: A. Nakamura, M. Nivat, A. Saoudi, P. S. P. Wang, K. Inoue (Eds.), Prallel Image Analysis. Proceedings, 1992.. VIII, 312 pages. 1992.

Vol. 655: M. Bidoit, C. Choppy (Eds.), Recent Trends in Data Type Specification. X, 344 pages. 1993.

Vol. 656: M. Rusinowitch, J. L. Rémy (Eds.), Conditional Term Rewriting Systems. Proceedings, 1992. XI, 501 pages. 1993.

Vol. 657: E. W. Mayr (Ed.), Graph-Theoretic Concepts in Computer Science. Proceedings, 1992. VIII, 350 pages. 1993.

Vol. 658: R. A. Rueppel (Ed.), Advances in Cryptology – EUROCRYPT '92. Proceedings, 1992. X, 493 pages. 1993.

Vol. 659: G. Brewka, K. P. Jantke, P. H. Schmitt (Eds.), Nonmonotonic and Inductive Logic. Proceedings, 1991. VIII, 332 pages. 1993. (Subseries LNAI).

Vol. 660: E. Lamma, P. Mello (Eds.), Extensions of Logic Programming. Proceedings, 1992. VIII, 417 pages. 1993. (Subseries LNAI).

Vol. 661: S. J. Hanson, W. Remmele, R. L. Rivest (Eds.), Machine Learning: From Theory to Applications. VIII, 271 pages. 1993.

Vol. 662: M. Nitzberg, D. Mumford, T. Shiota, Filtering, Segmentation and Depth. VIII, 143 pages. 1993.

Vol. 663: G. v. Bochmann, D. K. Probst (Eds.), Computer Aided Verification. Proceedings, 1992. IX, 422 pages. 1993.

Vol. 664: M. Bezem, J. F. Groote (Eds.), Typed Lambda Calculi and Applications. Proceedings, 1993. VIII, 433 pages. 1993.

Vol. 665: P. Enjalbert, A. Finkel, K. W. Wagner (Eds.), STACS 93. Proceedings, 1993. XIV, 724 pages. 1993.

Vol. 666: J. W. de Bakker, W.-P. de Roever, G. Rozenberg (Eds.), Semantics: Foundations and Applications. Proceedings, 1992. VIII, 659 pages. 1993.

Vol. 667: P. B. Brazdil (Ed.), Machine Learning: ECML – 93. Proceedings, 1993. XII, 471 pages. 1993. (Subseries LNAI).

Vol. 668: M.-C. Gaudel, J.-P. Jouannaud (Eds.), TAPSOFT '93: Theory and Practice of Software Development. Proceedings, 1993. XII, 762 pages. 1993.

Vol. 669: R. S. Bird, C. C. Morgan, J. C. P. Woodcock (Eds.), Mathematics of Program Construction. Proceedings, 1992. VIII, 378 pages. 1993.

Vol. 670: J. C. P. Woodcock, P. G. Larsen (Eds.), FME '93: Industrial-Strength Formal Methods. Proceedings, 1993. XI, 689 pages. 1993.

Vol. 671: H. J. Ohlbach (Ed.), GWAI-92: Advances in Artificial Intelligence. Proceedings, 1992. XI, 397 pages. 1993. (Subseries LNAI).

Vol. 672: A. Barak, S. Guday, R. G. Wheeler, The MOSIX Distributed Operating System. X, 221 pages. 1993.

Vol. 673: G. Cohen, T. Mora, O. Moreno (Eds.), Applied Algebra, Algebraic Algorithms and Error-Correcting Codes. Proceedings, 1993. X, 355 pages 1993.

Vol. 674: G. Rozenberg (Ed.), Advances in Petri Nets 1993. VII, 457 pages. 1993.

Vol. 675: A. Mulkers, Live Data Structures in Logic Programs. VIII, 220 pages. 1993.

Vol. 676: Th. H. Reiss, Recognizing Planar Objects Using Invariant Image Features. X, 180 pages. 1993.

Vol. 677: H. Abdulrab, J.-P. Pécuchet (Eds.), Word Equations and Related Topics. Proceedings, 1991. VII, 214 pages. 1993.

Vol. 678: F. Meyer auf der Heide, B. Monien, A. L. Rosenberg (Eds.), Parallel Architectures and Their Efficient Use. Proceedings, 1992. XII, 227 pages. 1993.

Vol. 683: G.J. Milne, L. Pierre (Eds.), Correct Hardware Design and Verification Methods. Proceedings, 1993. VIII, 270 Pages. 1993.

Vol. 684: A. Apostolico, M. Crochemore, Z. Galil, U. Manber (Eds.), Combinatorial Pattern Matching. Proceedings, 1993. VIII, 265 pages. 1993.

Vol. 685: C. Rolland, F. Bodart, C. Cauvet (Eds.), Advanced Information Systems Engineering. Proceedings, 1993. XI, 650 pages. 1993.

Vol. 686: J. Mira, J. Cabestany, A. Prieto (Eds.), New Trends in Neural Computation. Procidings, 1993. XVII, 746 pages. 1993.

Vol. 687: H. H. Barrett, A. F. Gmitro (Eds.), Information Processing in Medical Imaging. Proceedings, 1993. XVI, 567 pages. 1993.

Vol. 688: M. Gauthier (Ed.), Ada - Europe '93. Proceedings, 1993. VIII, 353 pages. 1993.

Vol. 689: J. Komorowski, Z. W. Ras (Eds.), Methodologies for Intelligent Systems. Proceedings, 1993. XI, 653 pages. 1993. (Subseries LNAI).

Vol. 690: C. Kirchner (Ed.), Rewriting Techniques and Applications. Proceedings, 1993. XI, 488 pages. 1993.

Vol. 691: M. A. Marsan (Ed.), Application and Theory of Petri Nets 1993. Proceedings, 1993. IX, 591 pages. 1993.

Vol. 692: D. Abel, B.C. Ooi (Eds.), Advances in Spatial Databases. Proceedings, 1993. XIII, 529 pages. 1993.